EARTH RISING

THE REVOLUTION

Toward a Thousand Years of Peace

Dr. Nick Begich

James Roderick

Earthpulse Press
P. O. Box 201393
Anchorage, Alaska
99520 U. S. A.

Special thanks to the
Anchorage Municipal Library system
for their research assistance.

ISBN 1-890693-43-X

Cover Art:
by Nick Begich and Shelah J. Slade
Earth image provided by NASA, Washington, D.C.

First Edition
Second Printing 2004

Printed in the United States of America

About the Authors

Dr. Nick Begich is the eldest son of the late United States Congressman from Alaska, Nick Begich Sr., and political activist Pegge Begich. He is well known in Alaska for his own political activities. He was twice elected President of both the Alaska Federation of Teachers and the Anchorage Council of Education. He has been pursuing independent research in the sciences and politics for most of his adult life. Begich received his doctorate in traditional medicine from The Open International University for Complementary Medicines in November, 1994.

He co-authored the book *Angels Don't Play This HAARP; Advances in Tesla Technology*, and wrote *Towards a New Alchemy: The Millennium Science.* His latest book *Earth Rising – The Revolution: Toward a Thousand Years of Peace* was co-authored with James Roderick in December, 1999. He is also the editor of Earthpulse Flashpoints, a continuing new-science book series. Begich has published articles in science, politics and education and is a well known lecturer, having presented throughout the United States and in nineteen countries. He has been featured as a guest on thousands of radio broadcasts reporting on his research activities including new technologies, health and earth science related issues. He has also appeared on dozens of television documentaries and other programs throughout the world including BBC-TV, CBC-TV, TeleMundo, and others.

Begich has served as an expert witness and speaker before the European Parliament. He has spoken on various issues for groups representing citizen concerns, statesmen and elected officials, scientists and others. He is the publisher and co-owner of Earthpulse Press and is also under contract as Tribal Administrator for the Chickaloon Village Council, a federally recognized American Indian Tribe of the Athabascan Indian Nation.

Dr. Begich is married to Shelah Begich-Slade and has five children. He resides just north of Anchorage in the community of Eagle River, Alaska, USA.

James Roderick was a 32-year resident of Alaska, working variously as a fisherman, gold-miner and trapper. His investigations into military toxics in Alaska led to stories in the press about HAARP, chemical weapons disposal in Alaska's waters, nuclear power plant contamin-ation in central Alaska and the military's illegal testing of Alaska natives and Eskimos with radioactive Iodine. His investigations continued until his death in August 2002 a few days after the Russian Duma objected to HAARP and its sinister applications.

Table Of Contents

Dedication

This collaborative work has been collected and written by Dr. Nick Begich and James Roderick with the support of hundreds of people over the last several years. We dedicate this work to our families, friends and fellow researchers who made this book possible.

We hope that this book serves as a catalyst for the future work of the Earthpulse Institute which is being founded with the release of this title. The Institute is being established in order to maintain, collect and release new information on these and other emerging technologies.

We wish to thank all of you who have supported our efforts in the past and those who are supporting us now by adding this book to your own collections, local library or passing it on to friends.

- *The new revolution in technology is vast and sweeping in every way. The technology can either serve the cause of mankind or be used against us.*

- *The revolution we are suggesting is one of personal empowerment and growth. We trust this book will contribute to your journey there.*

- *We are committed to the cause of peace and the day of the dawning of a thousand years of peace.*

Dr. Nick Begich **James Roderick**

Prologue

Earth Rising – The Revolution

"I believe, indeed, that overemphasis on the purely intellectual attitude, often directed solely to the practical and factual, in our education, has led directly to the impairment of ethical values."

Albert Einstein[1]

Earth Rising – The Revolution is about the dramatic changes which are taking place as a result of new technologies emerging from both military and civilian research. These technologies will increasingly impact every area of human activity. They will affect our lives in ways which challenge fundamental aspects of freedom, liberty, privacy and the very essence of our individual personhood.

The changes which are taking place will continue to increase in speed and impact. These technologies offer great potentials while at the same time presenting incredible risks to foundational democratic principles. This book provides the evidence of the new technologies which many have suspected but failed to prove. Our objective is to inform readers, create debate and see that the ethical concerns and risks are well defined. This work is an overview of the challenge, and, we believe, it will contribute to the kind of public discussion these innovations demand.

This book explores technologies which are revolutionary in their possibilities but even more importantly, technologies which if misused threaten our ability to advance to our highest individual potentials. This book presents the facts which demonstrate where many of the risks may be – while at the same time presenting a series of solutions which can change the way in which high-risk technologies are introduced into our societies, cultures and regional areas.

The pace of technological change is so pervasive, so rapid and so specialized that many of the advances escape the average person in society – yet those same average individuals become either the beneficiaries or victims of these new innovations.

1. Einstein, Albert. "The Need for Ethical Culture," Letter read on the occasion of the seventy-fifth anniversary of the Ethical Culture Society, New York, January 1951. Published in Mein Weltbild, Zurich: Europa Verlag, 1953. EPI1262

Technology advances under a set of rules different from other areas of human activity such as business, religion or politics, which have implicit rules which include discussion of what is ethical, reasonable and moral. With technological advances, which are the outgrowth of science, "objective results" are the measure of success and do not necessarily include ethical or moral considerations. The problem is in the pace of the advance of science, the lack of organized mechanisms to safeguard basic human rights and systems which ask and answer the ethical questions which these advances create.

Technology is doubling about every nine to ten months. What does this mean? Simply, all of the technology which has been created since the invention of the wheel is being doubled every nine to ten months. This doubling is also compounding as these new levels of advancement are again and again being doubled each cycle. Ten years ago this doubling process was occurring every five years and compressing into shorter and shorter intervals, eventually reaching the current nine to ten month window.

It is because of the speed of change, coupled with increasingly complex societal interactions, that the real issues regarding the direction of our technology are not presently debated – issues that require debate. They must be brought to the light of day for real discussion to take place with respect to their implications within the context of democratic societies.

The impact of our technological and scientific advances must be considered from the perspective of the individual first – as the central actor. Consideration of basic human rights and the impact of new technology, particularly military technology, on fundamental issues must be dealt with openly.

All authority – regional, national and eventually international, emerges from the consent of individual people comprising local populations. We believe that the right to develop any innovation, in any context, whether in science, business, religion or politics requires clear knowledge of both the risks and benefits associated with the technology. Informed consent also means informed debate when being asked to consider the issues which we face together.

In producing this book, we are seeking to engage public discussion of the technologies we identify, and, taking both a long and short view, evaluate their implications for each of us. We hope that this book provides solutions which will encourage individuals to recognize their own potentials in helping solve the problems surrounding the introduction of these new technologies. At the same time, we hope that people will gain strength in meeting the challenges we each face in our daily activity. We believe that this material will contribute to a shift in the way we see our individual and collective activities while also empowering us to continue to contribute to improving our regions and world. We believe that every individual has profound potential to contribute in ways which can create real solutions.

This book is divided into three primary sections. The first develops the context for change in terms of the authors' perspectives. This section describes the history of how the authors came to research these subjects and what has happened since 1994 when our research into these military technologies began. The emphasis of this section is on describing the path we followed to bring to the world's attention one of those technologies and, in the course of that effort, demonstrate the need for national and international debate to focus on these areas. The second section addresses specific technological areas with particular emphasis on mind control technologies, non-lethal weapon systems, geophysical (environmental) manipulation, anti-privacy technologies and other high-risk areas. The third section is centered on solutions. A set of solution possibilities is presented for consideration as starting points for discussions which may eventually lead to democratic solutions and public policy changes sufficient to maintain the basic freedoms democratically constituted governments require.

In the last five years a great deal has already been accomplished. Significant knowledge has been gained from our past activities which will contribute greatly to the presentation of this new research. A number of good friends have been made and methods developed to insure that these issues are brought to the forefront of public debate. We believe with absolute knowledge and conviction that a few people of good heart and honest intentions can still make a huge difference in shaping the world around us. We are committed to the effort and trust you will, by the end of the reading, join us in initiating *Earth Rising – The Revolution: Toward a Thousand Years of Peace.*

Chapter 1

Awakening to the Change

"Thus, it may be possible to 'talk' to selected adversaries in a fashion that would be most disturbing to them."[2]

In 1994, while reading an Australian journal, I came across a story about a new weapon technology being developed in my home state of Alaska – the High-frequency Active Auroral Research Program (HAARP). When I read this first account of the project I was concerned about the implications of the technology and, quite frankly, I did not believe what I was reading about what it could do. The claims regarding this project were outrageous. The article said that with this new technology the United States military could change the weather, manipulate all communications systems, transfer energy from one point on the planet to another, effect human behavior, view several kilometers of the subsurface on the earth and be used for over-the-horizon radar, among other things. When I read something this outrageous it is my habit to confirm the information by my own research – such was the case with HAARP. Now, almost six years later, I have traveled a very different path than the one I expected in order to answer my personal curiosity about these technologies and commit a portion of my life to awaken others to the concerns the issues raised.

I have always believed that a small group of committed and focused individuals could create large movements for change. I have always believed lasting change can occur when people of heart move in the direction of positive and lasting change. Our work is an example of the fact that even the most impossible obstacles and adverse circumstances can be overcome when motivations are toward the good, toward the truth and toward justice.

Six years ago when we began these efforts we had no external financial support. My wife and I sold some business property, took a loan on our house and published my first book, co-authored with Jeane Manning, called *Angels Don't Play This HAARP,* which has been translated into German under the title *Holes In Heaven*. Since

2. USAF Scientific Advisory Board. *New World Vistas: Air And Space Power For The 21st Century – Ancillary Volume.* 1996. EPI402

the release of that book, Earthpulse has published additional books written by ten different authors. We used our commercial activity to build an activist publishing organization with the mission of producing material which would have an impact on the direction of public policy. To this end, my company raised over $1,000,000US which we committed to the HAARP issue until we were able to achieve a high level of public awareness, independent confirmation of our work and political movement on the issue, which included a resolution from the European Parliament in January, 1999.

In the last six years we have continued our research into new technologies, issues of privacy and human rights, economics, health and politics. As a result, we have compiled very significant research materials and had the opportunity to share this material, to a limited degree, with the public.

When we originally researched HAARP we compiled about four hundred source documents and used two hundred of them in that book. The book created a great deal of controversy which continues even now. What we have learned since that time makes our earlier work seem quite limited in its revelations.

Since publication of *Angels Don't Play This HAARP,* we have acquired over 6,000 documents on new technology. In recent months, my colleague, James Roderick, has indexed approximately 1,200 of these documents for use in this book, from which over 600 have been used in these pages. We believe that this current work will stun even the most skeptical readers and researchers. The source documents have been selected because they prove many of the assertions made over the years regarding technologies that will have profound impacts and, most importantly, challenge the very essence of our humanity.

Where it all Began

I remember reading the first article about HAARP in *Nexus Magazine*. It was only a few column inches long and quite unbelievable. A United States Navy and Air Force project was being created in my home state of Alaska. After reading the article I just couldn't let go of it. I could not get the issues it had raised out of my mind. Should I ignore it? After all, it was only an article in an obscure Australian publication. In thinking about the story it seemed that it would be simple to confirm the sources which were referenced in the article. Within the next few days my search for the truth on the HAARP project began.

I made an initial visit to the library and began to find the referenced documents including a number of United States patents related to the program. These confirmed the *Nexus* article and in the course of looking those up additional material was discovered. I then went to the group Trustees for Alaska, an environmental organization, which was usually on top of environmental issues in Alaska. After a brief conversation, Trustees provided me a limited amount of material while explaining it really was not something they were interested in pursuing. They suggested that I try the Audubon Society because they believed that they had followed the issue.

It was a short walk down the street and around the corner to the building housing the Audubon Society. I found a small office tucked away on the second story of an old building looking out on the courthouse in downtown Anchorage. The folks running the office had a file with about fifty pages of information related to HAARP. Not very much, but a few promising leads. In one set of papers was a letter and short article by Clare Zickuhr expressing alarm and deep concern over the project. I asked the people in the office why they were not pursuing it, "Our interest is in the disturbance of wetlands and interference with migration patterns and this really is a small project." They were focused on the physical land disturbance and really had not given much consideration to what the facility would do – they trusted the military's public pronouncements. "After all, they were operating within the regulations."

A few days later I went back to the library trying to find the rest of the original materials which were in that first *Nexus* article I had read. I tracked down the writer, a fellow Alaskan who had followed HAARP since it first started. He had previously worked for the military and understood many of the technical aspects of the project and he provided us a number of pieces of information.

I took all of what had been collected and forwarded the materials to friends with the right scientific background. They reviewed the materials and at first suggested that the system could not work as it was designed. A few days later, I sent some additional information I had found – then came the call. Gael Flanagan said, "Nick, this is real important stuff - you have to write something. This story must get out." Her husband Patrick added, "It is so simple, it will work. It is dangerous and you must write something."

"I'm no writer," I said. "I can't write this story." At the end of a long and excited conversation I agreed to try and put something together. I would forward it to *Nexus,* considering they had published the earlier story.

In less than a week it was written. A couple dozen references, a few illustrations and ten or eleven typed pages made the story. As I got ready to mail it I thought; "Is this the right thing to do?" Would giving information to these people be contrary to my government's interests? I thought, Senator Ted Stevens was a friend of my family and it was his legislation which got the US Government to fund the project. Did he know the implications? I sent a copy of the draft article to his office in Washington, D.C., thinking that if there was a problem with the article or something I had missed his staff would let me know. Weeks passed and no response was received. It was odd because never in twenty-five years had a contact made to his office by any family member of mine gone unanswered. I sent the article to the magazine.

A week later I received a call from *Nexus* publisher Duncan Roads. He wanted to run the story in his next issue. He was excited about the content and the source material. We agreed that the story would run a few weeks later.

After the story was printed, I received a number of calls including one from James Roderick, who had been operating a small public interest research organization in Homer, Alaska, about a six hour drive from Anchorage. We exchanged some information and discussed what he had been doing to oppose the HAARP project. As it turned out, he and a small group of others had put together "No HAARP" in order to get the issue out. They had been at it for a couple of years by the time I became interested in the issue. We agreed to exchange information on HAARP as I was considering writing another article.

Over the next several months I met a number of other "No HAARPers." All were growing tired of the battle. They had tried for several years to get the attention of media, politicians and others in the attempt to get the issue out to the public. They were ready to quit but glad someone else had decided to take the issue on.

In October 1994, I was contacted by Jeane Manning, a Canadian writer who had also been following the issue. I agreed to meet Jeane a few weeks later on my way to Europe to exchange materials and discuss the possibility of working together on a follow-up article. It was clear that the material was too voluminous for only an article and we agreed to work together on a book about HAARP. Less than a year later the book was in print.

I am still amazed when I look back on those early days. I never imagined that my life would take such a different turn. In the years since, I have lectured and spoken in hundreds of forums throughout the world on the subject of HAARP. We made numerous contacts and gathered incredible amounts of additional material on HAARP as well as other technologies.

I was asked to provide materials for the media, elected officials and others in our effort to take this matter to the public. Our work was featured on hundreds of radio and television programs over the last six years and eventually reached the ear of political leaders. One of those individuals, Tom Spencer, was very interested in the technology and began to make a number of inquiries of his government in Great Britain and also the United States. Mr. Spencer was at that time the Chairman of the Environmental Committee of the European Parliament. Tom was committed to the issue once he had enough evidence to show what the United States was really up to with this project.

A few months later, Mr. Spencer received a new Chairmanship – he became Chairman of the Foreign Affairs Committee of the European Parliament. At that time the issues of military expansion, environmental impacts of the military and the larger issues of European security were being debated. A hearing was scheduled for

February 5, 1998, by the subcommittee of Foreign Affairs, The Committee on Security and Disarmament, in order to get to the bottom of the issue of HAARP.

The United States was invited to send their experts, as was the Secretary General of NATO, to debate the issues and answer the inquiries of Parliamentarians. We had also asked that the agenda for the meeting be broadened to include the issue of so called "non-lethal" weapons. NATO told the committee that they had no policy on either ionospheric manipulation for weapons development (the basis of HAARP) or non-lethal weapons. The hearing was held without the United States or NATO's participation and in opening our testimony we presented documents from NATO which clearly proved their involvement in both of these areas.

The hearing went well and capped what was our fifth trip to Europe on these issues. The representatives of the United States sat in the back of the hearing room, taking notes and registering the details of the meeting. They did not speak up until the hearings were over. The evidence was overwhelming and had been delivered over the previous visits to Europe. We had made it a point to provide as much documentation as could be amassed so that our research would be in the hands of those with the power to create real initiatives on these matters. When the hearing ended in Brussels the press responded:

> "The European Parliament is protesting a U.S. research project to investigate using the Earth's ionosphere for communications with deeply submerged submarines.
>
> Leaders of the parliament, which includes representatives from all 15 European Union nations, fear the project could have global impacts on weather, the environment and health.
>
> They have vowed to take their complaints directly to the U.S. Congress if they do not get the answers they want about the High Frequency Active Auroral Research Program (HAARP) from NATO and U.S. military officials."[3]

At the conclusion of the hearing the United States sent their "damage control" specialist in to make the rounds. He went from office to office of those who had been in attendance at the meeting attempting to discredit what was presented by attacking those of us who provided testimony. This effort was met with a much different response than the one the United States expected. Europeans wanted answers to their very serious questions on the HAARP project and on these other new weapon systems. These Parliamentarians understood that when people can't deal with real issues they attempt to deflect the issue by character assassination – shooting the messengers and forgetting the message. In this case it did not work. The evidence

3. Tigner, Brooks. "Europeans Protest U.S. Ionospheric Research." *Defense News*, Feb. 2, 1998. EPI760

presented deserved a response and these Parliamentarians were unmoved by the United States' and NATO's attempt to run from the issues. They wanted answers in response to the documentation that they now had, not rhetoric.

During the following months numerous attempts were made to gain input from the United States, all without success. Almost a year later the European Parliament passed a comprehensive resolution on military issues. Included in that resolution was the following:

> "24. Considers HAARP (High Frequency Active Auroral Research Project) by virtue of its far-reaching impact on the environment to be a global concern and calls for its legal, ecological and ethical implications to be examined by an international independent body before any further research and testing; regrets the repeated refusal of the United States Administration to send anyone in person to give evidence to the public hearing or any subsequent meeting held by its component committee into the environmental and public risks connected with the HAARP program currently being funded in Alaska;
>
> 25. Requests the Scientific and Technical Options Assessment (STOW) Panel to agree to examine the scientific and technical evidence provided in all existing research findings on HAARP to assess the exact nature and degree of risk that HAARP poses to both the local and global environment and to public health generally;
>
> 26. Calls on the Commission to examine if there are environmental and public health implications of the HAARP program for Arctic Europe and to report back to Parliament with its findings;
>
> 27. Calls for an international convention introducing a global ban on all developments and deployments of weapons which might enable any form of manipulation of human beings."[4]

This was a great step forward on this issue. It also demonstrated that a few people could bring forward complex issues in a manner which would get a fair hearing, at least in Europe.

Since our beginnings in 1994 one thing has become very clear – there was a great deal more out there than HAARP when it came to exotic new weapon systems. We had succeeded in getting the issue addressed and had been able to expand the issue in section "27" so that the other areas our research had uncovered could also be dealt with in the future.

4. European Parliament. Resolution on the environment, security and foreign policy. A4-0005/99. Jan. 28, 1999. EPI159

James Roderick has worked with me in our research efforts since our very first meeting in 1994. He has been there since the beginning in compiling some of the most incredible materials put forward in this book. We can say with great confidence that what appears in these pages is only the beginning of the next exposé. As each day passes new materials continue to arrive. We believe this current work to be our most important thus far.

Traditional Values & The Roots of Peace

Another important experience which contributed to our world view and current work began two years ago while visiting Eastern Europe. In my travels, I observed circumstances similar to my own region of the world. Alaska, you see, can best be viewed as a third world country lost in a first world nation. The trip to Eastern Europe had a significant impact on me and I wrote an essay on a concept I call cooperative capitalism. Having written the essay I was invited back to Europe at the invitation of Bartholomew I, the first Patriarch of the Orthodox Church in Istanbul, Turkey and Prince Philip the Duke of Edinburgh to present my ideas during a week-long conference on the *Environment and Poverty*. The concept presented in that paper was that values and beliefs are rooted in the individual person and that those values and beliefs form the foundation of regional culture. These beliefs can be translated into sustainable development models which amplify those values in opposition to outside influences which may otherwise be exploitive. This paper was shared with my friends around the world including Gary Harrison, Traditional Chief of Chickaloon Village, a recognized American Indian Tribe based 45 miles north of my home in Alaska.

The Chief was interested in the essay because it spoke about a long view, a view that was rooted in humanity in balance with the earth and sky. In native American traditions an overriding theme is to think in terms of seven generations into the future. In that culture, the decisions made today are contemplated in terms of their impact on the future generations. The earth in spiritual terms is mother and the sky is father – provided by the Creator to humans as caretakers and as a part of the whole of creation. This world view is shared by many in Alaska, which is one of the last places in the United States where traditional lifestyles are maintained in the current generation. Moreover, after incredible suppression of religious freedom and suppression of cultural expression, this group of first people in Alaska are looking to their past to establish their futures.

Over the last year the Chief and I have discussed the possibility of working together to help actualize their potential in creating a sustainable development model. In early August this year I was asked to consider working for the Tribe as their Tribal Administrator with primary responsibility for all project operations, facilitating the realization of the vision of the Traditional Tribal

Council and creating a sustainable economic base which will secure the health and welfare of tribal members into the future. My work with the Tribe began on August 11, 1999. The opportunity to create a sustainable model on a foundation of traditional values and individual cooperation is now taking real form as these concepts are being translated into specific projects. The initial funding for these projects has already been secured through United States government support combined with private support. This tribe is quite small, representing 200 remaining tribal members who control over 69,000 acres of land and all resources which are a part of the land.

As a part of my family history, indigenous issues have always been one of our central themes. My father, while serving in the United States Congress, was responsible for the largest indigenous land settlement in the history of the United States. This legislation transferred over $900,000,000US and 44 million acres of land in Alaska to thirteen native owned corporations set up to receive the resources. Shortly after this legislation was passed by the Congress and approved by the President, my father was lost on a plane flight from Anchorage to Juneau, Alaska. He was never found. Unfortunately, with his death, the follow-up activity related to this legislation in support of Alaska's first people was never completed. The structure of the land and cash transfers did not include Traditional Tribal Governments – the repository for native Alaskan values and beliefs. The established model was western in design and was a totally foreign concept to Traditional Tribal Elders who did not fully comprehend the western systems of economics and capitalism. Moreover, these western concepts resulted in profits being the focus rather that traditional values. The results in many instances have been tragic but not hopeless.

Working with this group has enhanced my own work. The ideas held by traditional cultures which have maintained their social order for thousands of years are worthy of serious consideration as working models for modern organizations. Traditional cultures around the world have significant contributions to make which are outside of economic interactions with western governments and corporations. Another realization, now shared by many international organizations, is that the best work towards change starts with the individual, then the local community and is best projected outward at the local level rather than have values and direction imposed by others from outside of the regions. The lesson learned is that "Traditional Values" held by people around the world are substantially the same – they are rooted in peace and cooperation and not in the waging of war.

The Tip of the Iceberg

All of what has been presented thus far is background which helps explain where our heart is in terms of this and our future work. A great deal of our effort has been centered on new military technologies which today represent the greatest challenges to our personal freedom and the integrity of the individual human being. Our present and future work is centered in the areas of peace activism, sustainable development and human rights.

Earth Rising – The Revolution is written with specific objectives in mind – to stimulate debate on technologies which will either dominate our lives to our detriment or be turned to the service of humanity. Our work involves four primary areas:

1. The development and use of so called "non-lethal" weapon systems and their implications in democratic societies;

2. A full discussion of technologies which are developed and being promoted by the military which will have a profound effect on the physiological, emotional and mental health of individuals;

3. An overview and full discussion of new technologies which will have serious impacts on the world's environment, and;

4. A review of unrelated, but potentially dangerous technologies, which are either already developed or expected soon by military planners.

These four areas are documented in a way which is irrefutable and clearly shows why we are concerned. The last section of the book is solution oriented. This section reviews the technology and applicable law where these new technologies and their uses can be challenged. Included in this section are specific areas where legislation can be enacted which could, in part, safeguard the public from the misuse of these new technologies.

"Non-lethal" technologies are not really non-lethal. In fact, even the United States government indicates that they can be lethal. The idea of damaging or debilitating humans while preserving property is the objective of some of these new systems. Expressing a chilling view, the father of the H-bomb, "...Edward Teller described mini-nukes, used to bombard enemy territory and destroy roads, bridges, and communication systems, as non-lethal...Teller told the conference that if civilian populations were warned before the bombs were dropped, 'a plan of this kind could work out without a single casualty.'"[5]

The underlying issue is one of controlling the behavior of humans in a way which serves the interests of the persons using such weapons. Many of the technologies discussed in public forums appear to be benign, at first glance. Things like nets and sticky foam used to subdue an adversary are often featured, as were, until recently, blinding lasers which would destroy the eyesight of an adversary. Chemicals and electromagnetic systems which dissolve rubber tires,

5. Rothstein, Linda. "The "soft-kill" solution." *The Bulletin of the Atomic Scientists*, March/April 1994. EPI1212

interfere with the operation of vehicles or cause malfunctions in computerized equipment are often mentioned as well. What has only been disclosed in vague terms by the military are weapons which interact with our living energetic systems – living systems which are responsible for sustaining our lives and mental processes. These new weapons are unlike anything ever contemplated by mankind. These are weapon systems which pierce the very integrity of the human being.

On the surface it is easy to see why some would desire to subdue rather than kill. On the surface it seems that some "minor" violation of civil rights might be overlooked in favor of maintaining existing social orders. On the surface it might even seem desirable to some people that controlling the behavior of others, in order to achieve political ends, becomes acceptable. The ethical line has not been drawn in terms of the use of this new technology. The reason this line has not been clearly drawn is because the science is advancing at ever-accelerating rates while the proper ethical discussion is lost in the march by militaries to their new revolution. We are suggesting in our work a counterrevolution, a revolution in thinking, a revolution rooted in looking back toward those traditional values from which all of the world's cultures and societies have emerged. We propose a revolution which will awaken each of us to the reality of our shared values and hopes for future generations. We propose a revolution which is focused in peace, which holds the individual human as the central actor in achieving that peace and stands firmly against violations of the rule of law and human dignity.

The issues uncovered in our research of new technology are frightening in their implications. We have uncovered over 40 United States patents dealing with the control and manipulation of the mental and emotional states of humans. We have culled from thousands of government, academic and major media reports the evidence of these new technologies in terms of their potential affects on us. We trace the roots of this research back to the early 1950s, when major initiatives began to be made by the military to control human behavior. The early attempts used chemicals and hallucinogenics to achieve some measure of control. Then in the early 1960s the interest changed to non-chemical means for affecting behavior. By the early 1970s, within certain military and academic circles, it became clear that human behavior could be modified by the use of subtle energetic manipulations. By the 1990s, the state of the technology had been perfected to the point where emotions, thoughts, memory and thinking could be manipulated through external means.

Stop for a moment and consider the impact of what this means – the idea that human thinking can be disrupted or manipulated in a way which can not be resisted. The ability to impact thinking in this way comes through, according to a leading Russian scientist, "as if it were a commandment from God, it can not be resisted." What else could be a greater violation of our own individual personhood? These new systems do not pierce the tissue – they violate the very essence of who we are – they violate the internal and private aspects of who we are as individuals. The idea that some external force can now disrupt not only our emotional states, but our health as well, should not come as such a great surprise. What is surprising is the volume of evidence now available to us that these technologies are here now.

One of the most revealing documents we found regarding these new technologies was produced by the Scientific Advisory Board of the Air Force. The Air Force initiated a significant study to look forward into the next century and see what was possible for new weapons. One of those forecasts shockingly revealed the following:

> "One can envision the development of electromagnetic energy sources, the output of which can be pulsed, shaped, and focused, that can couple with the human body in a fashion that will allow one to prevent voluntary muscular movements, control emotions (and thus actions), produce sleep, transmit suggestions, interfere with both short-term and long-term memory, produce an experience set, and delete an experience set."[6]

Think about this for a moment – a system which can manipulate emotions, control behavior, put you to sleep, create false memories and wipe old memories clean. Realizing this was supposed to be a forecast should not cause one to believe that it is not a current issue. These systems are far from speculative. In fact, a great deal of work has already been done in this area with many systems already in existence. The paper went on to say:

> "It would also appear possible to create high fidelity speech in the human body, raising the possibility of covert suggestion and psychological direction. When a high power microwave pulse in the gigahertz range strikes the human body, a very small temperature perturbation occurs. This is associated with a sudden expansion of the slightly heated tissue. This expansion is fast enough to produce an acoustic wave. If a pulse stream is used, it should be possible to create an internal acoustic field in the 5-15 kilohertz range, which is audible. Thus, it may be possible to "talk" to selected adversaries in a fashion that would be most disturbing to them."[7]

Is it possible to talk to a person remotely by projecting a voice into his head? The authors suggest that this would be "disturbing" to the victim – what an understatement, it would be pure terror. A weapon which could intrude into the brain of an individual represents

6. USAF Scientific Advisory Board. *New World Vistas: Air And Space Power For The 21st Century - Ancillary Volume.* 1996, pp. 89-90. EPI402
7. Ibid.

a gross invasion of one's private life. The idea that these new systems will be perfected in the next several years should be cause for significant discussion and public debate.

The Mind Has No Firewall

In another article published in the Spring 1998 edition of Parameters, US Army War College Quarterly, an article by Timothy L. Thomas appeared – "The Mind Has No Firewall." The article was perhaps the most revealing in terms of what can be expected in the future.

For decades, the United States, the former Soviet Union, and others have been involved in developing sophisticated new systems for influencing human physical and mental health. The desire and focus of this research has been to discover ways of manipulating the behavior of humans to meet political ends in the context of war-making and defense. What is interesting in all of this is the sophistication of external devices which can alter our very nature. In the article "The Mind has No Firewall" the author states:

> "A recent Russian military article offered a slightly different slant to the problem, declaring that 'humanity stands on the brink of a psychotronic war' with mind and body as the focus. That article discussed Russian and international attempts to control the psycho-physical condition of man and his decision-making processes by the use of VHF-generators, 'noiseless cassettes,' and other technologies."

The article goes on to describe that the aim of these new weapons is to control or alter the psyche or interfere with the various parts of the body in such a way as to confuse or destroy the inner-body signals which keep the living system operational. The article describes the way "Information Warfare Theory" neglects the most important factor in information warfare – the human being. Militaries publicly focus on hardware and software, neglecting the human "data-processor." In the information warfare theories put forth in the past, discussion was limited to manmade systems and not the human operator. Humans were considered in information warfare scenarios only in that they could be impacted by propaganda, deceit and deception – tools recognized as part of the military mindset and arsenal. This book publicly explores a more sinister approach, an approach which must be considered in the context of basic human rights and values which are fundamentally and foundationally based on our right to think freely. The article went on:

> "Yet the body is capable not only of being deceived, manipulated, or misinformed but also shut down or destroyed – just as any other data-processing system. The data the body

> receives from external sources – such as electromagnetic, vortex, or acoustic energy waves – or creates through its own electrical or chemical stimuli can be manipulated or changed just as the data (information) in any hardware system can be altered."

The aim of any information war ultimately deals with human beings. The policy of the United States is to target all information dependent systems "whether human or automated" and the definition extends the use of these new technologies to people – as if they were just data-processing hardware.

The Parameters article went on to discuss the work of Dr. Victor Solntsev of the Baumann Technical Institute in Moscow. He insists that the human body must be viewed as an open system instead of simply as an organism or closed system. This "open system" approach has been held by many Russian researchers and others going back to at least the early 1970s, according to documents held by Earthpulse. What is interesting is that it has taken thirty years to be seen in the open literature as a credible view of reality. Dr. Solntsev goes on to suggest that a person's physical environment can cause changes within the body and mind whether stimulated by electromagnetic, gravitational, acoustic, or other stimuli.

The same Russian researcher examined the issue of "information noise" which can create a dense shield between a person and external reality. The "noise" could be created as signals, messages, images or other information with the target population the consciousness of the group or individuals. The purpose would be to overload a person so that he no longer reacts to the external stimulus or information. The overloading would serve to destabilize judgment or modify behavior.

According to Solntsev, at least one computer virus has been created which will affect a person's psyche – Russian Virus 666. This virus appears in every 25th frame of a computer's visual display where a mix of color, pulse and patterns are reported to put computer operators into trance. The subconscious perception of the display can be used to induce a heart attack or to subtly manage or change a computer operator's perceptions. This same system could be used in any television or visual broadcast.

In a July 7, 1997, U.S. News and World Report article, it was revealed that scientists were seeking specific energy patterns which could be externally applied to the body of individuals for the purpose of modifying their behavior. The article addressed some of the important public revelations about these new systems. These "revelations" represent but a small part of the story.

Since we first discovered these materials we have pieced together the bits and pieces of information that weave together a more complete story. What we now know is that the most powerful weapon systems will be those which subdue populations through the use of subtle energy fields which trick the body and mind into reacting as if the signals were normal and natural. These systems can also be used to create total disorientation in people, cause "mystery" illnesses as well as be used to induce, at a distance, heart failure or significant respiratory distress, among other things.

One of the other important areas of our current work deals with the issues of privacy. Under the guise of such thinking as, "if you don't have anything to hide, why are you concerned? This is for your own safety and protection" – comes a theme voiced by every repressive regime ever to emerge out of human interactions. Fear is being used to dismantle our system of laws and civility. International drug traffickers and terrorists are the first targets of these new technologies, based on the idea that the average person will look the other way when the rights of these suspects are considered. However, except in the most extreme situations, the first questions which should be raised deal with the idea of innocence until guilt is proven, concepts of due process of law, the right to a reasonable defense and a fair trial. Who is the guiding hand that will otherwise decide the guilt or innocence of suspected persons? Should we look the other way or insist that the values on which all democratic constitutional law is based are equally enforced? Developments on the privacy front are summarized below and dealt with in greater detail later. They include:

- The United States government, in cooperation with several other governments, now monitors all forms of electronic communications including telephone, fax transmissions and internet communications. This is done through huge computer systems that pull key phrases or words out of all of these communications and then deliver the information to numerous security agencies in the participating countries. To get around each country's domestic privacy laws another country does the spying and transfers the data to interested parties. Part of this system is called Echelon.

- It is now possible to follow the movements of any individual anywhere through their cell phone. In fact, in the United States, under the guise of locating people in emergency situations, it will soon be required that all cell phones phones have this tracking ability. It is also possible, with this new technology, to use microcircuits in these phones just like a hidden microphone to monitor the conversations of cell phone owners even when the device is switched off.

- In many urban areas micro-cameras are being installed inclusive of microphone pick-ups in order to monitor "criminal activity." These cameras are being installed in numerous major urban areas and, based on miniaturization and decreasing technology costs, it will soon be possible to literally monitor entire countries.

- Sophisticated computer systems are under development which are designed to read body language in order to interpret potential violent or other behaviors and dispatch authorities to the scene.

• The ability to look into the homes of individual people, using very high tech infrared cameras and other technology combinations, is now available to law enforcement. These technologies allow people to literally look into the bedrooms and private activity of anyone.

• The ability of televisions, through cable connections, to be both transmitter and receiver is not far away and is being tested and introduced as new burglar alarms and home safety devices.

• The technology to look through peoples clothing to find hidden weapons or drugs is now in use by some policing and military organizations.

• Roving wiretaps which involve the idea that whole areas of cities can be tapped in order to find a single person who might be engaged in criminal activity. It is the same logic behind collecting DNA samples to prove an individual's innocence or getting fingerprinted at the bank in order to cash a check – all are gradual intrusions into privacy.

• Genetic predisposition-to-violence studies which involve the concept that certain genetic markers can be identified which show who is likely to exhibit criminal behavior are now ongoing. Once identified, the individual can be medicated with mood altering drugs in advance of such anticipated behavior. During the Bush administration, this program was almost implemented in urban school systems in the United States. The $400,000,000US initiative was stopped when those engaged in real science began to object. Now, nine years later, this issue is again surfacing, the interesting point being that the same indicators in terms of personality for criminals are the same for non-conforming elements of society which produce the leadership within countries.

• Mandatory bank transaction reports including "reporting suspicious activity" are also being initiated in a number of countries around the world.

These are but a few of the things that are happening now and are presented here only to give indications of the general direction of these invasive technologies. A great deal more will be presented in succeeding chapters.

Earth War?

In other areas of our work we have discovered that manipulation of the environment is becoming increasingly an area of focus for military organizations. There has been a good deal of speculation about the possibilities of creating artificial weather and of controlling the weather. This it not new and has been the subject of ongoing military research for decades. Moreover, in 1977, the United States signed a treaty with over sixty other countries calling for a ban on "geophysical warfare." This refers to manipulation of the environment for military advantage.

The use of new weapons is not limited to governments and sophisticated science laboratories. In April 1997, the United States Secretary of Defense, William Cohen, made the following comment:

> "Others are engaging even in an eco-type of terrorism whereby they can alter climate, set off earthquakes, volcanoes remotely through the use of electromagnetic waves."[8]

This is not new either but has its roots in 1960-70s era research by American scientists and continues to appear in numerous articles and reports. The idea of creating artificial weather, including cyclones, is being explored. In a recent article in the Wall Street Journal it was reported that "a Malaysian company, BioCure Sdn. Bhd., will sign a memorandum of understanding soon with a government-owned Russian party to produce the Cyclone."[9] The deal with the Russians was set up so that if the technology did not work the Malaysians did not have to pay for the attempt. There have been other reports of Russian research in this area.

More recently, Dr. Bernard Eastlund, the inventor of the HAARP technology, has completed work sponsored by the European Space Agency which demonstrates the ability of HAARP-type instruments to control and manipulate weather systems including tornadoes and hurricanes. We have compiled a large amount of evidence about the state of this technology and can demonstrate that these capabilities exist. Unfortunately, the 1977 treaty allows for domestic testing and use of these technologies. These treaties are now in need of significant revision to take into consideration the advances in technology and the fact that one can not change weather in one location without profound effects around the world.

Star Trek & Star Wars

The development of new energy weapons has occupied the imaginations and resources of our national and private laboratories. In a 1989 patent a most interesting bit of this science is revealed. One such weapon idea is owned by the United States Department of Energy. It is a new kind of weapon which allows electromagnetic or acoustic energy to be focused into a tight package of energy which can be projected over great distances without dissipating. When scientists think of this energy being projected through the air it is

8. DoD News Briefing, Secretary of Defense William S. Cohen. Cohen's keystone address at the Conference on Terrorism, Weapons of Mass Destruction, and U.S. Strategy. April 28, 1997. EPI317
9. *Wall Street Journal.* "Malaysia to Battle Smog With Cyclones." Nov. 13, 1997, p. A19 EPI322

always assumed that the energy would dissipate, dispersing at such a rapid rate that no weapon's effect could be realized. What has been discovered is that there is a way to create this system. In a U. S. patent the following summary appears:

> "The invention relates generally to transmission of pulses of energy, and more particularly to the propagation of localized pulses of electromagnetic or acoustic energy over long distances without divergence."

> "As the Klingon battle cruiser attacks the Starship Enterprise, Captain Kirk commands "Fire photon torpedoes." Two darts or blobs of light speed toward their target to destroy the enemy spaceship. Stardate 1989, Star Trek reruns, or 3189, somewhere in intergalactic space. Fantasy or reality. The ability to launch localized packets of light or other energy which do not diverge as they travel great distances through space may incredibly be at hand."[10]

The patent describes the energy effect as "electromagnetic missiles or bullets" which could destroy almost any object in their path. These emanations would travel near the speed of light destroying anything in their way.

Disturbing Advances

New energy weapons have been described as being capable of creating symptoms of sea sickness, can be used to resonate the inner organs to cause pain and spasms, induce epileptic-like seizures or cause cardiac arrest. Other weapons include, according to our research, those which cause or prevent sleep, override voluntary muscle movements or otherwise affect the brain. For example, 100,000 units of the "Black Widow," which overrides muscle movement, were added to the Russian government's arsenal in recent years.

The term 'psycho-terrorism' was created by Russian writer N. Anisimov of the Moscow Anti-Psychotronic Center. He indicates that psychotronic weapons can be used to take away part of the information which is stored in a person's brain and send it to a computer which reworks it to the level needed to control the person. The modified information is then reinserted into the person's brain and thought by them to be their own information. These systems are then able to induce hallucinations, sickness, mutations in human cells, zombification or even death. These technologies include VHF generators, X-rays, ultrasound and radio waves. Russian army Major I. Chernishev described in the military journal *Orienteer* (February 1997), how "psy" weapons are under development all over the globe.

10. US Patent #4,959,559, Sept. 25, 1990. Electromagnetic or Other Directed Energy Pulse Launcher. Inventor: Ziolkowski, Richard W. Assignee: United States of America as represented by the United States Department of Energy. EPI2

Specific types of weapons he noted were:

• A psychotronic generator which produces a powerful electromagnetic emanation capable of being sent through telephone lines, TV, radio networks, supply pipes and incandescent lamps. This signal would manipulate behavior of those in contact with the signal.

• A signal generator that operates in the 10-150 Hertz band which when operating in the 10-20 Hertz range creates an infrasonic oscillation that is destructive to all living organisms.

• A nervous system generator which is designed to paralyze the central nervous systems of insects. This same system is being refined to have the same effect on humans.

• Ultrasonic signals of very specific design have been created. These devices are supposedly capable of carrying out bloodless internal operations without leaving a mark on the skin. They can also be used to kill.

• Noiseless cassettes have been developed by the Japanese which has given them the ability to place infra-low frequency voice patterns over music, patterns that are detected by the subconscious. The Russians claim to be using similar "bombardments" with computer programming to treat alcoholism and smoking.

• The 25th-frame effect discussed above is a technique where every 25th frame of a movie reel or video footage contains a message that is picked up by the subconscious so as to alter the conscious mind.

• Psychotropics are defined as medical preparations used to induce a trance, euphoria, or depression. These are referred to as "slow-acting mines." Symptoms could include headaches, noises, voices or commands in the brain, dizziness, pain in the abdominal cavities, cardiac arrhythmia, or even the destruction of the cardiovascular system.

What is written here is the tip of a very large iceberg. These bits of information are intended to draw your attention to the state of the technology and where we see it going. Our conclusions are not based on speculation but, rather, are based on the facts presented by military and academic researchers from the United States and around the world. It is our sincere hope that our current work will represent the necessary stimulus to keep the momentum of debate fresh and alive so that we, together, can begin to create the changes which are necessary to assure that human values are at the center of all technological developments and deployments. We believe that we can maintain the momentum for change.

One of the roots of the problem deals with free flowing debate of weapons concepts. We believe that the greatest issue moving against humanity is the "secrecy syndrome" that our governments have succumbed to since the end of World War II. This government paranoia is the enemy of freedom. We need to recognize that within democratic institutions the requirement to disclose weapons and technology concepts should be a matter for public debate. We do not have to tell people how to build these systems but we must be able to debate these new concepts.

The Revolution Begins

Sometimes the word "revolution" brings to mind acts of violence and social disorder. I suggest a different vision rooted in human expression that in our lifetimes we are just now beginning to see. We have seen the spirit of humanity rise to its highest level in the largely peaceful revolutions of India, eastern Europe, the Philippines and numerous other regions where human beings have begun to recognize their mutual obligations to one another, the natural world and our individual concepts of the Creator and creation itself. In democracies throughout the world, individuals are expressing their discontent with existing governments which are viewed as serving interests which are not shared by the general population. These western governments have, in many instances, missed the mark in translating human values into political outcomes. As a result, the governments of Japan, Germany, France, Great Britain and countless others are being transformed and reshaped. Throughout the world the influence of the United States has interfered with the established cultures by exporting to them those elements most civilized people find objectionable.

The revolution was never formally announced but it has begun. The new thinking which is developing challenges existing power systems and seeks greater human consideration. I am pleased to be a small part of this revolution because embedded within it is a great opportunity to change the essence of our living experience.

In all of our travels we find it quite remarkable that so many very different people, in very diverse places, are seeing the same problems and seeking similar solutions. Communication between individuals is becoming increasingly possible as each of us now possesses the ability to reach millions of others with our ideas through the Internet as well as other communications systems. It is interesting to note that the Internet itself was designed by the military to further its own ends and it has now become one of the leading tools for a global revolution in thinking and for the democratization of the world by empowering individuals in ways never anticipated by military planners. The revolution begins with each of us translating our ideas into action and recognizing that we are not alone in this effort.

We often hear the cry that apathy is our enemy, the cry that we are alone and there is no leadership to make the changes which many of us see as necessary.

This is a leaderless revolution. One of the things we have learned from studying the military approaches to controlling populations is the concept of "leaderless revolution." The greatest fear of any established order is that some revolutions appear to be spontaneous expressions, as occured, to a large extent, throughout eastern Europe. Normally, the way revolutions are squashed by those in power is through access to the leadership of those challenging movements. Those in power neutralize the leadership through economic isolation, social isolation via scandal, misinformation or even worse. On the other hand, a "leaderless revolution" can not be stopped as there is no head or tail – there are only the individual actions of many people with similar ideas expressed in their daily lives and activities. We, through our efforts in the HAARP debate, have demonstrated that individuals can achieve a large measure of success even beginning with limited resources. We have always moved in the knowing, in the absolute understanding that through peaceful means great change is possible. In each challenge which we have faced we have come to an even greater conviction that there is incredible power in action itself. Action is the mysterious step of faith.

Faith is taking what is known internally and acting on this internal knowledge in our outward expressions of action in the material world. We externalize our inner ideas into action which creates other interactions which flow toward the creation of new realities. Faith is the actualization of what we know to be right and true. Trying, attempting and working toward change is all that is really required. We do not need grand plans or great detailed effort. We need only to take those steps which are in front of each of us, trusting that millions of others are doing the same. We are not alone but are fully united in the direction of what is true and worthy of human action. We are not required to win every battle. It is not a matter of winning and losing but, rather, of trying. The real power is in actions themselves – It is in the doing.

We know that there are no challenges too great. There are no obstacles too large. There are no thresholds of peace which we cannot cross. To the good, and toward the light of truth, we invite each to follow his own conscience in creating a world that seven generations from now will reflect the true vitality of the human spirit. We are each created in the image of the Creator, with but one objective, to reclaim our birthright as created beings holding the responsibility of stewardship of the planet. We are not passive participants. We are co-creators of the reality around us. Let us each make sure that the reality we create reflects the highest values of the human spirit.

Chapter 2

HAARP & The ABM Treaty

"Wherever the military has been in Alaska it has fouled the nest."

Dan O'Neill[11]

In February 1998, while testifying before the European Parliament, I indicated that the United States would violate the Anti-Ballistics Missile (ABM) Treaty with the former Soviet Union. I asserted that the United States would take the position that because the Soviet Union had dissolved, the agreement was no longer valid. On this point, the Members of the Committee could not agree with us because most believed that the United States would honor the agreement, recognizing the Treaty as a stabilizing document for the entire world. In less than ten months from our pronouncement, the United States announced their new initiative for a missile defense system outside of the Treaty. United States Senator Ted Stevens made the arguments just as predicted.

The original ABM Treaty allowed only one defense system in each country – the United States and the Soviet Union. It had been our contention that HAARP represented research which would lead to the eventual deployment of a new missile defense system based in Alaska. We also made clear that the United States was moving forward with new technologies while giving the world, and our allies, the impression that there was a real effort underway toward reductions in arms. The reality, time would demonstrate, was a new race toward the new technology.

What Ever Happened to Star Wars?

Remember Star Wars, the concept which would move the theater of war to space? In 1995, the funding for Star Wars was widely reported as a dead issue when full funding was defeated by the United States Congress. However, Star Wars did not end as many unpopular programs do – they just get new names.

11. AP. "Missile hearing draws support, some critics." *Anchorage Daily News,* Nov. 3, 1999. EPI1162

> "This year the Ballistic Missile Defense Organization (once called the Strategic Defense Initiative) got $3.7 billion. That's up from $2.8 billion in 1995, and is very near the peak level spent during the Cold War."12

The problem was that the United States was not fully covered under its single ABM system, leaving Alaska and Hawaii unprotected. Alaska produces about 22% of U.S. oil requirements, holds all 33 strategic minerals in commercial quantities and represents tremendous natural resource wealth. Hawaii is a strategic point for the entire Pacific presence of the United States military. These two areas, in the estimation of military planners, had to be protected. The need for a system to protect Alaska and Hawaii is acknowledged by the authors of *Earth Rising – The Revolution* and we agree that these regions should be protected. We believe that this could be accomplished in a more open and honest way, particularly when it affects our allies and the American public.

What is interesting is that the billions spent on Star Wars systems, which these became known as, were only for "research," according to the military's mission statements. The technology is being advanced in the hope that a system might be developed early in the next century. The external threats are now being characterized as "rogue states and terrorist organizations" which might gain nuclear weapon delivery technologies. While the threats are not imagined and need to be addressed, it is not responsible to create word games which end public debate and allow systems thought to be discontinued the latitude to proceed.

In another "offshoot of the Reagan administration's Strategic Defense Initiative" satellite-disabling lasers have been developed. A test, at less than full power, was performed at the end of 1997 to demonstrate the ability of the system to hit its target. The demonstration was a success and now many are concerned that this may provoke an arms race in space."13

What has become clear is that what the United States says to its citizens, friends and adversaries tends to be variations of the same misinformation and manipulations of the facts.

HAARP

This is where our story began. HAARP – The High Frequency Active Auroral Research Program is a joint project of the United States Air Force and Navy, based in Alaska. The project is a research program designed to study the ionosphere in order to develop new weapons technology.

The HAARP system is designed to manipulate the ionosphere, a layer which begins about thirty miles above the earth. The transmitter or HAARP device on the ground is a phased array antenna system – a large field of antennas designed to work together in

12. Fox, Adrienne. "Star Wars: Force Not With Us, US Remains Defenseless Against Missile Attack." *Investor's Business Daily*, Aug. 25, 1997. EPI150
13. Richter, Paul. "Army laser zaps satellite." *Anchorage Daily News*, Oct. 21, 1997. EPI426

focusing radio-frequency energy for manipulating the ionosphere. The radio frequency energy can be pulsed, shaped, and altered in ways never possible with other transmitters or at the power levels of this new facility.

The device will have an effective-radiated-power of 1 billion watts when completed in the first phase of the project. It will be used for "earth penetrating tomography" (looking through layers of the earth to locate underground facilities or minerals), communications with submarines, to manipulate communication of others, over-the-horizon radar, energy transfers from one part of the world to another, creating artificial plasma (energy) layers or patches in the ionosphere, to alter weather and may be used as an anti-satellite weapon. The story of HAARP and its related technology is a part of the puzzle of a "Revolution in Military Affairs" being pursued by the United States government, but it is only a small part.

My earlier book, which was co-authored with Jeane Manning, *Angels Don't Play This HAARP: Advances in Tesla Technology,* provides greater detail on the HAARP program. This current work grew out of our interest in these new technologies and the need to put some of the pieces together for greater consideration by the public.

"Millions of dollars have gone to Alaska universities to harness the power of the aurora borealis, the electrical energy shimmering in the northern skies... The earmarks also financed, among other things, 'a supercomputer north of the Arctic Circle' without naming the University of Alaska Fairbanks as the beneficiary."[14] Early on in the HAARP project there was confusion about "harnessing the energy of the aurora" (the northern lights) for use by the military. The real use appears to be for over-the-horizon effects, communications and ABM defense-related research. The University of Alaska again is the recipient of a supercomputer for "academic" research. The reality once again is that the University is being used to conduct military experiments for advanced weapon systems without full public disclosure or debate.

The National Science Foundation has initiated a number of studies including the Ionospheric Interactions Initiative. This initiative is intended to explore the following:

> "Processes of interest include (but are not limited to): Langmuir turbulence; electron acceleration, including the production of optical and IR emissions; the generation, maintenance and/or suppression of field-aligned plasma structures; the modulation of ionospheric currents to generate ULF/ELF/VLF radio waves in the ionosphere; and other nonlinear processes such as the production of stimulated electromagnetic emissions, plasma waves, etc., in the ionosphere."[15]

14. Weiner, Tim. "Universities tap Congress' deep pockets." *Anchorage Daily News*, Aug. 24, 1999. EPI563
15. National Science Foundation. Proposals for the Ionospheric Interactions Initiative, NSF 97-61. Feb. 17, 1997. EPI726

What is being sought in this initiative is what is taking place at HAARP. These experiments are providing the information needed to create systems which can gain control of the ionosphere. This knowledge is critical to the advance of communications technology, missile defense and other important technologies. "HAARP is a scientific endeavor aimed at studying the properties and behavior of the ionosphere, with particular emphasis on being able to understand and use it to enhance communications and surveillance systems (e.g., over-the-horizon) for both civilian and defense purposes."[16] The military summarizes their work through 1997 as follows:

"High Frequency Active Auroral Research program: HAARP – USAF
FY 90 Special Interest Congressional Program

• '...to support projects examining the exploitation of auroral ionospheric phenomena for improved military and submarine communications programs, and high energy technology applications.'

- Year-to-Year Funding Continuous Since Inception
- FY90 ($10M) funds to Director, Defense Research and Engineering (DDR&E)
- FY91–FY97 funding various AF Program Elements/Counterproliferation

- Program Jointly Managed by the Air Force and Navy
- AF included in program with Navy and ARPA per direction of DDR&E
- Execution assigned to Phillips Laboratory and Office of Naval Research
- PL/ONR Memorandum of Understanding established (Mar 90)
- Program status and needs provided annually at the request of Congress

HAARP Objectives – USAF

- Characterize the processes Triggered in the atmosphere and Space by High Power Radio Waves
- Understand the physical mechanisms that drive the processes
- Develop techniques to control / exploit selected processes
- Conduct Pioneering Experiments to Assess the Potential of Emerging Ionospheric Technology for DoD Uses
- Insure technology superiority
- Avoid technological surprises
- Support Other National Research Programs of Interest to the DoD, Such As the National Space Weather Initiative

16. U.S. Army (Tecom). HAARP. http://www.tecom.army.mil/tts/1997/proceed/abarnes/sid006.htm
EPI440

HAARP: Assessment of the Viability of New Concepts for Potential DoD Application – USAF

- **ULF/ELF/VLF Wave Generation in the Ionosphere**
- Communications
- Imaging of underground structures/tunnels/etc.
- Radiation belt control

- **Generation of Ionospheric Irregularities**
- Controlled scattering of radio waves (HF ducting, signal deflection)
- Artificial generation/suppression of radio scintillation effects

- **Creation of Energetic Electron Effects in Space**
- IR / Optical emissions to calibrate/confuse space sensors
- Spacecraft charging/safety

HAARP Appropriations SUMMARY: FY90 - FY97

FY90:	**OSD**	**$ 10.0M**
FY91:	**AF/STP**	**9.0M**
FY92:	**AF/STP**	**5.0M**
	PL/GP	**24.0M**
FY93	**PL/GP**	**2.5M**
FY94	**PL/GP**	**2.5M**
FY95	**PL/GP**	**5.0M**
FY96	**OSD**	**10.0M**
	PL/GP	**5.0M**
	AFOSR	**3.0M**
FY97	**OSD**	**7.5M**
	PL/GP	**7.5M**
FY90-97	**TOTAL:**	**$91.0M**[17]

Dr. Bernard Eastlund developed a number of innovations which were used in creating the HAARP program which required, "A large source of energy, such as available in a large oil or gas field, or from nuclear power, could be used to produce electricity which would then be used to generate electromagnetic waves in the RF (radio frequency) region (from 1.5 to 7 MHZ in this study). The RF waves were then to be focused by a large phased array antenna at points in the ionosphere at altitudes of 150 km and above to produce local field strengths high enough to accelerate electrons to relativistic energies."[18] This vision of Dr. Eastlund's was realized by the

17. Presentation for Conference on Business and Economic Development in Alaska. Air Force Acquisition in the 21st Century. Art Money, Assistant Secretary of the Air Force (Acquisition), 27 June, 1997. EPI16
18. Eastlund, Dr. Bernard J. "Applications Of In Situ Generated Relativistic Electrons In The Ionosphere." Eastlund Scientific Enterprises Corporation. Dec. 13, 1990. EPI1141

development of the HAARP transmitter in Alaska. Building on this concept, Dr. Dennis Papadopoulos, one of those brought into the project by Dr. Eastlund, invented "A method and apparatus for causing interruption in the ionospheric electrojet to produce ULF/ELF/VLF signals enhances low frequency communications capabilities. A high power transmitter heats ionospheric electrons to enhance the electron-neutron collision rate with an antenna beam that can be modulated in the ELF and VLF frequency ranges."[19] It was this invention which allowed Dr. Papadopoulos to prove that the "earth-penetrating tomography" uses of HAARP would work and be effective. Tests were successfully completed on this application in 1996.

DARPA didn't waste any time taking advantage of the experimental results from HAARP. The following appeared in 1997:

> "We propose to provide a numerical capability to map the ELF/VLF electric and magnetic fields above the ground, as perturbed by threat underground structures of interest to the DoD. MATLAB source code will be delivered that will produce these perturbed fields as a function of radio frequency, as well as values for the unperturbed background field and the external noise. The latter two signals are the primary limiting factors to the sensitivity of any low-frequency surveillance system. Two sources will be employed in the calculations of the perturbed fields; 1) distant HAARP/HIPAS low-frequency modulated signals; 2) a local source, for example from a coil near the surveillance area. This phenomenological capability is sorely lacking at present, as various sensor and surveillance concepts are being designed and offered for subsurface monitoring. There is little basis for assessing feasibility or performance effectiveness among proposed candidates. The MATLAB package we will deliver will be comprised of extensive tools our company already has in place, modified and adapted to handle subsurface DoD threat objects and the above two radio signal sources. Example calculations will be provided to demonstrate how the code is used, and to give feasibility estimates for the most obvious system implementations of such sensors. The output of Phase I will form the basis of the Phase II effort to: a) design, model, prototype, and test and ELF/VLF sensor, and b) use ELF/VLF field maps to examine the methodology for inverting/extracting information about underground structures."[20]

In earlier work conducted by the Navy a patent was developed for creating enhanced communications. "A natural gigantic loop antenna is created by the earth's magnetosphere in combination with

19. US Patent #5,053,783, Oct. 1, 1991. High Power Low Frequency Communications By Ionospheric Modification. Inventor: Papadopoulos, Dennis. EPI1117
20. CODAR Ocean Sensors, Ltd. ELF/VLF Electromagnetic Detection and Characterization of Deeply Buried targets. Topic# DARPA 97-088. EPI1200

a satellite having long insulated conductors ending in a metallic plate. The satellite may be equipped with a transmitter and relay equipment capable of transmitting in the Extra Low and Ultra Low frequency range. These waves are transmitted efficiently by the huge antenna and may be received above, on, or penetrate below the surface of the earth. Such waves may be used to carry communications or for other purposes such as transmission between a satellite and a submarine."[21] These kinds of mechanisms have been replaced by the more versatile ground-based technology available through HAARP.

Additional tests were planned for HAARP to explore a number of applications in 1999, including communications and earth penetrating tomography. "The HAARP project will be conducting a "winter campaign" from February 8th to 28th. According to the project, "Emphasis will be placed on the joint or complementary operation of the two ionospheric modification facilities and optical auroral radio frequency measurements.' HAARP will have an expected operating power of 960 KW for the campaign. A series of ELF measurements are also planned for the spring."[22] This test was later moved into March.

What are the Risks?

Several significant risks were identified in the course of our research on HAARP, including the possibility that these experiments could have devastating effects on our environment and health. One of the most significant risks was identified by Brooks Agnew, a specialist in earth-penetrating tomography. He wrote that:

> "The high energy Active Auroral Research Project (HAARP) is designed to test the effects of billions of watts of concentrated radio energy on the ionosphere of our planet. Extremely high altitude lenses, formed with billions of cubic yards of ionospheric particles, are being used to focus radio carrier waves onto the surface of our planet for military purposes. From my research I know that if the correct frequency harmonic for that carrier is chosen randomly, the result will be an absolutely catastrophic release of pure energy. The sky would literally appear to burn. What cannot be calculated or imagined is how hot the fire would be, or how long it would burn, or what elements or isotopes would remain after the reaction is complete."[23]

What Mr. Agnew suggests is a worst-case scenario, based on his research into these issues and his field tests of similar technology

21. US Patent #3,866,231, Feb. 11, 1975. Satellite Transmitter Of ULF Electromagnetic Waves. Inventor: Kelly, Francis J. Assignee: The United States of America as represented by the Secretary of the Navy. EPI1118
22. *Copper River Country Journal.* "HAARP Project Has Campaign Coming Up." Jan. 7, 1999. EPI362
23. Agnew, Brooks. "Radio Tomography Of Geological Strata." EPI789

on a much smaller scale than that which is anticipated by the HAARP project planners.

In the course of the ongoing debate with the military about HAARP, one of the major issues in dispute has to do with the amount of power needed to create some of the various effects anticipated by the military. A very important aspect of the power debate has to do with the "pulsing or bursting" of energy from the transmitters. We have maintained that short, high-power, high-frequency bursts of energy can be used to trigger larger releases of energy. In the case of HAARP, remember that the first phase seeks an effective-radiated-power level of one billion watts. One of the applications of the HAARP program has been the creation of what is known as an electromagnetic pulse. This is a energy pulse which is capable of interfering with electronic devices such as computers, airplane navigation systems, communications systems, power grids and virtually all systems dependent on electronics to operate. In other words, just about everything modern societies rely on. What is possible with pulsed energy of this type is illustrated below:

> "It could take the energy contained in two car batteries, compress it and release it in 150 billionths of a second, thus exceeding (in those 150 nanoseconds) the electrical generating capacity of the United States by about 20 times.
>
> In so doing, Aurora simulated the high-energy gamma rays that could scramble the electronic controls of everything from toaster ovens to guided missiles.[24]

Since the beginning of the controversy surrounding this technology, a great deal of information has continued to come forth from scientists around the world. The implication that this technology is to be used in conjunction with other systems is a logical deduction for those familiar with the science. Dr. Rosalie Bertell, a medical doctor and physicist, has put forward the following with respect to this technology:

> "It would be rash to assume that HAARP is an isolated experiment which would not be expanded. It is related to fifty years of intensive and increasingly destructive programs to understand and control the upper atmosphere.
>
> It would be rash not to associate HAARP with the space laboratory construction which is separately being planned by the United States. HAARP is an integral part of a long history of space research and development of a deliberate military nature.
>
> The military implications of combining these projects are alarming. Basic to this project is control of communications, both disruption and reliability in hostile environments. The power wielded by such control is obvious.

24. Hampson, Rick. "Cold War Doomsday Machine Meets Own Doom." *Los Angeles Times*, March 24, 1996. EPI433

> The ability of the HAARP/Spacelab combination to deliver a very large amount of energy, comparable to a nuclear bomb, anywhere on earth via laser and particle beam, are frightening.
>
> The project is likely to be "sold" to the public as a space shield against incoming weapons, or, for the more gullible, a device for repairing the ozone layer."25

Other risks which have been identified are enumerated in our earlier work, *Angels Don't Play This HAARP.*26 These include disruption of communications, electronics, human health, migration patterns of animals, weather changes, runaway effects on the ionosphere and several others.

In the beginning of our work in opposition to HAARP we suggested that the government would exceed the power levels and other perimeters established by the regulatory authorities. We suggested that when this occurred the contractor would likely be blamed and forgiveness would be asked. This is what did happen during the summer of 1998. One of the leading HAARP scientists issued the following statement when this situation actually transpired:

> "We have established specific procedures by which the contractor is to notify the government, in advance, when such on-the-air engineering tests are required. On this occasion, not only was such notification not received, but the contractor chose frequencies, modulation types and power levels that would not have been permitted under any circumstances. (In no case are transmissions within the amateur radio allocations ever established)."27

This also serves to illustrate one of the other points we have made from the beginning. We had suggested that the capability of HAARP was well beyond the regulatory perimeters which had been established. We said that when the military felt it was necessary, or it served "national security" interests, or some "accident" happened, these limits would be exceeded and the device tested. How many more accidents or tests thought to be disallowed will occur before people realize that the government will test this device's full capabilities regardless of the risks to the public?

Fly by Wire

"The world has changed significantly for air travellers in the 1990s. New generation aircraft, such as the Airbus A319/320/321/330/340 series and the Boeing B777, are now in service. These aircraft are 'fly-by-wire' – their primary flight control is achieved through

25. Bertell, Rosalie, PhD, GNSH. Background of the HAARP Project. July 1, 1996. EPI725
26. Begich, Nick, M.D. and Jeane Manning. *Angels Don't Play This HAARP.* Earthpulse Press Incorporated, Sept. 1995. EPI1229
27. Kennedy, Edward J. "The Interference Was Unintentional." Letter to the Editor, *Copper River Country Journal*, June 18th, 1998. EPI717

computers. The basic maneuvering commands which a pilot gives to one of these aircraft through the flight controls is transformed into a series of inputs to a computer, which calculates how the physical flight controls shall be displaced to achieve a maneuver, in the context of the current flight environment."28 These kinds of fly-by-wire systems are very susceptible to radio interference as well as other wireless interference, which is one of the reasons we can not use radios, televisions, cell phones and the like on aircraft and no electronic devices at all during take-offs and landings.

"In Alaska, pilots are concerned with EMI possibly created by a proposed High-Frequency Active Auroral research Program (HAARP). HAARP research would use high-powered transmissions to study ionospheric processes. The fear is that the transmissions could adversely affect aircraft navigation and communication, in addition to fly-by-wire flight controls."29 This particular concern was acknowledged by the military and some measures were instituted for the obvious safety reasons. Nonetheless, a part of our earliest discovery of the HAARP system was its potential use as an electromagnetic pulse weapon (EMP). These weapons are intended to deliberately cause this effect on electronic systems such as these. By their own press releases the military acknowledges their interest in these matters. As was reported in *Aviation Week & Space Technology*, "The U.S. Army, like the Air Force, is developing weapons that generate electromagnetic pulses (EMP) to disable aircraft and vehicles. One set of targets is the 'fly-by-wire' systems of electronic sensors and computers that keep aircraft stable."30 Since that article was written, these systems have been perfected.

Liquid Mirrors

The addition of what are called liquid mirror telescopes is an interesting addition to the HAARP mix of technology. In this case the so-called "telescope" is used in conjunction with the earlier pulsing activities. These pulsations cause the ionosphere to vibrate like a string on a piano. Through pulsing the high-frequency signal, what occurs is that the ionosphere begins to vibrate at the pulse rate. This causes the ionosphere to release energy in the form of a returning signal which emanates from the ionosphere at a frequency which is the same as the pulse rate. Another way of seeing this is that the instrument on the ground stimulates the ionosphere in a way which makes it functionally part of the machine on the ground. It is a component of the machine connected by the energy beam of the device instead of some other physical connection. The use of the mirrors in conjunction with HAARP is best described by their own personnel, who said: "Although astronomy and upper-atmospheric research are the

28. Ladkin, Peter. "Computer-Related Incidents and Accidents with Commercial Airplanes." Technische Fakultat, Universitat Bielefeld, Germany, March 6, 1996. EPI460
29. Herskovitz, Don. "Killing Them Softly." *Journal of Electronic Defense*, Aug. 1993. EPI697
30. *Aviation Week & Space Technology*. "Army Prepares For Non-Lethal Combat." May 24, 1993. EPI699

immediate beneficiaries of a new liquid-mirror telescope in Alaska, experiments using the telescope may lead to new military techniques for signaling submarines and probing battlefields for secret tunnels and bunkers."[31] This particular device was again provided as only a "research tool" as if this somehow separated it from how the research would most likely be used in creating technological applications – weapons.

Wireless Energy

One of the other uses of HAARP deals with the transfer of energy from the ground to space platforms and other low orbiting devices. The possibility of using this type of instrument was illustrated in the following press reports:

> "An experimental airplane fueled by microwaves transmitted from Earth will take off in Canada this summer: in theory, it could stay aloft for months. By the next century, spacecraft could routinely rise to orbit on engines powered by energy beams, lifting large payloads at a tiny fraction of current cost."[32]

> "A High Altitude Long Endurance (HALE) platform hovering more than 20,000 meters (66,000 feet) over the city of Boston could send and receive millions of telecom signals simultaneously for the metropolitan area. It is one of several types of systems that could provide the wireless infrastructure of the next century."[33]

Wireless energy transmission has been a long sought goal in science. One might wonder what happens to those objects which get in the way of the beam. If the objects contain any sophisticated electronics the results could be disastrous. In the case of living things, we assert, based on the evidence presented in later chapters, that humans and other living things can be severely impacted as well.

Totally Open Project?

In Alaska there was a high degree of openness when the project began. This openness has gradually decreased as greater world scrutiny is focused on the project. Early on, in Alaska, the State legislature held a hearing on HAARP. The following was reported in Alaska's press:

> "Now the project is attracting legislative scrutiny.
> Rep. Gene Kubina, D-Valdez, who asked for the hearing after receiving more than 40 calls and letters about the project, said he wants to know more about oversight.

31. Browne, Malcom W. "Scope System Also Offers a Tool For Submarines and Soldiers." *New York Times*, Nov. 21, 1995. EPI719
32. Broad, William J. "Beam me up, Orville." *Anchorage Daily News*, Oct. 9, 1999. EPI324
33. Evans, John V. "New Satellites for Personal Communications." *Scientific American*, Vol. 278, No. 4, April 1998. EPI769

> 'Right now, there's so much fear,' he said. "People are scared of the government lying to them. I don't want to play into anyone's hysteria. I just want people to have a place to get questions answered."34

The hearing was interesting and effective in that it provided an opportunity to discover the major points of disagreement with the military and academics on the other side of the issue. The military had restrictions placed on their testimony which appeared to be imposed by Senator Steven's office. We had been lead to believe that there would be a series of hearings so that the full story could be told. After the hearing was held, the Chairperson of the committee, Representative James, was advised and pressured into not holding additional hearings. Unfortunately, this prevented us from answering the points in disagreement until the issue eventually reached the European Parliament.

Presently, the project managers are reluctant to provide interviews for media organizations. In the last year both Fuji TV and Spiegel TV have been refused interviews at the site, being told that there was no one available to meet them at the facility. Yet, when representatives from these organizations went to the site, there were full scientific teams available who could have answered their questions but were forbidden from doing so.

SuperDARN

In related developments around Alaska and the world, a new system of phased array antennas are being erected, with the official story being quite similar to the stories the military was releasing about HAARP. "These antennas have the capability of changing their frequencies and their pulse rates, she said. 'And it will be open to other scientists around the world...anyone doing ionospheric studies can use the antennas.'"35 They said that, "SuperDARN is an international HF radar network designed to measure global-scale magnetospheric convection by observing plasma motion in the Earth's upper atmosphere."36 The scientists were insisting that the facility would be used to measure natural phenomena in the ionosphere while failing to acknowledge that it would also be used to measure man-made effects or perhaps even be involved in creating them. What they were saying was that, "Weather in space affects modern communications technology. Solar activity causes turbulence in the ionosphere that can cause disruptions in satellite and communication systems, Bristow says. SuperDARN's radars will help scientists understand the activity in the ionosphere, a region roughly

34. Komarnitsky, S.J. "Antenna fears grab attention of lawmakers." *Anchorage Daily News,* April 3, 1996. EPI803
35. Jeffrey, Sue. "SuperDARN Critics Appeal for More Time To Study Project." Kodiak Daily Mirror, Vol. 58, No. 38, Feb. 24, 1998, p.1. EPI14
36. NSF Award Abstract - #9704717. Arctic Research Initiative Expansion of the SuperDARN Radar Network, Sept. 4, 1997. EPI134

60 to 200 miles above the surface of the earth."[37] Although we acknowledge that SuperDARN is a separate project, its use in conjunction with HAARP is clear from the materials we have collected. This system will have additional installations built in Alaska and across Canada.

The ABM Unraveling

The Anti-Ballistics Missile (ABM) Treaty was destined to dissolve when new technology was developed and the Soviet Union fell. Many inside the military might even have seen this as an opportunity to advance the array of technologies which had evolved from the bloated defense budgets of the Reagan and Bush Administrations.

Interest by military planners in a viable "Star Wars" weapon system did not stop when Clinton became President, even though he voiced his opposition to this area of defense spending. The military believed that even though the use of technology was politically opposed they could still advance the concepts in research projects knowing that the political climate would change. They suggested, "The future of SAT systems technology seems assured. Unless a major policy effort is made to ban all research on SAT and ABM systems, the importance of space-borne reconnaissance and military communications demands the ability to strike at space vehicles."[38] Applying this rationale, the research continued.

One of the ideas which began to surface was the idea of hitting a "bullet with a bullet" or basically trying to hit an incoming missile with a missile. The following information began to reveal the intentions of the military:

> "The Minuteman National Missile Defense (NMD) option consists of the following major components: satellites with infrared sensors...early warning radars...X-band radars...and interceptor missiles, based at Grand Forks, that consist of three Minutemen booster stages and a fourth stage that includes a kinetic kill vehicle (KKV)."[39]

Even though the world was being sold on a "peace dividend," the idea that peace would release national budgets for things of greater importance to human advancement than creating the best killing systems was slowly slipping away as the technological advances accelerated during the 1980s and 1990s.

37. Jeffrey, Sue. "Scientists propose Kodiak site for ionosphere study." *Kodiak Daily Mirror*, Jan. 9, 1998. EPI705
38. Nordlie, John. "The Rise, Fall, and Rebirth Of Anti-Satellite Weapons." http://people.cs.und.edu/~nordlie/papers/asat.html EPI335.
39. Rand Research Brief, RB-47. "The Minuteman National Missile Defense Option." Published 1997. http://www.rand.org/publications/RB/RB47/index.html EPI27

Even though the cold war was over, tensions would remain high, as was demonstrated when U.S. security experts believed the world was on the brink of nuclear war when Russia mistook a harmless research rocket for a NATO missile, a Norwegian newspaper reported in the summer of 1997.[40] Even though the Cold War ended years before, the United States remained hard at work on new and modified designs for nuclear arms, a formerly secret federal document revealed in 1997.[41] Even with all of the efforts toward disarmament, our country and our adversaries are continuing to spend billions in weapons related research. It is acknowledged by these authors that the move toward peace is not an easy journey or without risks. Does this mean that research should be stopped? Does this mean that we should honor the treaties we have already signed? Does this mean that we will begin a new arms race with new weapons?

The issue of the ABM treaty began to be raised again in 1997 because of the threats posed by emerging adversaries. In 1997 it was reported that "Within the next three to five years, North Korea and China will have long-range missiles that could threaten Alaska...For that reason, the United States should withdraw from the Anti-Ballistic Missile Treaty of 1972 and renegotiate its position, including better protection for Alaska..."[42] This was the beginning of the public awareness campaign to gain support for the unilateral abandonment of the ABM Treaty by the United States.

As a part of the expansion of the defense systems, Alaska continued to be the place of greatest interest. "...the Ballistic Missile Early Warning facility at Clear Air Force Station...continues serving the nation as an important element in America's defense shield. Ground was broken on April 16 for a $106 million project to replace the existing mechanical radar with what is known as a PAVE PAWS system."[43] These new systems were substantially more powerful and efficient and would mix well with the other technologies being developed here.

"The current plan for defending against long-range missiles is descended from the Reagan-era Star Wars program that hoped to use satellites to detect and knock out incoming missiles. Some $45 billion was spent on developing such a system without success."[44] The lack of success was not total. A great deal was learned in terms of detection technology as well as the limits of space-borne systems. Once the limits were identified these became the targets for finding solutions which would lead to a powerful new integrated set of technologies. We have asserted for over six years that these would include the technological innovations developed through HAARP and other new systems in various stages of development.

---

40. AP. "Newspaper says error nearly caused nuclear war." *Anchorage Daily News*, July 6, 1997. EPI616
41. Broad, William J. "Critics attack U.S. nuclear weapons work." *Anchorage Daily News*, Aug. 18, 1997. EPI608
42. Phillips, Natalie. "Expert warns of Asian missile threat to Alaska." *Anchorage Daily News*, Oct. 24, 1997. EPI151
43. *Anchorage Times*. "Martians Beware." May 7, 1998. EPI13
44. Kizzia, Tom. "New defense plan on tap." *Anchorage Daily News*, Nov. 17, 1998. EPI152

A press report declared, "...top defense officials say they are prepared to deploy a $6.6 billion system that would blow up an incoming missile with another, provided they can overcome technical problems that have plagued their long research effort. In light of those stubborn technical hurdles, however, they announced they were delaying the target date of deployment from 2003 to 2005."45 Mixed messages were being sent by the United States, creating a good bit of uncertainty on the part of both our traditional allies and adversaries. "China criticized a U.S. plan to deploy an anti-missile defense system as destabilizing...and urged Washington not to sell missile defenses to rival Taiwan."46 The following day another press report appeared which said, "The United States assured Russia Friday it wanted to stay within the Anti-ballistic Missile (ABM) Treaty, despite plans to spend more on developing a defense umbrella against rogue missiles."47

By March 1999, the uncertainty was becoming reality when it became clear that a new system was the priority of the Congress. Reuters reported, "Still, the bipartisan vote...does place considerable pressure on President Clinton, who is scheduled to decide in June of next year whether to build a network of radar and missile interceptors designed to knock down any nuclear warheads intentionally fired by rogue nations or accidently launched from Russia or China."48 What is interesting in all of this is that although publicly Clinton was waffling on the issue, he had already appropriated billions of dollars on research into new anti-missile systems. The military wasn't waiting around for the politics to get caught up with reality and made clear their intentions when the head of the initiative said, "As the committee is aware, on January 20, 1999, Secretary Cohen announced our revised NMD program approach, which included adding some $6.6 billion to the program through fiscal year 2005. For the first time, the department has allocated the funding necessary to fully develop and deploy an NMD system. The other key element that the secretary announced was our expectation that the likely deployment date would be 2005 based on meeting key technology milestones. However, we have preserved the option to field the system earlier than 2005, if the threat warrants it and we have made good technical progress."49

The pressure to not move forward with the system continued to mount during 1999 and "Clinton and the Pentagon have recommended waiting until June 2000 before making a decision on whether to deploy, saying that to rush uncertain technology could create a false sense of security and weaken U.S. security."50 The

45. Richter, Paul. "Missile shield will be built, Pentagon says." *Anchorage Daily News,* Jan. 21, 1999, p. A-3. EPI299
46. AP. "U.S. anti-missile plan draws China's condemnation." Jan. 22, 1999. EPI154
47. Wright, Jonathon. "U.S. Assures Russia It Wants To Keep Missile Pact." Reuters. Jan. 22, 1999. EPI365
48. McDonald, Greg. "Missile defense favored." *Anchorage Daily News*, March 19, 1999. EPI157
49. American Forces Information Service, *Defense Viewpoint*, April 14, 1999. Prepared Statement by Director, Ballistic Missile Defense Organization, before the Senate Appropriations Committee Subcommittee on Defense. http://www.defenselink.mil/cgi-bin/dlprint EPI216
50. Roll Call Report Syndicate. "How Alaska's Congressional Delegation Voted." *Anchorage Daily News*, May 31, 1999. EPI130

public posturing still did not line up with what was actually happening. By June we had tested and proven that a system could be built with the official report being, "After six straight failures, a $3.8 billion experimental missile-defense system scored its first hit Thursday, shooting down a rocket in a test that left a puff of smoke and a twisted white trail of vapor across the desert sky."[51] The tests were far from definitive in the view of critics. The military's own people were "casting new doubt on the progress of the government's leading missile defense technology, saying two recent test flight successes haven't established how well the controversial system could handle an actual attack."[52]

Testing continued through the year with another test successfully completed. "The latest triumph came Saturday night when the most advanced of these models – the one that would be used to defend all 50 states against missile attacks – flew its first intercept attempt and pulverized a dummy warhead about 140 miles above the central Pacific Ocean."[53] The project was moving forward even though the official line was that the project decisions were on hold.

Throughout the process the military made clear that they would continue to advance the technology and that what was eventually developed might in fact be quite different from the current plans. We believe that this is more in line with the truth in that the system once begun will continue to expand and technically advance, incorporating all research in the field.

Russia and China Scream

"If the United States builds a missile defense system, 'Russia will be forced to raise the effectiveness of its strategic nuclear arms forces and carry out several other military and political steps to guarantee its national security.'"[54] The August 1999 story continued to amplify the complaints of the Russians.

"The Clinton administration has begun negotiations with Russia on treaty amendments that could allow it to place missiles in Alaska for its proposed anti-ballistic-missile defense network...the administration won't decide whether to deploy a missile defense until next summer."[55] This continues to be the "official" position of the administration despite the fact that the system is now being prepared.

The concerns continued to mount. The press reported that, "Russia is not the only country to see the ABM treaty as a cornerstone of strategic stability which, if kicked away, could bring

51. AP. "Experimental anti-missile system finally scores a hit after 6 failures." June 11, 1999. EPI153

52. Richter, Paul. "Missile defense tests raise questions." *Anchorage Daily News*, Aug. 24, 1999. EPI561

53. Graham, Bradley. "Test's success boosts missile defense odds." *Anchorage Daily News*, Oct. 4, 1999. EPI1176

54. Wadhams, Nick. "U.S., Russia agree on missile defense, reducing nuclear arsenals." *Anchorage Daily News*, Aug. 20, 1999. EPI315

55. Daily News wire services. "Missile defense talks on." *Anchorage Daily News*, Sept. 9, 1999. EPI475

the rest of the arms control edifice crashing down. 'What would this [scrapping the ABM treaty] do to missile control regimes, and what would its effects be on the third world?' asks a senior British official."56 The implications for the third world were there and had formed the United States rationale for developing the national missile Defense (NMD) system. It was also becoming part of the United States Presidential debate. The NMD was "designed to protect the U.S. from a limited missile attack from a rogue state or an accidental launch from a nuclear power, is already an issue in the presidential election. Texas Gov. George Bush, the Republican front-runner, is committed to deploying a missile protection system, as are all other GOP contenders."57

The Russians did have concerns about the problem of "rogue" states but they have no interest in changing the ABM Treaty. Grigory Berdennikof, a senior arms control specialist at Russia's Foreign Ministry, was reported to have said, "We are open to cooperation," he told the Times. "But if our cooperation means changing the ABM treaty our answer is 'thanks but no thanks.'"58

The Treaty was being changed or would be abandoned if this is what was required to gain a state-of-the-art defense system. "The accord, which has been a cornerstone of arms control for more than 25 years, sharply restricts the testing and deployment of antimissile defenses. The Clinton administration is now trying to amend it so that the United States can build a new 'battle management' radar system in Alaska by 2005 and deploy 100 antimissile interceptors there."59 What must be understood is that the entire system integrates a number of separate pieces. Each of these pieces, which fell outside of the agreement, would be addressed as separate issues. As separate items in the overall plan, work was already underway in Alaska with upgrades to our early-warning systems.

Despite the ABM Treaty the United States was not waiting for anyone's approval so "the talks began as Russia and China introduced a U.N. resolution Thursday demanding strict compliance with the ABM treaty, signaling their strong opposition to the U.S. plan."60 In addition, "The Russian military warned the United States Monday that it has enough weaponry to overwhelm any anti-ballistic missile system, and it threatened to deploy more atomic warheads if the United States builds a national missile defense system."61 In other words, there is no peace dividend, but, rather, the beginning of a new race to arms.

56. Buchan, David and Fidler, Stephen. "Star Wars strikes back." *Financial Times,* Oct. 6, 1999. EPI1174
57. Buchan, David and Fidler, Stephen. "Missile defense could spur arms race." *Anchorage Daily News,* Oct. 11, 1999. EPI373
58. AP. "U.S. seeks revision of ABM treaty." *Anchorage Daily News,* Oct. 17, 1999. EPI1161
59. Gordon, Michael R. "Russia says no deal on ABM treaty." *Anchorage Daily News,* Oct. 21, 1999. EPI1173
60. Wadhams, Nick. "Arms talks begin with Russia resisting changes to ABM treaty." *Anchorage Daily News,* Oct. 22, 1999. EPI1175
61. Hoffman, David. "Russia warns U.S. against building missile defense." *Anchorage Daily News,* Oct. 26, 1999. EPI376

Saber rattling increased with each step the United States took, with the following report appearing in November 1999: "Not only has Russia resorted to lobbing deadly missiles at civilian targets in downtown Grozny (as discussed above), but its military chiefs say they will use atomic warheads against the United states if it deploys a ballistic defense."62 Followed by the Russians demonstrating their capibility when it was reported that the Russian Navy's Arctic Fleet test-fired two ballistic missiles "from a nuclear cruiser submarine, underlining the country's combat-readiness amid heightened tensions with the West. The missiles were launched...from a site in the Barents Sea, and the warheads hit their targets..."63 Launches off the shores of the United States are indefensible by interceptor missiles. "The system currently envisioned would protect against a limited long-range missile attack...NMD is not being designed to defend against large scale mass attacks or attacks by short-range missiles launched near U.S. territory."64

It was not just our traditional adversaries but our allies who were objecting as well. "The American campaign to develop a protective shield against ballistic nuclear missiles is provoking serious alarm among the European allies, who fear that it could weaken the political and military links between the United States and Europe and trigger a dangerous arms race with Russia and China.65 Yet, at the end of 1999, Russia was showing signs of their willingness to negotiate changes to the ABM Treaty in exchange for U.S. aid in a number of areas.

The Plan Advances Regardless

During late 1998 and throughout 1999, the Environmental Impact Statement was being readied in order to meet the United States regulatory requirements for building the system. Despite the official pronouncements the engagement of the process clearly demonstrated which direction we were moving in. The United States was moving forward regardless of the opposition. What is also interesting to note is that the design of the system had materially shifted from the satellite based "Star Wars" concepts of the Reagan/Bush Administrations to a largely ground based system. During President Clinton's last election, his opponent, Bob Dole, had made the point that a missile defense system was achievable as a ground based system at much lower cost than the older concepts. Despite the rhetoric, this is exactly where the project is and it did not matter who won the election. On this point they would have both ended up in the same place – A ground-based star wars system was well underway by the end of 1999.

The following comments were taken from the "Executive Summary" of the Environmental Impact Statement:

62. *Anchorage Times*. "Missile duplicity." Nov. 2, 1999. EPI1171
63. *Anchorage Daily News*. "Russian Navy test-fires missiles." Nov. 18, 1999. EPI1219
64. Boeing. Boeing Missile Defense: The Boeing role in National Missile Defense and Theater Missile Defense. 1999. http://www.boeing.com/news/feature/nmd/main.html EPI1220
65. Drozdiak, William. "Plans to alter ABM worry allies." *Anchorage Daily News*, Nov. 7, 1999. EPI1191

"ES.1.1 Introduction. The NMD Joint Program Office of the Ballistic Missile Defense Organization is responsible for developing and deploying the NMD system. In the year 2000, there will be a Department of Defense (DOD) Deployment Readiness Review to review the technical readiness of NMD elements. Thereafter, the United States Government will determine whether the threat, developed capability, and other pertinent factors justify deploying an operational NMD system.

The NMD system would be a fixed, land-based, non-nuclear missile defense system with a land –and space-based detection system capable of responding to limited strategic ballistic missile threats to the United States. The NMD system would consist of five elements:

- Battle Management, Command, Control, and communications, which includes the Battle Management, Command and Control (BMC2), the communication lines, and the In-Flight Interceptor Communications System (IFICS) Data Terminal as subelements.

- Ground-Based Interceptor (GBI)

- X-Band Radar (XBR)

- Upgraded Early Warning Radar (UEWR)

- Space-based detection System

This EIS examines the land-based NMD elements. The space-based detection system, Defense Support Program Satellites, is an existing system that is being replaced by the Air Force independent of an NMD decision.

ES.1.2 Purpose And Need For The Proposed Action. The proliferation of weapons of mass destruction and technology of long-range missiles is increasing the threat to our national security. The purpose of the NMD program is defense of the United states against a threat of a limited strategic ballistic missile attack from a rogue nation.

ES.1.3 No-Action Alternative And proposed Action. This section describes the Proposed Action and the No-action alternative. The No-action alternative is not to deploy the NMD system. If the initial decision made is not to deploy, the NMD program would use the time to enhance the existing technologies of the various system elements. The NMD program would also have the option to add new elements if and as they are developed. For the potential sites being considered for NMD deployment, the No-action Alternative would be a continuation of activities currently occurring or planned at those locations.

Ground-Based Interceptor (GBI). The GBI would be dormant and would remain in the underground launch silo until launch. Launches would occur only in defense of the United States from a ballistic missile attack. There would be no flight testing of the missiles at the NMD deployment site. The GBI site would contain launch silos and related support facilities. Under the Proposed Action, up to 100 GBI silos could be located at one of the locations shown in Figure ES-1, or up to 100 silos could be deployed at both one site in Alaska and one site in North Dakota. When the GBI site becomes fully operational, the total site-related employment would be 250 to 360 direct jobs."[66]

The Debate Continues

The usual issues continue to be addressed as the military moved from rural community to community throughout Alaska in their attempt to sell the project to citizens. The assurances made were standard in appealing to areas where unemployment was high and the economic needs of individuals could be exploited. "The Defense Department organization planning for a national missile defense system said locating the launch site in Alaska raises few environmental concerns and would create hundreds of short and long-term jobs."[67] The environmental issues also became part of the debate with the military following the standard line while opponents raised the reality of what Alaska has historically experienced. "The project's draft environmental study said the environmental footprint of the system would be minimal. 'Wherever the military has been in Alaska it has fouled the nest,' countered Dan O'Neill."[68]

66. National Missile Defense (NMD) Deployment Draft Environmental Impact Statement, September 1999. U.S. Army Space and Missile Defense Command. Huntsville, Alabama. EPI1189
67. Whitney, David. "Report favors Alaska missile site." *Anchorage Daily News*, Sept. 30, 1999. EPI830
68. AP. "Missile hearing draws support, some critics." *Anchorage Daily News*, Nov. 3, 1999. EPI1162

Chapter 3

Beam Wars Speed of Light Weapons

***"You could stop an elephant with a voltage on par of having your hand on a static ball."*[69]**

The rush to disarmament, although it appears substantial, is really about dropping old technology in favor of new systems. The previous technologies which are now the subject of disarmament debate are being replaced with more efficient and more controllable technology. The more well known weapons of mass destruction are being replaced with precision systems which use new principles of design in order to create speed-of-light weapons with incredible implications. The Electronic Industries Association (EIA) believes that it is this new technology which will shape the future. "The EIA's most dramatic – and perhaps most controversial – prediction concerned the impact that speed-of-light weapons would have on the anticipated 'Revolution in Military Affairs' in the period 2010 and beyond. The EIA predicts that 'whichever side owns the speed-of-light weapons will dominate the battlespace in almost every respect.'"[70] It is believed, "In the eyes of some experts, beam weapons could change the nature of war as dramatically in the near future as the machine gun did earlier in the 20th century. Beam weapons range from shoulder-held laser guns that fire a concentrated, destructive ray of light to large high-tech devices that could destroy satellites by knocking out their electronics with powerful radio transmissions or laser beams."[71]

"Directed Energy Warfare (DEW) is a major emerging military technology. It involves the effects of laser, microwave, charged particle or neutral particle beams on any tactical or strategic military system on the face of the earth, in the air or in space. The assessment of the susceptibility of military components, subsystems, systems and platforms to DEW, and the hardening of these

69. Johnson, Christina. "Laser Puts More Authority Behind The Command, 'FREEZE!'" *San Diego Daily Transcript,* April 21, 1998. EPI680
70. Green, Gerald. EIA Forecast: "Speed-of-Light Weapons Will Dominate the Future Battle space." *Journal of Electronic Defense,* Dec. 1996. EPI421
71. Reuters News Agency. "Superpowers' new 'beam weapons' may revolutionize warfare." *Washington Times,* March 27, 1990. EPI435

components against such weapons, is necessary to the development of offensive and defensive capabilities in this area."72 This new direction represents one of the major shifts in weapon system development and another thrust point in the military's "Revolution in Military Affairs."

The Defense Intelligence Agency (DIA) and the Central Intelligence Agency (CIA) have expressed concern about this new direction of warfare. The former head of the DIA, Air Force Lieutenant-General Leonard Perroots indicated as far back as 1991 that the former Soviet Union would have developed these capabilities by the early 1990s. "The DIA, unlike the CIA, is allowed to use 'mirror imaging' in its reports – that is attributing one's own motives and weapons capabilities to the 'other side'. Perroots wrote in a magazine in March last year that he thought the Soviets would begin deploying battlefield beam weapons within the next two or three years."73

In 1998, the Russians offered to cooperate with the United States in developing Strategic Defense Initiative (SDI) concepts. The Russian proposal, and the article referenced below, spoke of a system which would create a new defense against intercontinental ballistic missiles. This system's components were essentially the same as HAARP. The offer "highlights the fact that the Russians are strong in many of the key scientific areas in which the SDI has been weak – especially the frontier area of nonlinear processes involved in the generation, propagation, and absorption of powerful pulses of electromagnetic and particle-beam radiation."74 The United States turned down the offer of the Russians and our research indicates that the United States did not need Russian help in advancing technology already being developed in the United States.

Power for the Lasers

"An extraordinary plan to beam more power to orbiting communications satellites than they can get from the sun may make use of a free-electron laser under development at the Department of Energy's Lawrence Berkeley National Laboratory's Accelerator and Fusion Research Division. 'If you want to do better than the sun, you need a powerful source,' says Alexander Zholents of the Center for Beam Physics, who has helped design a laser called IFRA (Ignition Feedback Regenerative Amplifier). At 200 kilowatts peak power, IFRA would be the most powerful free-electron laser in the world."75 This device would provide a ground source for powering satellites or low orbit space platforms, reducing the fuel components needed to keep these objects in orbit over long periods. Avoiding the need to

72. Naval Research Lab. Directed Energy Warfare Susceptibility and Hardening. (BAA 742) http://www.nrl.navy.mil/BAA/baa742.html EPI796
73. Kennard, Peter. "Field of Nightmares." *Weekend Guardian*, Feb. 2-3, 1991. EPI691
74. Tennenbaum, Jonathan. "Behind the Russian SDI Offer: A Scientific, Technological, and Strategic Revolution." *21st Century*, Summer 1993, pp. 36-43. EPI17
75. DOE. Beaming Power to Communication Satellites from the Ground. August/September 1999. http://home.doe.gov/lasers.htm EPI808

refuel, and lengthening the time that these objects can stay aloft, would represent significant improvements.

One of the other new technologies "is a device known as a collective accelerator, a large tube surrounded by very powerful magnets. The magnets accelerate orbits (small, invisible elements of atoms) to the velocity of light. These streams of protons are pulsed, hundreds of times per second. Each pulse produces tens of billions of electron volts, enough to virtually disintegrate anything it strikes."76 This again gets back to the idea of "pulsing" energy in order to magnify the impact of otherwise limited power sources. The pulsing of energy gives the device the ability to not only disintegrate the object but to do so at the speed-of-light. So why isn't this part of the ABM debate? Perhaps because this technology represents the real direction the United States intends to go, to a weapon which would stop any incoming object whether sent from Russia, China or anywhere else, destroying everything in its path.

One of the other issues surrounding these new weapons is related to the power supplies. The need for enormous amounts of power for the operation of these new lasers was partially met, interestingly enough, with the help of the Russians. "A 15-million-watt, self-contained, transportable Russian generator has completed two months of acceptance testing and is being delivered to the Air Force's Phillips Laboratory. This $4 million Pamir-3U magneto-hydrodynamic generator, which uses modified solid-rocket fuel to create a plasma that is processed to create electricity, will be used for laboratory research that may lead to a power source for high-energy lasers or advanced high-power microwave weapons."77 The ideal source for not just lasers, but also HAARP, are magnetohydrodynamic generators, which are portable and create the amounts of power needed to operate these systems. Phillips Laboratory is where HAARP research is based.

Particle Beams

"Neutral particle beams can probe satellites' interiors, which enable them to discriminate on mass and determine the satellite's function with confidence. That capability, now absent, is essential for enforcing existing space agreements, defining the verifiability of future agreements, and determining the capabilities of unknown space objects. It is complemented by their ability to image at close ranges and measure spectral information at intermediate ranges. The platforms would be of very low current and moderate energies, which would give to them no significant anti-satellite capability."78 This low power system could be used to determine if the space object is

76. Major General USAF (retired) George J. Keegan, Jr., "The New 'Zap' Guns, A look At The New Revolution In Beam Weapons," *Gung Ho*, Jan. 1982 Issue, pp. 35-49, 54. EPI1
77. Phillips Laboratory Public Affairs Office. Phillips Lab Getting Russian Generator. Press release, 18 April, 1995. PL Release No. 95-32. Contact: Rich Garcia. EPI6
78. Los Alamos National Laboratory. Collaborative, Remote, In-Depth Inspection and Verification of Satellites with Neutral Particle Beams. LA-11804-MS. May 1992. EPI875

carrying nuclear payloads or other materials which might be forbidden by treaty agreements. "Unlike lasers, neutral particle beams (NPB) travel through the atmosphere very poorly. That characteristic dictates that NPB weapons will have to be stationed in space. But it also means that such weapons will not be a threat to Earth targets, the beam would dissipate more than 100 kilometers above the Earth's surface."79 This is the belief, at present, but like many other systems identified in these pages, the technology changes very rapidly. What is thought of as impossible today becomes the science of tomorrow.

The Big Lasers

AirBorne Laser (ABL) technology is reaching maturity. "Aircraft platforms now exist that can carry the necessary crews, fuels, and equipment constituting a laser weapon system with potentially high operational effectiveness. This means ABLs could be used in a variety of missions, including TBM (Theater Ballistic Missile) kills at ranges of 400+ kilometers (km), counterair and anti-cruise-missile kills at 100+ km range, and defense of airborne high-value assets against air-to-air and surface-to-air missiles (SAM). They could also perform surveillance, command and control (C2), and battle management tasks yet maintain an effective self-defense capability. These missions could cover wide areas by capitalizing on the flexibility and responsiveness inherent to air power, while leveraging ABL's precision into a potent force multiplier for boost-phase missile intercepts."80 These systems are not something futuristic but were available in 1994 when this article was originally written.

Experiments in these areas continued through the 1990s. "The High Altitude Balloon Experiment (HABE) is a research program to validate acquisition, tracking, and pointing technologies for directed energy weapons (lasers) that could be used for ballistic missile defense. The program is managed by the Air Force's Phillips Laboratory at Kirtland Air Force Base, New Mexico, for the Ballistic Missile Defense Organization. Under the program, the Laboratory will design and fly a complex electro-optical payload aboard a high-altitude balloon to validate technologies that could be deployed from an airplane or spacecraft. This four-year program will consist of a series of high-altitude balloon experiments which began in June 1993, from Clovis, New Mexico."81 Three years later it was reported, "In a controversial test that offers new proof of the vulnerability of the world's growing fleet of government and commercial satellites, a ground-based laser has struck a satellite 260 miles above Earth."82 These technologies are no longer theoretical but real and being improved at a very rapid rate. The near-term expectation is that this technology will be perfected within a few years.

79. Fulghum, David. "Key Contract Near in Developing Particle Beam for SDI." *Navy Times*, Oct. 5, 1987. EPI64
80. Coulombe, Lt. Col. Stephan A. (USAF). "The Airborne Laser: Pie in the Sky or Vision of Future Theater Missile Defense?" *Air Chronicles*, Fall 1994. EPI226
81. Phillips Laboratory Computation Services Division. HABE, the High Altitude Balloon Experiment. April 15, 1996. EPI432
82. Richter, Paul. "Army laser zaps satellite." *Anchorage Daily News*, Oct. 21, 1997. EPI426

High-Power Microwave

High-power microwaves are being explored for use on both hardware and humans. Potentials of this new technology have been explored in an International Red Cross document. One of the uses described is an Electromagnetic Pulse (EMP) weapon which gives an operator the same ability to wipe out electronic circuits as a nuclear blast would provide. The main difference is that this new technology is controllable, and can be used without violating nuclear weapons treaties.

The report described some of the effects of these types of systems as follows:

* "Overheats and damages animal tissue."

* "Possibly affects nervous system."

* "Threshold for microwave hearing."

* "Causes bit errors in unshielded computers."

* "Burns out unprotected receiver diodes in antennas."

The effects are based on microwaves being pulsed "between 10 and 100 pulses per second." The report confirmed that non-thermal effects were being researched. These non-thermal effects included damage to human health when the effects occurred "within so-called modulation frequency windows or power density windows."[83] This research is also not particularly new, "Scientists have been studying the possibly of developing electromagnetic weaponry for at least a quarter of a century. In 1965, the CIA and the state department began a program called, appropriately enough, Pandora project. Its mandate was to study the health effects of a microwave beam which U.S. officials had discovered zapping their embassy in Moscow."[84]

The way these weapons work was clearly described when the report noted their effect on machines:

> "A HPM (High Power Microwave) weapon employs a high power, rapidly pulsating microwave beam that penetrates electronic components. The pulsing action internally excites the components, rapidly generating intense heat which causes them to fuse or melt, thus destroying the circuit...HPM (weapons) attack at the speed-of-light thus making avoidance of the beam impossible, consequently

83. "Expert Meeting on Certain Weapon Systems and on Implementation Mechanisms in International Law", Report of the International Committee of the Red Cross, Geneva, Switzerland, May 30 - June 1, 1994. Issued July 1994. EPI1232
84. Goldenthal, Howard. "Oppression in an age of high technology." *Canada's Toronto Star*, June 27, 1989. EPI685

negating the advantage of weapon systems such as high velocity tactical missiles."

In other words, with this kind of weapon there is no machine which could get by this invisible wall of directed energy. Once again "Phillips Laboratory is the Air Force High Power Microwave Program (HPM) manager and national leader in the development of HPM sources. The objectives of the Air Force HPM technology program are to determine the effects of high power microwaves on military systems, develop hardening technologies as needed, and develop the technologies and data base for military application of HPMs."85 These new areas also require that the U.S. military devise protections against the new technology they are creating, expecting eventual proliferation. To this end, they are also working on the "hardening technologies" which would insulate U.S. systems from these types of new weapons.

Canadian government concerns regarding the availability and ease in which this technology could be built by garage amateurs was another issue which was beginning to be raised. "The use of High Power Microwaves as a directed energy weapon is a significant emerging threat. There is concern because of the availability of off-the-shelf systems and the possibility of having considerable damage done to electronic systems, perhaps without clear evidence of the cause. Canada responded to this threat by developing expertise in the HPM analysis, measurement and hardening methods."86

Blinding Lasers

One of the early "non-lethal" weapons was the blinding laser, which was intended to blind combatants rather than kill them. The introduction of this weapon was met with strong public opposition because of the inhumanity of this new technology. The outcry was a clear signal to military planners that if they wanted to develop such weapons it could only happen out of the public's view. In this case the program was stopped, a move representing a step forward in clear thinking and fair public discussion where human values were at the center of the decisions. "The adoption in Vienna of a new fourth protocol prohibiting blinding with laser weapons represents a significant breakthrough in humanitarian law. The prohibition, in advance, of an abhorrent new weapon the production and proliferation of which appeared imminent is a historic step for humanity. It represents the first time since 1868, when the use of exploding bullets was banned, that a weapon of military interest has been banned before its use on the battlefield and before a stream of

85. Office of Public Affairs, Phillips Laboratory. "High Energy Microwave Laboratory." Fact Sheet, Jan. 1994. EPI428
86. Defense Research Establishment Ottawa. Electronic Warfare: High Power Microwaves. http://www.dreo.dnd.ca/pages/factsheet/ew/ew0003_e.html EPI797

victims gave visible proof of its tragic effects."[87] This project stopped and the Clinton administration "ordered the Pentagon to withdraw from active service new laser guns which are designed to blind enemies permanently. Two prototypes of the "Dazer" laser gun were taken to Saudi Arabia during the Gulf War but were not used."[88] We believe that the fate of the blinding laser, when it was relegated to the scrap heap of inhumanity, is the same fate to which many of these new technologies should be condemned.

Electromagnetic Pulse Guns

Several variations of electromagnetic guns were under development and were tested in the last decade. These systems followed several different principles of design, with the underlying energy being the same. The objective of using various systems to create a maximum delivery speed for the killing blows delivered by the technology seemed to drive the research teams. In one application a new gun was used to launch an object, like a bullet, only it would move several kilometers per second. One report reads in part, "Electromagnetic guns use high current pulses (on the order of a few MEGA Amperes) to electromagnetically launch objects at tremendous velocities (several Kilometers per second). Electromagnetic guns can launch projectiles to velocities well beyond today's conventional guns."[89]

In other developments, real life was catching up with Star Trek when Eric Herr from San Diego, California, was "granted a patent for a 'phaser' that uses laser light to stun or kill. Herr's invention uses lasers to generate intense beams of ultra-violet light...The currents can be manipulated to cause painful contractions, stun a victim painlessly, or induce a heart attack."[90] Set phasers on stun. Where had we heard that before? Modern science fiction, Star Trek in this case, was proving to be the precursor to the reality. "Unlike tasers, tetanizers send a precisely modulated high-frequency pulse that overloads the body's nerve channels, shorting them out temporarily with no sensation, Herr said. Because of this, 'You could stop an elephant' with a voltage on par of having your hand on a static ball."[91] The tetanizers invented by Herr once again capitalized on the power of pulsing the energy, thereby increasing the power of the impact. The same principles for creating enormous pulse weapons for manipulating the ionosphere were being applied on a very small scale to these personal sized weapons.

87. International Committee of the Red Cross. "Vienna's Historic Success: Blinding Laser Weapons." Nov. 1995. EPI434
88. Wilson, Peter. "Pentagon Prohibited From Using Laser Gun." *The Weekend Australian*, Sept. 23-24, 1995, p.14. EPI8
89. IAP. Electric Armaments. http://www.iap.com/eleclaun.html EPI806
90. *New Scientist*. "Set phasers to shock." Nov. 1, 1997. EPI679
91. Johnson, Christina. "Laser Puts More Authority Behind The Command, 'FREEZE!'" *San Diego Daily Transcript*, April 21, 1998. EPI680

The reality of speed-of-light weapons is already upon us, the research will continue, and the use of these systems will increase as rapidly as the research advances. Although some of these systems are likely needed in a very unstable world, many, like the blinding laser, humanity can do without. The loss of access to some of these technologies will not have an impact on the balance of power in the world. Forbidding the use of some of these systems will, however, have a tremendous impact on the balance of humanity and justice within civil society.

Chapter 4

Lies and Experimentation

"At least 500,000 people were used as subjects in Cold War-era radiation, biological and chemical experiments sponsored by the federal government" [92]

When is it right for a democratically constituted government to lie? Is it ever right to use people as experimental guinea pigs to test new technology without clear informed consent? How many men, after World War II, were sent to the gallows to hang by their necks for believing that human experimentation was the right thing to do? The country largely responsible for those trials, the United States, would, after World War II, repeat these violations of human rights by experimenting on thousands of citizens without either their informed consent, or, for that matter, even their knowledge that the experiments were going on. This chapter revisits some of these atrocities which will continue to be repeated if measures are not taken to end these crimes against humanity.

In the last six or seven years much information has been released on these subjects and a few points stand out in all of it. One of the major points is that some of the experiments were conducted within the official safety limits of the technology at the time. Those official limits did not match experimental results – in other words the safety limits were not safe and the government knew it. One of the other points was that in all instances the sacrifice of a few disenfranchised individuals was judged acceptable for the greater good of the country. In other words, unnamed someones decided who would be sacrificed to the experimenter's tools.

This was by far the most difficult chapter to write. Periodically, I had to stop and clear my eyes as each injustice touched me. I was raised in politics surrounded by statesmen and women who were fully committed to service to humanity, believing that they were protecting us from the very things I am now writing about. It grieves me tremendously as I think about the millions of Americans who have stood against injustice, dishonesty and tyranny only to be faced with

92. MacPherson, Karen. "500,000 endangered by tests since '40." *Washington Times*, Sept. 29, 1994. EPI620

the reality of what happens even in the greatest democracies when government is permitted to hide in corners of secrecy. We must rise to the truth, and open wide the doors of justice, exposing the evil we have allowed by our trusting natures to fester under the guise of "national security." We suggest that there is no "national security" worth preserving unless the integrity of every human being has the same equal standing. There is no security for any person unless that same security extends to every other person. What is freedom without principles? What is justice without truth? What is right when the government lies?

The Basic Rationale for the Big Lie

Public safety is the usual logic which is applied to extreme violations of civil law by governing authorities. It is the trade-off for peace, in their eyes at least. In December 1998, the United States Secretary of Defense made some interesting observations, observations of the kind that have always formed the basic rationale for repression. He said:

> "The best deterrent that we have against acts of terrorism is to find out who is conspiring, who has the material, where they are getting it, who they are talking to, what are their plans. In order to do that, in order to interdict the terrorists before they set off their weapon, you have to have that kind of intelligence-gathering capability, but it runs smack into our constitutional protections of privacy. And it's a tension which will continue to exist in every free society – the reconciliation of the need for liberty and the need for law and order.
>
> And there's going to be a constant balance that we all have to engage in. Because once the bombs go off – this is a personal view, this is not a governmental view of the United States, but it's my personal view – that once these weapons start to be exploded people will say protect us. We're willing to give up some of our liberties and some of our freedoms, but you must protect us. And that is what will lead us into this 21st century, this kind of Constitutional tension of how much protection can we provide and still preserve essential liberties."[93]

Although Secretary of Defense Cohen indicates that this is his personal view and not the view of the U.S. Government, what does this say about the position of the second in command of the most powerful military organization in the world? We believe that this explains how lies can be told without a blink of the eye or the twitch of a muscle. We believe that this explains how people in high position

93. Cohen, Secretary of Defense William S. Hemispheric Cooperation In Combating Terrorism, Defense Ministerial of the Americas III. *Defense Viewpoint*, Dec. 1, 1998. EPI660

rationalize their behavior when the very rights of those they are charged to protect are violated. Is this the democracy our fathers fought for – one in which lies are traded for freedom – one in which a dubious security is traded for liberty? The basic human rights defined by all enlightened constitutional democracies and democratic republics do not provide any license for dishonest government, violations of civil rights or intrusions into the very essence of our individuality and bodies.

Medical Ethics?

The World Health Organization (WHO) has published some interesting standards concerning the ethics of medical experiments. They said:

> "Medical progress demands that in research the 'benefit of the patient' should not be interpreted in a narrow sense, since this could hamper progress and deprive future patients of benefits. When a research project is of direct diagnostic or therapeutic relevance to the individual patient, the ethical problems involved tend to be simple ones. When, however, a research project is intended to extend medical and scientific knowledge in general, without specific benefit to the subject, the situation is different and the principles of medical ethics need to be applied in a broad sense, in relation to the benefit expected to accrue in the future to patients or human beings in general."[94]

What did they say? It sounds like a very nice way to say that we need a few humans to use as lab rats in order for the rest of us to improve our health and advance science. For most people this is an unacceptable standard. This is the same kind of thinking that was applied in the death camps of Germany by Nazi scientists experimenting on humans. Who will decide which people will be the modern lab rats? Will they be incarcerated prisoners like the victims of the some earlier medical holocausts, black men, American Indians, children, military personnel or just ordinary citizens? The truth is, that from time-to-time it has been all of these groups which have been the victims of the United States' big lie in the face of an illusion of freedom.

Human Experiments

The United States has engaged in experiments on its own citizens since the end of World war II. These experiments have crossed the line and are unjustified and inhumane. During the 1990s several news reports appeared around the world which drew attention to these tests. As each report was issued people were shocked at what our

94. WHO Expert Committee on the Use of Ionizing Radiation and Radionuclides On Human Beings For Medical Research, Training, and Nonmedical Purposes. Geneva, 1-8 March, 1977. EPI1180

country had done and the government's lack of any real justification for its actions. Many Americans reacted to each new revelation with a sickened feeling. However, when a number of these reports are assembled in one place, the scope of the abuses becomes apparent and have a heavy emotional impact on the reader.

"Government researchers appear to have had a pattern of choosing 'vulnerable' people – minorities, the poor, prisoners and retarded children – for Cold War-era radiation experiments, Energy Secretary Hazel O'Leary said Tuesday. 'This appearance of treating some citizens as 'expendable' is especially repugnant.'"[95] The groups which were targeted for these experiments were those with the least opportunity to object, the uneducated or those easily victimized. "Children in orphanages were used to test experimental vaccines for diphtheria and whooping cough for several decades after WWII,"[96] according to a newspaper story appearing in 1994. "At least 500,000 people were used as subjects in Cold War-era radiation, biological and chemical experiments sponsored by the federal government, a congressional agency said yesterday...the tests conducted from 1940 through 1974, ranged from radiation to biological and chemical agents like mustard gas and LSD."[97]

These experiments involved every area of the United States. "Stockyards in Kansas City, wheat fields in North Dakota and sites from central Alaska to the Georgia coast were used for Cold War-era research on biological warfare, according to Army documents."[98] "People as far away as the East Coast may have been exposed to as much radiation fallout from nuclear tests in the 1950s as residents directly downwind from the Nevada blasts, according to preliminary information from a National Cancer Institute study."[99]

In Alaska it was reported that "The North Slope Borough has filed a claim with the federal government seeking $428 million on behalf of 79 Inupiaq villagers who were given radioactive iodine tablets during 1950s medical research."[100] The following material surfaced during the release of records of this experiment:

> "During the last two years we have conducted a series of I131 studies in Eskimos and Indians in Alaska, using different groups of Whites as controls.
>
> Tracer doses of up to 50 microcuries were used to determine the thyroid uptake, urinary elimination, total plasma and protein bound I131, using sensitive scintillation counters as part of a complete mobile isotope laboratory which could be transported to remote areas of arctic Alaska.

95. MacPherson, Karen. "Radiation researchers went after 'vulnerable' subjects." *Anchorage Daily News*, Jan. 26, 1994. EPI1183
96. AP. "Vaccines were tested on orphans." *Anchorage Daily News*, June 11, 1997. EPI75
97. MacPherson, Karen. "500,000 endangered by tests since '40." *Washington Times*, Sept, 29, 1994. EPI620
98. Brasher, Philip. "Documents show Army tested biological agents from Arctic to Georgia." *Anchorage Daily News*, June 19, 1994. EPI624
99. Neergaard, Lauran. "Research finds '50's radiation had long reach." *Anchorage Daily News*, July 26, 1997. EPI583
100. Hulen, David. "Borough sues over '50's study: 79 villagers given doses of radioactive iodine." *Anchorage Daily News*. EPI577

The subjects included both coastal and inland Eskimos, and two groups of Athabascan Indians of Fort Yukon and Arctic Village."[101]

"Officials with the National Academy of Sciences say they plan to begin looking into whether radioactive iodine experiments done nearly 40 years ago on Alaska Natives were safe and ethical."[102] The government will investigate the government in determining the standards for this experimentation, rather than an independent body with full investigative power. In a country where tens of millions of dollars are spent to root out the truth behind the sexual activities of President Clinton, pennies are being spent to bring justice to the victims of government experimental abuses.

The government also experimented on prison inmates using chemicals designed to disable adversaries. They used ill-trained personnel to carry out many of the tests, some of which included radioactive substances. "The Army tested an incapacitating agent, EA 3167, which it hoped to add to its chemical warfare stock. Inmates suffered hallucinations and confusion for up to three weeks. Kligman tested radioactive isotopes despite having little training in radioactive medicine."[103]

Along with the experiments came the denials by government organizations which again attempted to justify themselves under the guise of national security. Even worse was the thought of respected members of the scientific community being involved in this type of experimentation. Sandra Marlow, the founder of the Center for Atomic Radiation Studies in Boston, Massachusetts, "calls the U.S. government's use of human subjects in radiation experiments, and the alleged cover-up of those experiments, 'the Holocaust of American scientific integrity.'"[104] This Holocaust is shocking in its size and implications. The fact that as many, perhaps more, than 500,000 Americans were victimized by these experiments implies the cooperation of thousands of knowledgeable scientists and unwitting participants. Where were the whistle-blowers? Where were the ethical individuals in these programs? Were these scientists participating to keep their jobs? Were they afraid to speak the truth, going through the motions of their individual tasks each day without even a thought for their victims or, perhaps, with burning guilt waiting for release?

The term "weapons of mass destruction" was coined by the United States in pointing out the biological threats posed by Iraq during the Gulf War. The United States was the major supplier of this technology along with some European countries. These and other biological weapons had been developed to provide a weapon which

101. Rodahl, Kaare. *Human Acclimatization To Cold.* Arctic Aeromedical Laboratory. October 1957. EPI1185
102. AP. "Experts to Review '50s radioactive tests on Natives." *Anchorage Daily News,* Jan. 26, 1999. EPI11991
103. Kinney, David. "Inmates used as drug guinea pigs." *Anchorage Daily News,* June 8, 1998. EPI617
104. Dunham, Mike. "Researcher wants feds to come clean." *Anchorage Daily News,* Aug. 11, 1997. EPI622

could create a nuclear-equivalent level of death.105 "For example, the transmissible agents smallpox and pneumonic plague were intended for use in 'nuclear-equivalent situations' requiring a strategic strike against the enemy's home territory. Other agents were developed for tactical use, even individual assassination."106

Somewhere along the line. information begins to leak from behind the door of secrecy on programs such as MKULTRA, radiation experiments or biological agent tests. Investigations are launched, some details are disclosed, but how many files have been shredded to preserve the careers of the perpetrators? The U.S. Department of Energy has begun releasing some materials on just their own projects:

> "'We started out with a universe of 3.2 million cubic feet of records scattered all over the country,' she says. From these, OHRE compiled a database of the most critical 150,000 pages of photos and text, describing slightly more than 150 experiments, Weiss says. But, she adds, 'we know about and are trying to piece together information about at least another 150.' Those additional data should be on-line later this year.
>
> Files on the first 150 experiments are now being photographed and loaded onto Internet's World Wide Web. The electronic address for the 'home page,' which allows computer users to browse through attached files, is http://www.eh.doe.gov/ohre/home.htm. DOE also offers electronic mail (E-mail) support services for accessing the files at ohre-support@hq.doe.gov."107

Some victims were able to sue for damages which resulted from their unwitting victimization. "Mary Jean Connell will receive $400,000 – $8,000 for each year the woman survived after being injected with radioactive uranium in medical experiments that were kept hidden even from her."108 While some victim's cases were settled, others were denied based on the nature of the allegation being too vague, particularly in the case of those asserting victimization by electromagnetic weapons. "United States government attorneys say that a class action suit, allegedly on behalf of persons who had been unwitting subjects of government-sponsored radiation and electric and magnetic field exposure tests, should be dismissed because its 'vague allegations are insufficient to confer organizational standing.'"109 The problem with the victims of these technologies is that their stories sound unbelievable, yet the evidence shows that their claims could be

105. Thomas, William. *Bringing the War Home*. Published by Earthpulse Press, Inc., 1997. EPI1236
106. Davis, Christopher. "Bioterrorism Special Report: Inside Out." *New Scientist*, May 29, 1999. http://www.newscientist.com/nsplus/insight/bioterrorism/insideout.html EPI881
107. *Science News*. "Human irradiation data now on Internet." Vol. 147, p. 102. EPI757
108. Warrick, Joby. "Unwitting guinea pigs compensated." *Anchorage Daily News*. Nov. 20, 1996. EPI578
109. *Andrews Electromagnetic Field*. "Human Testing Claims 'Too vague to Confer Standing', U.S. Argues." October 1998. http://www.andrewspub.com EPI625

quite real. In the section of this book which deals with the subject of mind control, the evidence of the possibilities is even more shocking in its implications for abuse and the most vile of manipulation.

The Big Test Pool - The U.S. Military

"During World War II and the Cold War era, DOD and other national security agencies conducted or sponsored extensive radiological, chemical and biological research programs. Precise information on the number of tests, experiments, and participants is not available, and the exact numbers may never be known. However, we have identified hundreds of radiological, chemical, and biological tests and experiments in which thousands of people were used as test subjects. These tests and experiments often involved hazardous substances such as radiation, blister and nerve agents, biological agents, and lysergic acid diethylamide (LSD). In some cases, basic safeguards to protect people were either not in place or not followed. For example, some tests and experiments were conducted in secret; others involved the use of people without their knowledge or consent or their full knowledge of the risks involved."[110]

"A former soldier given LSD nearly 40 years ago in experiments by the Army and CIA won more than $400,000 from arbitrators to settle his lawsuit against the government. He was then a young soldier who thought he was participating in a test of equipment and clothing. Instead, the Army and CIA gave Stanley and 740 other volunteer soldiers LSD to observe the drug's effects."[111]

These experiments continue even today and are being openly resisted by service personnel who believe in defending their country with their lives but object to the use of their bodies for experimentation. The issue is one of truth and privacy. The right of individuals to determine what goes on in their mind and body is, and should be, beyond the reach of the government. Yet the game continues, in mass, with the standard Orwellian line of "we know what is good for you." As recently as the Gulf War, experiments were conducted which have lead to a number of casualties. "The use of the drug pyridostigmine bromide (PB) by 250,000 soldiers during the Persian gulf war 'cannot be ruled out' as a cause of lingering illnesses in some veterans, according to a new report prepared for the Defense Department."[112]

Under a military order, the use of experimental and unapproved drugs on service personnel is permitted. The soldier is asked to risk his life twice – once at the hands of his own government and once when facing the enemy. The order is as follows:

> "Sec. 2. (b) It is the expectation that the United states government will administer products approved for their

110. GAO. Human Experimentation: An Overview on Cold War Era Programs. GAO/T-NSIAD-94-266, Sept. 28, 1994. EPI618
111. *Honolulu Star Bulletin*. "GI given LSD wins $400,577 settlement." March 5, 1996. EPI576
112. *Anchorage Daily News*. "Drug considered possible cause of gulf war syndrome." Oct. 19, 1999. EPI190

intended use by the Food and Drug Administration (FDA). However, in the event that the Secretary considers a product to represent the most appropriate countermeasure for diseases endemic to the area of operations or to protect against possible chemical, biological or radiological weapons, but the product may, under certain circumstances and strict controls, be administered to provide potential protection for the health and well-being of deployed military personnel in order to ensure the success of the military operation."[113]

Now, at the end of 1999, military personnel and reservists are voicing their objections to being either used in experiments or forced, under threat of court martial, to allow the government to invade their bodies any way they choose. After a mass exodus out of one reserve squadron, "Gen. Ralph E. Eberhart, the vice chief of staff, ordered the creation of the anthrax task force" after one-quarter of an Air Force Reserve squadron in California quit rather than take the six-shot vaccine.[114] This is occurring around the country and being widely reported. Some are being discharged, with others quitting rather than allowing the government to violate their bodies.

From Investigator to Aggregator

Earthpulse has, since our beginnings, searched for information in all mediums, including internet, print, audio, video, direct interview and experience in putting together our research. We, and many others, take a strong position on the right to publish from the perspective of fulfilling the responsibilities associated with the right of free speech. This requires that the media work to maintain an informed public which engages in vigorous public debate. What was once good investigative journalism is now being named, in the interests of political-correctness, as something akin to "information-terrorist." The idea that thoughts and information would be so threatening to existing orders that denial, subversion and lying to the public becomes institutionalized and routine is shocking. It is at this point that the very essence of democracy is being attacked as a threat to national security. The military's concerns are expressed below:

> "The World Wide Web provides the Department of Defense with a powerful tool to convey information quickly and efficiently on a broad range of topics relating to its activities, objectives, policies, and programs. It is at the heart of the Defense Reform Initiative and is the key to the reengineering and streamlining of our business practices. Similarly, fundamental to the American democratic process is the right of our citizens to know what government is doing, and the corresponding ability to judge its performance.

113. Executive Order 13139. Improving Health Protection of Military Personnel Participating in Particular Military Operations. Sept. 30, 1999. EPI371

114. Bowman, Tom. "Resignations lead Air Force to reassess vaccine policy." *Anchorage Daily News*, March 18, 1999, p, A-6. EPI199

> At the same time, however, the web can also provide our adversaries with a potent instrument to obtain, correlate and evaluate an unprecedented volume of aggregated information regarding DoD capabilities, infrastructure, personnel and operational procedures. Such information, especially when combined with information from other sources, increases the vulnerability of DoD systems and may endanger DoD personnel and their families.
>
> Component heads must enforce the application of comprehensive risk management procedures to ensure that the considerable mission benefits gained by using the web are carefully balanced against the potential security and privacy risks created by having aggregated DoD information more readily accessible to a worldwide audience."115

> "'The biggest mistake people make is they don't understand how easy it is to aggregate information,' Walsh said. The lesson from this is that even though what is posted on the net is perfectly innocent in and by itself, when combined with other existing information, a larger and more complete picture might be put together that was neither intended nor desired."116

While the points raised by the military in their reports are well taken, an overreaction can also take place and information that should be open to the public becomes unavailable. Public debate is central to democracy and the maintenance of freedom. Free flow of information is essential to this process.

Creating Public Prejudice through Media

"Certain events over the past 20 years, particularly the Oklahoma City bombing, have shown that domestic terrorism is a legitimate concern for U.S. law enforcement officials. According to Joseph Roy Sr. of the Southern Poverty Law Center in Montgomery, Ala., which publishes 'Klan Watch' and 'Intelligence Report,' there exists a large underground movement orchestrated by extremists with little or no regard for law enforcement officials. Roy says that since the Oklahoma City bombing, investigations into subversive groups have grown from about 100 a year to about 1,000 per year. Potential terrorists could include loners, or those within cults or other extremist groups. Bombings, shootings, kidnappings, extortion, and even chemical or biological attacks make the list for possible threats. An especially vulnerable area is the public transportation infrastructure. Many people with anti-government and anti-tax sentiments have joined militias. These groups, protected by constitutional rights, may

115. Deputy Secretary of Defense. Information Vulnerability and the World Wide Web. Memorandum, Sept. 24, 1998. EPI217
116. Stone, Paul. American Forces Press Service. "Internet Presents Web of Security Issues." Sept 1998. EPI104

be stockpiling weapons and planning life-threatening events. Roy notes that use by radicals of the right to free speech as provided by the First Amendment poses problems for law enforcement officials. Gun shows, patriot rallies, the Internet, and other media facilitate the spread of these organizations. In its Spring 1998 edition, Klan Watch reported 179 patriot Web sites. Roy says another trend worth watching is racial unrest – racial polarization is increasing, and many believe a race war is imminent. Law enforcement officials need to prepare and train now for a wide variety of potential domestic terrorist threats."117

Overreaction by government is a greater threat to democracy than those posed by real or imagined threats. We have personally met many people associated with the "patriot" movement and the vast majority of these individuals are law abiding citizens with no motives other than to live under the rules of our constitutionally constituted democratic republic and nothing more. The vilification of the "patriot" movement proceeds, with media assistance, under principles of "guilt by association." Their methods, in reality, would be no different than polling axe murderers for political affiliation and claiming the results support the notion that most are either democrats or republicans and then drawing the conclusion that all democrats or republicans are guilty of axe murder because of their politics. Most of the patriot movement is about regional politics within the constraints of our form of government and does not represent any threat to anyone.

When Government Abuses Secrecy Hiding Behind Power

One of the most notorious examples of the U.S. government's abuse of its citizens was under the cloak of secrecy in a program operated by the CIA known as MKULTRA. The words of the CIA themselves are the most contemptible in their implication of the Agency for their knowledgeable, unethical and illegal use of Americans in these experiments. A memorandum for the director reads as follows:

"Memorandum For: Director of Central Intelligence
Subject: Report of Inspection of MKULTRA

1. In connection with our survey of Technical Services Division, DD/P, it was deemed advisable to prepare the report of the MKULTRA program in one copy only, in view of its unusual sensitivity.

2. This report is forwarded herewith.

3. The MKULTRA activity is concerned with research and development of chemical, biological, and radiological materials

241. Morrison, Richard. "Domestic Terrorism: How Real Is the Threat?" *Law Enforcement technology*, Jan. 1999. Vol. 26, No. 1. Source: NLECTC *Law Enforcement & technology News Summary*, Feb. 4, 1999. EPI1038

capable of employment in clandestine operations to control human behavior. The end products of such research are subject to very strict controls including a requirement for the personal approval of the Deputy Director/Plans for any operational use made of these end products.

4. The cryptonym MKULTRA encompasses the R&D phase and a second cryptonym MKDELTA denotes the DD/P system for control of the operational employment of such materials. The provisions of the MKULTRA authority also cover [deleted]. The administration and control of this latter activity were found to be generally satisfactory and are discussed in greater detail in the main body of the report on TSD.

5. MKULTRA was authorized by the then Director of Central Intelligence, Mr. Allen W. Dulles, in 1953. The TSD was assigned responsibility thereby to employ a portion of its R&D budget, eventually set at 20%, for research in behavioral materials and [deleted] under purely internal and compartmented controls, (further details are provided in paragraph 3 of the attached report). Normal procedures for project approval, funding, and accounting were waived. However, special arrangements for audit of expenditures have been evolved in subsequent years.

6. The scope of MKULTRA is comprehensive and ranges from the search for and procurement of botanical and chemical substances, through programs for their analysis in scientific laboratories, to progressive testing for effect on animals and human beings. The testing on individuals begins under laboratory conditions employing every safeguard and progresses gradually to more and more realistic operational simulations. The program requires and obtains the services of a number of highly specialized authorities in many fields of the natural sciences.

7. The concepts involved in manipulating human behavior are found by many people both within and outside the Agency to be distasteful and unethical. There is considerable evidence that opposition intelligence services are active and highly proficient in this field. The experience of TSD to date indicates that both the research and the employment of the materials are expensive and often unpredictable in results. Nevertheless, there have been major accomplishments both in research and operational employment.

8. The principal conclusions of the inspection are that the structure and operational controls over this activity need strengthening; improvements are needed in the administration of the research projects; and some of the testing of substances under simulated operational conditions was judged to involve excessive risk to the agency.

Report of Inspection of MKULTRA/TSD

1. Introduction

2. The MKULTRA charter provides only a brief presentation of the rationale of the authorized activities. The sensitive aspects of the program as it has evolved over the ensuing years are the following:

a. Research in the manipulation of human behavior is considered by many authorities in medicine and related fields to be professionally unethical, therefore the reputations of professional participants in the MKULTRA program are on occasion in jeopardy.

b. Some MKULTRA activities raise questions of legality implicit in the original charter.

c. A final phase of the testing of MKULTRA products places the rights and interests of U.S. citizens in jeopardy.

d. Public disclosure of some aspects of MKULTRA activity could induce serious adverse reaction in U.S. public opinion, as well as stimulate offensive and defensive action in this field on the part of foreign intelligence services.

5. The inspection of MKULTRA projects in biochemical controls of human behavior raised questions in the following area of policy and management which are dealt with in the balance of this report:

a. Scope of the MKULTRA charter

(1) Over the ten year life of this program many additional avenues to the control of human behavior have been designated by the TSD management as appropriate to investigation under MKULTRA charter, including radiation, electro-shock, various fields of psychology, psychiatry, sociology, and anthropology, graphology, harassment substances, and paramilitary devices and materials.

(2) Various projects do not appear to have been sufficiently sensitive to warrant waiver of normal Agency procedures for authorization and control.

c. Advanced testing of MKULTRA materials:

It is the firm doctrine in TSD that testing of materials under accepted scientific procedures fails to disclose the full pattern of reactions and attributions that may occur in operational situations.

> TSD initiated a program for covert testing of materials on unwitting U.S. citizens in 1955. The present report reviews the rationale and risks attending this activity and recommends termination of such testing in the United States...

II. Modus Operandi

6. The research and development of materials capable of producing behavioral or physiological change in humans is now performed within a highly elaborated and stabilized MKULTRA structure. The search for new materials; e.g., psilocybin from Mexican mushrooms, or a fungi occurring in agricultural crops, is conducted through standing arrangements with specialists in universities, pharmaceutical houses, hospitals, state and federal institutions, and private research organizations who are authorities in the given field of investigation in their own right. Annual grants of funds are made under ostensible research foundation auspices to the specialists located in the public or quasi-public institutions. This approach conceals from the institution the interest of CIA and permits the recipient to proceed with his investigation, publish his findings (excluding intelligence implications), and account for his expenditures in a manner normal to his institution.

10. The final phase of testing of MKULTRA materials involves their application to unwitting subjects in normal life settings. It was noted earlier that the capabilities of MKULTRA substances to produce disabling or discrediting effects or to increase the effectiveness of interrogation of hostile subjects cannot be established solely through testing on volunteer populations. Reaction and attribution patterns are clearly affected when the testing is conducted in an atmosphere of confidence under skilled medical supervision.

11. TSD, therefore, entered into an informal arrangement with certain cleared and witting individuals in the Bureau of Narcotics in 1955 which provided for the release of MKULTRA materials for such testing as those individuals deemed desirable and feasible...

12. The particular advantage of these arrangements with the Bureau of Narcotics officials has been that test subjects could be sought and cultivated within the setting of narcotics control. Some subjects have been informers or members of suspect criminal elements from whom the Bureau had obtained results of operational value through the tests. On the other hand, the effectiveness of the substances on individuals at all social levels, high and low, native American and Foreign, is of great significance and testing has been performed on a variety of individuals within these categories.

14. The MKULTRA program officer stated that the objectives of covert testing concern the field of toxicology rather than medicine; further, that the program is not intended to harm test individuals, and that the medical consultation and assistance is obtained when appropriate through separate MKULTRA arrangements. The risk of compromise of the program through correct diagnosis of an illness by an unwitting medical specialist is regularly considered and is stated to be a governing factor in the decision to conduct the given test. The Bureau officials also maintain close working relations with local police authorities which could be utilized to protect the activity in critical situations.

17. The final stage of covert testing of materials on unwitting subjects is clearly the most sensitive aspect of MKULTRA. No effective cover story appears to be available. TSD officials state that responsibility for covert testing is transferred to the Bureau of Narcotics. Yet they also predict that the Chief of the Bureau would disclaim any knowledge of the activity. Present practice is to maintain no records of the planning and approval of test programs...

30. TSD has initiated 144 projects relating to the control of human behavior; i.e. [deleted] during the ten years of operation of the MKULTRA program. Twenty-five (25) of these projects remain in existence at the present time, while a number of others are in various stages of termination.

31. Active projects may be grouped under the following arbitrary headings. Many projects involve activity in two or more of the areas listed.

a. basic research in materials and processes
b. procurement of research materials
c. testing of substances on animals and human beings
d. development of delivery techniques
e. projects in offensive/defensive BW, CW, and radiation
f. miscellaneous projects; e.g., (1) petroleum sabotage, (2) defoliants, (3) devices for remote measurement of physiological processes."[118]

What more can be said for July 26, 1963, when this document was written? Keep in mind that throughout the period of the program numerous others were underway and hundreds of documents were shredded to protect the guilty and hide many aspects of the technology. After 1963, the evidence shows a significant interest in radio frequency (RF) weapons and direct control of human behavior at a distance. These aspects of the research are where the greatest modern emphasis has been, rather than chemical or biological agents which violate existing treaties. These new technologies are exempt

118. Report of Inspection of MKULTRA. Memorandum For: Director of Central Intelligence. July 26, 1963. MORI DocID: 17748. EPI1201

from regulation because they are of low, "safe," power levels and their effects are not widely understood. These technologies will be discussed in detail in another chapter.

In another memorandum released on these projects the following is revealed:

"Memorandum For: The Record
Subject: MKULTRA, Subproject 119

1. The purpose of this subproject is to provide funds for a study conducted by [deleted] to make a critical review of the literature and scientific developments related to the recording, analysis and interpretation of bioelectric signals from the human organism, and activation of human behavior by remote means. When initiated this study was being done on a consultant basis by [deleted]. The reason for converting this into a subproject is to provide more flexibility in the disbursal of funds for various kinds of assistance and equipment needed.

2. As indicated in the attached proposal this study is to provide an annotated bibliography and an interpretive survey of work being done in psychophysiological research and instrumentation. The survey encompasses five main areas:

a. Bioelectric sensors: sources of significant electrical potential and methods of pick-up.

b. Recording: amplification, electronic tape and other multi-channel recording.

c. Analysis: autocorrelators, spectrum analyzers, etc. and coordination with automatic data processing equipment.

d. Standardization of data for correlation with biochemical, physiological and behavioral indices.

e. Technique of activation of the human organism by remote electronic means.[119]

The electromagnetic revolution was underway, according to this memorandum, as far back as 1960, which makes sense in light of the patents and inventions being developed at that time. Mind Control was becoming mechanically possible and not just theoretically possible.

The Commission on CIA Activities Within the United States, *Report to the President by the Commission On CIA Activities Within The United States* was issued in June, 1975:

119. Memorandum For: The Record. Subject: MKULTRA, Subproject 119. Draft, Aug. 17, 1960. Doc. 119.18.1 EPI1204

"While the research and development of new CIA scientific and technical devices is naturally undertaken within the United States, the evidence before this Commission shows that with a few exceptions, the actual devices and systems developed have not been used operationally within this country.

However, the Agency has tested some of its new scientific and technological developments in the United States. One such program included the testing of certain behavior-influencing drugs. Several others involved the testing of equipment for monitoring conversations. In all of the programs described, some tests were directed against unsuspecting subjects, most of whom were U.S. citizens.

1. The Testing of Behavior-Influencing Drugs on Unsuspecting Subjects Within the United States.

In the late 1940's, the CIA began to study the properties of certain behavior-influencing drugs (such as LSD) and how such drugs might be put to intelligent use. This interest was prompted by reports that the Soviet Union was experimenting with such drugs and by speculation that the confessions introduced during trials in the Soviet Union and other Soviet Bloc countries might have been elicited by the use of drugs or hypnosis. Great concern over Soviet and North Korean techniques in "brainwashing" continued to be manifested into the early 1950's.

The drug program was part of a much larger CIA program to study possible means for controlling human behavior. Other studies explored the effects of radiation, electric-shock, psychology, psychiatry, sociology and harassment substances.

The primary purpose of the drug program was to counter the use of behavior-influencing drugs clandestinely administered by an enemy, although several operational uses outside the United States were also considered.

Unfortunately, only limited records of the testing conducted in these drug programs are now available. All the records concerning the program were ordered destroyed in 1973, including a total of 152 separate files.

In addition, all persons directly involved in the early phases of the program were either out of the country and not available for interview, or were deceased.

The Commission did learn, however, that on one occasion during the early phases of this program (in 1953), LSD was administered to an employee of the Department of the Army without his knowledge while he was attending a meeting with CIA personnel working on the project.

Prior to receiving the LSD, the subject had participated in discussions where the testing of such substances on unsuspecting subjects was agreed to in principle. However, this individual was not made aware that he had been given LSD until about 20 minutes after it had been administered. He developed serious side effects and was sent to New York with a CIA escort for psychiatric treatment. Several days later, he jumped from a tenth floor window of his room and died as a result.

The General Counsel ruled that the death resulted from 'circumstances arising out of an experiment undertaken in the course of his official duties for the United States Government,' thus ensuring his survivors of receiving certain death benefits. Reprimands were issued by the Director of Central Intelligence to two CIA employees responsible for the incident."[120]

It is again time to revisit this program and those that ran parallel to it. The new science of applying subtly modulated energy to manipulate brain activity or in order to alter other physiological functions should be investigated again across all agencies and departments of the government. It can also be seen in terms of why there is such an increase in interest in surveillance and other invasive technologies, resulting in "national security" interests being valued more than foundational human values.

WACO & Civil Rights?

The siege of the Branch Davidians by federal law enforcement and military authorities was a travesty of American law. The evidence has shown that the issue belonged in the hands of Texas justice and policing authorities and that federal intervention could have been avoided, sparing the lives of both Branch Davidians and law enforcement officials. Local law enforcement officials had had reasonable nonconfrontational relations with these individuals and had no reason to expect anything different.

Attorney General Janet Reno, in September 1999, "ordered U.S. Marshals to the FBI's headquarters to seize a previously undisclosed tape recording of voice communications between FBI commanders and field agents during the ill-fated tear-gas assault at the Branch Davidian compound, Justice Department officials said."[121] Senate Judiciary Committee Chairman Orrin Hatch, R-Utah, "said he too is launching an inquiry. Reno and other leaders at Justice, he said don't 'have much credibility at this point.'" The FBI

120. Commission on CIA Activities Within the United States. *Report to the President by the Commission On CIA Activities Within The United States.* June 1975. EPI854
121. *New York Times.* "Marshals seize Waco tape." *Anchorage Daily News*, Sept. 2, 1999. EPI1072

had revealed the week before "that 'a very limited number' of incendiary tear gas grenades were shot at a bunker in the Waco, Texas, compound in the early hours of April 19, 1993. Later, the main wooden structure where the Branch Davidians had barricaded themselves burst into flames."[122] A few days after these reports ran, "former Republican Sen. John Danforth agreed Wednesday to oversee an independent review of the 1993 government standoff with the Branch Davidians, government officials said. Reno's decision came as the top Republican in the Senate said he now has doubts about who started the fire that ended the fatal siege in Waco, Texas, and believes it is time for Reno to step down."[123]

The sad story is that this was not the only time that the government had trouble with the truth when it came to abusing its own citizens and "it is not just those on the radical right who are angry at learning that the FBI failed to admit for six years that it used military tear gas cartridges that can ignite.'The government has an unfortunate long history of not coming clean with the mistakes it makes which dates back through history,' said John Lunsford, a researcher with the Coalition for Human Dignity, a Seattle-based group that monitors white-supremacy organizations."[124]

Looking deeper into the controversy what had surfaced was ...a report, subpoenaed by the House Government Reform Committee, which "confirmed that a 40mm shell casing reexamined by the rangers was a type of military tear gas round that 'burns at 500 to 700 degrees Fahrenheit, and is capable of igniting flammable items.'"[125]

The gas used at Waco was banned by international convention and yet was used on U.S. citizens, a type of situation which was pointed out in an International Red Cross document on the subject of chemical weapons. The report looked at the ramifications of international law regarding use of these technologies. It pointed out weaknesses in the international conventions regarding the use of chemical weapons:

> "Therefore, when the Convention (Chemical Weapons Convention) comes into force next year, activities involving them – activities such as development, production, stockpiling and use – will become illegal, unless their purpose is a purpose that is expressly not prohibited under the Convention. One such purpose is 'law enforcement including domestic riot control purposes.'[126] Unfortunately, the Convention does not define what it means by 'law enforcement' (whose law? what law? enforcement where? by whom?), though it does define what it means by

122. Abrams, Jim. "Republicans pledge probe of Waco siege." *Anchorage Daily News*, Aug. 30, 1999. EPI841
123. Yost, Pete. "Danforth to Head Waco Inquiry." *Newsday.com*, Sept. 8, 1999. http://www.newsday.com/ap/rnmpwh1u.htm EPI1143
124. Hedges, Michael. "Radical groups abuzz over Waco revelations." *Anchorage Daily News*, Sept. 4, 1999. EPI1100
125. Schwartz, Jerry. "Resurrection of Waco." *Anchorage Daily News*, Sept. 26, 1999. EPI381
126. Chemical Weapons Convention, Article II.9(d). EPI1261

'riot control agent', namely 'any chemical...which can produce rapidly in humans sensory irritation or disabling physical effects which disappear within a short time following termination of exposure.' State parties are enjoined 'not to use riot control agents as a method of warfare.'"[127]

The lines between federal and local police are disappearing as increased federalization and control are fostered on local police agencies. The need for local control with a potential for federal assistance does make sense, but not the other way around. Police activity needs to remain in the hands of state and local authorities.

Civil Rights and Harmless Errors

"...During a search for evidence in the assault case, specifically Heflin's bloodied shirt, cowboy hat and some cigarette packs, SWAT team members pointed guns at small children, including Heflin's 4-year-old granddaughter. Her mother said the little girl was chased down a hall with a red laser beam from an assault rifle positioned in the middle of her back."[128] Aggressive police actions are resulting in increased risks to innocent bystanders while frustration continues to build because of the backlash of civil libertarians.

The U.S. Supreme Court, voting 5 to 4, "said some convictions may be allowed to stand despite the use of confessions obtained in violation of the defendant's constitutional rights. In an opinion by Chief Justice William H. Rehnquist, the court said there may be so much other evidence of guilt that the use of an involuntary confession could be considered 'harmless error.'"[129] Another step taken on a path best left untraveled.

The Rule of Law
The Ethics of the Unethical

One of the things which has always bothered me as a researcher is how the "little guy" is always held to a high standard of accountability while big organizations get away with murder. I am not suggesting that individuals should be held to a lesser standard – quite to the contrary. Organizations responsible for the security of the nation should be held to the highest standards. We must ask ourselves what these agencies are charged with protecting and whether their actions follow the values expressed in law. Are there reasons that the government should be excused from meeting the requirements of the law? Is there good cause for hiding behind laws which allow for the exploitation of other laws? The following illustrates the point:

127. "Expert Meeting on Certain Weapon Systems and on Implementation Mechanisms in International Law", Report of the International Committee of the Red Cross, Geneva, Switzerland, May 30 - June 1, 1994. Issued July 1994. EPI1232
128. Draper, Electra. "Judge OKs suing SWAT, sheriff." *Denver Post,* Aug. 19, 1999. http://www.denverpost.com/news/news0819h.htm EPI384
129. Marcus, Ruth. "Court Splits On Coerced Confessions." *Washington Post,* March 27, 1991. EPI604

> "A former CIA officer from the agency's top secret 'black bag' unit that breaks into foreign embassies to steal code books was charged with espionage Friday for tipping off two countries about the CIA's success in compromising their communications."[130]

Douglas Groat was fired in 1996 from the CIA's Science and Technology Directorate and could have faced the death penalty. These super secret teams are sent around the world to break into embassies and other locations to steal codes and other information so that the National Security Agency (NSA) can intercept another country's classified communications and know their contents. The article concluded:

> "The CIA has never publicly acknowledged the existence of its black-bag teams because their operations are by their nature illegal. And they not only target America's adversaries but embassies of friendly powers."[131]

Consider the contents of this article from the perspective of one of our allies. Remember a few years ago the outrage of our government when we discovered that the State of Israel was using its intelligence gathering resources in the U.S. It was an outrage – or was it just the game we all play? Why should we expect anything less of our allies then we expect of ourselves?

Murder rather than trial was reported in 1989 to be one of the options for the "war on drugs" which was being used to increase fear and assert greater levels of control over the population by security agencies. "The assassination of major international drug traffickers is under consideration by senior U.S. officials devising aggressive new measures against them. Encouraged by the President, Congress and a new drug czar demanding stepped-up efforts to deal with narcotics control, a National Security Council-convened team is seriously considering 'military operations' including such actions against 'a number of high-value targets,' according to a senior administration official."[132]

---

130. *Anchorage Daily News.* "Ex-CIA Officer Faces Charges of Espionage." James Risen (*Los Angeles Times*) p. A-3, April 4, 1998. EPI589
131. Ibid.
132. Greve, Frank. "U.S. weighs assassination of foreign drug traffickers." *Philadelphia Enquirer*, June 10, 1989. EPI572

Special Access Programs

"Special access programs (SAPs) employ a variety of security measures that are far more restrictive than those used in 'ordinary' classified programs and that largely shield them from independent oversight.

'DOE is not necessarily aware of what SAPs are being worked on in DOE facilities...We found this condition to be true at DOE headquarters, a DOE operations office and a national laboratory'...

The Inspector General also found that cost accounting for SAPs and other classified programs was often wildly inaccurate, if not deliberately fraudulent. For example, some program costs were charged to overhead and paid for parasitically by other accounts. Several classified programs reviewed by the IG falsely reported a total estimated program cost of zero dollars."133

"When normal security methods are insufficient, special access controls safeguard operational and technological advantages from potential enemies by limiting access to information about, or observation of, certain weapons, weapons systems, techniques and operations. DoD, SAF, HQ USAF, and many commands, agencies, and program offices work together in SAPs to create, maintain, and discard special access controls.

These controls provide extraordinary protection by: keeping personnel access to the minimum needed to meet program goals; setting investigative or adjudicative criteria for persons seeking access; naming officials who determine whether cleared people have a need-to-know; using access lists and registered unclassified nicknames (and, in some cases, classified code words) to identify information needing additional protection; security guides and procedures specifically tailored for certain information and equipment; and supporting and overseeing infrastructures.

Only 'core secrets' have special access controls. Normal security controls, classification, and handling systems protect other information. Examples of activities protected with special access controls might include a technology breakthrough or exploitation of an enemy's weakness. A SAP could also fix a US weakness or protect extremely sensitive operations."134

"Cover stories may be established for unacknowledged programs in order to protect the integrity of the program from individuals who do not have a need-to-know. Cover stories must be believable and cannot reveal any information regarding the true nature of the contract. Cover stories for Special Access Programs must have the approval of the PSO (Program Security Officer) prior to dissemination."135 One such example of a cover story: "In the darkest days of the cold war, the military lied to the American public about the true nature of many unidentified flying objects in an effort

133. *Secrecy & Government Bulletin*. "Mystery Programs Thrive at DOE." Issue No. 37, July 1994. EPI872

134. USAF. Special Access Programs, Air Force Instruction 16-701. Jan. 28, 1994. EPI389

135. Horgan, John. "Lying by the Book." *Scientific American*, Oct. 1992. EPI706

to hide its growing fleets of spy planes. The deceptions were made in the 1950s and '60s amid a wave of UFO sightings that alarmed the public and parts of official Washington."[136]

The lies, experiments and cover stories continue in every field. The lies have been institutionalized and are as intricate and as devastating as a missile. Breakthroughs in health, energy, propulsion, communications and information technologies are classified and hidden from view. Some of these technologies were discovered applying the same ethical rules used on death camp victims during World War II, only this time it was Americans experimenting on Americans – hiding behind their badges of rank and the sacrifices of our forefathers.

136. Broad, William J. "Air Force lied about UFO's to hide spy planes." *Anchorage Daily News*, Aug. 3, 1997. EPI637

Chapter 5

Non–ionizing Radiation The Debate

"...the density of radio waves around us is now 100 million or 200 million times the natural level reaching us from the sun..."[137]

"The human species has changed its electromagnetic background more than any other aspect of the environment. For example, the density of radio waves around us is now 100 million or 200 million times the natural level reaching us from the sun. Nor is there any end in sight."[138] "Intense exposures – the kind a person standing next to a powerful radar antenna would get – can burn the skin and underlying tissue. According to some scientists, a high dose can also cloud the eyes with cataracts, much as a hot frying pan turns the transparent part of an egg opaque. As for the effect of lower doses, however, scientific opinion on the question has split along national boundaries."[139] "In the USSR and other eastern European countries, thermal effects were not used as criteria for safety standards. Instead, so-called nonthermal effects, effects reported after exposure to low levels insufficient to cause a measurable rise in temperature, were the criteria in setting maximum permissible limits."[140] In the United States two key recommendations for the EPA emerged from a recent conference "1) develop RF radiation exposure guidance as soon as possible, and (2) conduct additional research in a number of areas, particularly with respect to the potential for 'nonthermal' effects."[141]

"The principal electromagnetic biological effects of greatest concern are behavioral aberrations, neural network perturbations, fetal (embryonic) tissue damage (including birth defect), cataractogenesis, altered blood chemistry, metabolic changes and suppression of the

137. Becker, Robert and Selden, Gary. *The Body Electric: Electromagnetism And The Foundation Of Life.* William Morrow, New York. 1985. ISBN 0-688-06791-1. EPI857
138. Ibid.
139. Gold, Michael. "The Radiowave Syndrome." *Science.* EPI494
140. Silverman, Charlotte. "Nervous and Behavioral Effects of Microwave Radiation in Humans." *Journal of Epidemiology,* Vol. 97, No. 4. April, 1973. EPI369
141. U.S.EPA. *Summary and Results of the April 26-27, 1993 Radiofrequency Radiation Conference, Volume 2: Papers.* 402-R-95-011, March 1995. EPI728

endocrine and immune systems."[142] The military emphasis on the disabling and killing effects of this knowledge is creating the environment for unregulated experimentation in this area. "There is also a fundamental ethical dilemma for doctors. The development of this new generation of weapons incorporates knowledge from the remarkable advances made in medical science; two examples are calmatives and eye attack lasers. The ultimate expression of this dilemma is the potential development of race specific weapons based on knowledge of genetic engineering and human genome diversity."[143] These are the same types of medical experiments which ended on the gallows in the Nuremberg trials.

European interest in these issues continues to grow and by December 1999 resulted in the release of a draft request for proposals which included the following language:

> "'Physiological and environmental effects of electromagnetic radiation' is a project being conducted by the STOA Unit in the Directorate-General for Research of the European Parliament.
>
> **Scientific research into the biological and health effects of electromagnetic fields.**
>
> A survey of the findings of research into the biological and health effects of non-ionising radiation has already been submitted to Parliament in the aforementioned 1994 Lannoye report, which states that 'all these results undoubtedly help to provide a reliable scientific basis on which the decision-makers must rely in defining standards and regulations'. It also points out that even if the mechanisms causing biological injury have not clearly been elucidated, we have today sufficient information to adopt the standards and regulations on the basis of two guiding principles:
>
> 1) the precautionary principle.
>
> 2) the World health Organisation's ALARA (As Low As Reasonably Achievable) principle, under which exposure to radiation must be as low as reasonably possible, which excludes avoidable exposure to radiation.
>
> Since 1994 the views set out in the Lannoye report have been endorsed by authoritative, well publicized scientific studies, which, however, appear to have been ignored both by the author of the proposal for a recommendation and by the experts on DG XXIV's Scientific Committee (as is apparent

142. *Microwave News.* "Biological Effects of Microwave Radiation: A White Paper." September/October 1993. EPI729
143. Coupland, Robin M., Surgeon. "Non-lethal" weapons: precipitating a new arms race. *British Medical Journal*, Vol. 315, July 12, 1997. EPI565.

from the bibliography which appears in the committee's opinion).

No mention is made of the effects of low-frequency EMFs (generated by electricity transmission lines) on cell membrane receptors, which pass on into the cell itself and trigger off enzymatic activity and the production of chemical messages which can activate genetic transcription. The relevant data are, nonetheless, to be found in the conclusions of the European Community symposium on 'Electromagnetic transmissions: the latest scientific evidence, potential threats and strategies to reduce risk' held in London on 27 October 1994 and in the collective work published by Springer Verlag in 1995 under the title 'On the nature of electromagnetic field interactions with biological systems' (edited by A.H. Frey).

The 1999 Tamino Report also points out that these effects are of fundamental importance in understanding exactly how EMFs may be involved in the process of carcinogenesis, which is considered to involve two stages: *Initiation,* when the initial genetic damage occurs (to DNA) and *promotion* of the proliferation of cancerous cells.

Normally, the agents involved in the process of initiation (ionising radiation, alkylating agents, etc.) are not active in the subsequent promotion stage, which is triggered by agents which may either interact with membrane receptors or inhibit natural mechanisms designed to eliminate cancerous cells (for example, the immune system). A large number of laboratory studies indicate that EMFs are instrumental in the promotion of tumors (see inter alia the research by W. Loescher and others, referred to in the Scientific Committee's opinion, which concludes that there is limited evidence from laboratory studies in support of the theory that EMFs promote tumors). Furthermore, research has shown that 50 Hz EMFs have the effect of depressing the immune system and reducing melatonin secretion, which are of vital importance in understanding how EMFs might promote tumors. In this connection, the work carried out by Liburdy (which is described in the aforementioned book edited by Frey) is of particular interest, in that it demonstrates that melatonin continues to have an oncoststic effect at an exposure level of 0.2 micro tesla, while that effect is blocked at 1.2 micro tesla.144

What this suggests is that the classical theories of how the mind and body interact with electromagnetic fields are much different than originally thought. Subtle energy can be modulated to interact with the human body as easily as a radio wave interacts with a radio receiver. In other words, the classical theories are partially wrong in

144. European Parliament. STOA Project Brief. *Physiological and environmental effects of electromagnetic radiation.* October 1999. EPI1209

that they do not fully explain all of the reactions which are observed in the body. The Navy has abstracted over a thousand international professional papers by private and government scientists which explore these issues. Capt. Paul Tyler writes:

> "Even though the body is basically an electrochemical system, modern science has almost exclusively studied the chemical aspects of the body and to date has largely neglected the electrical aspects. However, over the past decade researchers have devised many mathematical models to approximate the internal fields in animals and humans. Some of the later models have shown general agreement with experimental measurements made with the phantom models and animals. Presently most scientists in the field use the concept of specific absorption rate for determining the Dosimetry (dosages) of electromagnetic radiation. Specific absorption rate is the intensity of the internal electric field or quantity of energy absorbed... However, the use of these classical concepts of electrodynamics does not explain some experimental results and clinical findings. For example, according to classical physics, the frequency of visible light would indicate that it is reflected or totally absorbed within the first few millimeters of tissue and thus no light should pass through significant amounts of tissue. But it does. Also, classical theory indicates that the body should be completely invisible to extremely low frequencies of light where a single wave length is thousands of miles long. However, visible light has been used in clinical medicine to transilluminate various body tissues."

> "A second area where classical theory fails to provide an adequate explanation for observed effects is in the clinical use of extremely low frequency (ELF) electromagnetic fields. Researchers have found that pulsed external magnetic fields at frequencies below 100 Hertz (pulses/cycles per second) will stimulate the healing of nonunion fractures, congenital pseudarthroses, and failed arthroses. The effects of these pulsed magnetic fields have been extremely impressive, and their use in orthopedic conditions has been approved by the Food and Drug Administration."

> "Recently, pulsed electromagnetic fields have been reported to induce cellular transcription (this has to do with the duplication or copying of information from DNA, a process important to life). At the other end of the non-ionizing spectrum, research reports are also showing biological effects that are not predicted in classical theories.

> For example, Kremer and others have published several papers showing that low intensity millimeter waves produce biological effects. They have also shown that not only are the effects seen at very low power, but they are also frequency-specific."[145]

The development of these technologies and research projects represent the beginning of the research which had started as far back as the late 1950s and had significantly progressed by 1984 when the above report was written. The European Parliament's study will surely point the direction for the next century as it is hopefully completed sometime in 2000 or 2001.

Gene Crime

The direction of science and increase in our understanding has placed man in a position to dictate and manipulate genetics and DNA in all ways – from creating cones to selecting specific genetic attributes. The research has also shown that subtle energy fields, when manipulated, pulsed, shaped or modulated on specific carrier waves, can alter life itself. The ethical debate of the new technology can not be avoided.

In one research laboratory "...scientists are developing a technique to weed out embryos at high risk of developing cancer. The development is to be evaluated by the Human Fertilization and Embryology Authority, but moves to weed out 'unacceptable' embryos have been described as the 'Nazification' of medicine by a leading expert on medical ethics."[146]

"In February of 1992, psychiatrist Frederick Goodwin announced a government plan to identify at least 100,000 inner city children whose alleged biochemical and genetic defects will make them violent in later life. Treatments...include screening programs through the schools, special 'day camps' for children, behavior modification for the family, and psychiatric treatment including drugs."[147] "It should be noted that projects aimed at discovering or measuring 'cognitive dysfunction' or 'neurophysiological deficits' ultimately attempt to link brain dysfunction to violent or criminal behavior."[148] "Now the basic purpose of what NIMH is trying to do is on this light: to design and evaluate psychosocial, psychological and medical interventions for at-risk children before they become labeled

145. Low-Intensity Conflict and Modern Technology, Lt. Col. David J. Dean USAF, Editor, Air University Press, Center for Aerospace Doctrine, Research, and Education, Maxwell Air Force Base, Alabama, June 1986. EPI709
146. *BBC News*. "Warning over 'Nazi' genetic screening." Aug. 9, 1999. EPI1217
147. Bregin, Peter R., M.D. and Breggin, Ginger. The Center For The Study of Psychiatry Update Report. "The Federal Violence Initiative: Threats to Black Children (and others)." *Psych Discourse*, Vol. 24, No. 4., April 1993. EPI115
148. Breggin, Peter R., M.D. and Breggin, Ginger Ross. "The Hazards of Biologizing Social Problems." Center For The Study Of Psychiatry, Inc. June 2, 1993. EPI106

as delinquent or criminal."149 "Researchers have already begun to study the genetic regulation of violent and impulsive behavior and to search for genetic markers associated with criminal conduct. Their work is motivated in part by the early successes of research on the genetics of behavioral and psychiatric conditions like alcoholism and schizophrenia."150 "The authors say there is now widespread evidence that children who show aggressive behavior at about age 8 'are more likely than others to exhibit delinquent, criminal or violent behavior in adolescence and adulthood.' This pattern of behavior in aggressive 8-year olds is likely to persist as they get older even if they leave their families, schools or neighborhoods, the report found."151 "But violence research at NIH, as the panel heard, is very much alive. Nine institutes and two centers have funded about 300 research projects totaling $42 million for 1993, with the bulk of work being done at the National Institute of Mental health (NIMH). Much of the NIH violence research is aimed at determining biological factors that may underlie violent behavior."152 This issue was firmly resisted by a committed group within the profession and the initiative was stopped. The idea of race or genetic markers for behavior were the stuff of science fiction and Nazi science, yet here it was masquerading as ethical science in the United States of America.

By 1997, the theme was raised again: "The view of mental illness as a brain disease has been crucial to the effort to destigmatize illnesses such as schizophrenia and depression. But it's just one example of a much broader biologizing of American culture that's been going on for more than a decade...everything from criminality to addictive disorders to sexual orientation is seen today less as a matter of choice than of genetic destiny."153 "No one believes there is a single 'criminal gene' that programs people to maim or murder. Rather, a person's genetic makeup may give a subtle nudge toward violent actions."154 In another government-funded study, "poor black and Hispanic boys were given a now-recalled diet drug to test for violent tendencies is being criticized as risky and racist...The researchers defended their efforts as a legitimate attempt to understand the roots of violence."155 Perhaps there are markers but what else do these combinations show in terms of creativity, leadership or other characteristics? Who will decide what characteristics to cultivate and which to breed out? What will be the results over time? Will we create a reaction, a disease or some other unsuspecting effect as we manipulate the genetics of our own species? Will we create human

149. Goodwin, Frederick K. "Conduct Disorder as a Precursor to Adult Violence and Substance Abuse: Can the Progression be Halted?" American Psychiatric Association Annual Convention. May 1992. EPI107
150. A Conference sponsored by The Institute for Philosophy and Public Policy and The National Institutes of Health. "Genetic Factors in Crime: Findings, Uses & Implications." Oct. 9-11, 1992. EPI110.
151. Butterfield, Fox. "Study Cites Biology's Role in Violent Behavior." *New York Times*, Nov. 13, 1992. p. A-12. EPI111
152. *Science*. "Panel Finds Gaps in Violence Studies." Vol. 260. June 11, 1993. EPI114
153. Herbert, Wray. "Politics of Biology." *U.S. News & World Report*, April 21, 1997, pp. 72-80. EPI39
154. Toufexis, Anastasia. "Seeking the Roots of Violence." *Time*, April 19, 1998. EPI112
155. AP. "Psychiatric experiments on boys draw complaints." *Star Tribune*, April 18, 1998. EPI626

"killers" as we created killer bees or will we contribute to new breakthroughs in science?

"Italian scientists have genetically engineered mice to live up to 35 percent longer than normal, an experiment that offers the strongest evidence yet that aging in mammals is controlled by a genetic switch. The researchers...deleted a gene in the mice that makes them vulnerable to cell damage."156 While some are lengthening the life of mice others are stealing the genetic material of isolated groups of people and running to the patent office to claim the latest unique material. "Indigenous germplasm could provide the key to massive profits. Earlier this year Genetic Engineering news reported that 30 citizens of Limone, an isolated Italian community, were found to carry a unique gene that protects against cardiovascular disease. Scientists from Swedish and Swiss pharmaceutical companies, as well as from the University of Milan, have since swarmed over the townspeople, taking blood and other samples, and applying for patents."157

The DNA Switch

Experiments with lab rats demonstrated that subtle energy fields do have a direct effect on the brain activity of mammals. "During the probe trial, magnetic field-exposed animals spent significantly less time in the quadrant that contained the platform, and their swim patterns were different from those of the controls. These results indicate that magnetic field exposure causes a deficit in spatial 'reference' memory in the rat."158 Additional "studies on DNA have shown that large electron flows are possible within the stacked base pairs of the double helix. Therefore, gene activation by magnetic fields could be due to direct interaction with moving electrons within DNA. Electric fields as well as magnetic fields stimulate transcription, and both fields could interact with DNA directly."159 What this means is that energy below the ionizing level can impact the very structure of our genetic makeup provided that the energy is manipulated correctly. The risk of exposure in a world with between 100 and 200 million times more artificial radio frequency energy alone than what nature produces is cause for alarm. The complex interactions of these various fields create standing energy fields and other energetic effects which can have profound impacts on mind and body.

"An increase in single-strand DNA breaks was observed after exposure to magnetic fields of 0.1, 0.25, and 0.5 mT, whereas an increase in double-strand DNA breaks was observed at 0.25 and 0.5 mT. Because DNA strand breaks may affect cellular functions, lead to carcinogenesis and cell death... our data may have important

156. McCall, William. "Genetically manipulated rodents live longer, suffer no side effects." *Anchorage Daily News,* Nov. 18, 1999. EPI1218
157. *Earth Island Journal.* "Patenting Indigenous Peoples." Spring 1993. EPI57
158. Lai, Henry et al. "Acute Exposure to a 60 HZ Magnetic Field Affects Rats' Water-Maze Performance (introduction)." *Bioelectromagnetics,* 19:117 -122(1998). EPI192
159. Blank, Martin and Goodman, Reba. "Do Electromagnetic Fields Interact Directly With DNA?" *Bioelectromagnetics* 18:111 - 115 (1997). EPI191

implications for the possible health effects of exposure to 60Hz magnetic fields."160 Using this knowledge to manipulate DNA presents its own risks. "Scientists expect to be able to produce the first genetically targeted drugs in five years. The drugs would repair faulty DNA within the cell and might be used to treat conditions such as diabetes and cystic fibrosis Dr. Nathanson said: 'No one has been able to tell me why if we can produce genetically targeted drugs with a good effect, we won't be able to produce similar drugs with a bad effect in the same time-scale.'"161

Today, in 1999, the greatest breakthroughs are just beginning to be announced. What their long term consequences will be remains to be seen. "Recent advances in the generation of ultrashort high electrical power pulses have opened new venues in the field of bioelectrics. Electrical pulses with duration down to less than one billionth of a second but at voltages exceeding ten thousand volts allow us to explore and utilize electrical interactions with biological cells without heating the tissue. The high frequency components in the ultrashort pulses have been shown to provide a pathway to the interior of cells. Pulsed, high power microwave and millimeter wave sources allow us to explore and utilize nonlinear processes on the molecular level, with the potential to modify molecular structures, such as DNA, selectively."162 In other words, the mystery is solved: non-ionizing radiation does cause profound changes beyond the "safety levels" of classical models of human life.

The Pulse Of Life

"A wide range of biological effects from exposure to microwave fields, both in animals and in vitro, have been reported in the literature...These studies indicate that modulation plays an important role in eliciting biological responses with weak microwave fields."163 The pulsing or modulation of the energy carrier, whether it is a radio carrier, microwave, electric current or even a sound wave can result in profound changes. These changes occur when the energy is in sync, harmonized or resonating with the electromagnetic component of the tissue or at the right brain frequency to affect behavior. "The essential molecular functions appear in fact to be determined by electromagnetic mechanisms. A possible role of molecular structures would be the carrying of electric charges which generate, in the aqueous environment, a field specific to each molecule. Those exhibiting such corresonating or opposed fields ("electroconformational coupling") could thus communicate, even at

160. Lai, Henry and Singh, Narendra P. "Acute Exposure to a 60 HZ Magnetic Field Increases DNA Strand Breaks in Rat Brain Cells (introduction)." *Bioelectromagnetics* 18:156 - 165 (1997). EPI193
161. Laurance, Jeremy. "Genetic weapons to provide force for high-tech ethnic war." *The Independent*, July 2, 1997. EPI489
162. Old Dominion University. First International Symposium on Nonthermal Medical/Biological Treatments using Electromagnetic Fields and Ionized gases, April 12-14, 1999. Announcement. http://www.ece.odu.edu/~emed99 EPI778
163. Litovitz, T.A. et al. "Bioeffects Induced by Exposure to Microwaves are Mitigated by Superposition of ELF Noise (introduction)." *Bioelectromagnetics*, 18:422 - 1997. EPI195

a distance. Therefore, a minute variation in the structure of molecules (plus or minus an atom, or a rearrangement of an amino acid, for example), which even slightly modifies their radiating field would allow their message to be received or not by a receptor, as in the FM waveband."[164] In other words, the body communicates based upon electromagnetic principles which operate at tremendous speeds and can be influenced or overridden by other energy sources. Again it is a matter of harmonics or resonance in order to create the most significant effects.

"Mystery high frequency radio impulses have been bombarding the Eugene-Springfield area for as long as six years and may be affecting people's health...The paper said the source of the signals is unknown...They say it is being broadcast at 4.75 megahertz and is pulsating about 1.100 times per second."[165] The mechanisms for understanding the effects of these energies is being recorded in several diverse laboratories with the mounting evidence of the proofs open science requires. One of observations shows that, "At the core of observed sensitivities to low-level EMF fields are a series of cooperative processes. One such series involves calcium ion binding and release. Available evidence points to their occurrence at cell membranes and on cell surfaces in the essential first steps of detecting EM fields. Also, attention is now directed to newly defined roles for free radicals, that may also participate in highly cooperative detection of weak magnetic fields, 'even at levels below thermal (kT) noise.'"[166] One of the other effects which has been observed shows that interaction of specific fields with chemicals present in the environment or body can also contribute to significant changes. "This 'increase in genomic instability,' they suggested, could mean that chronic exposure to very strong EMFs 'may result in an increased incidence of congenital malformations and cancer. We propose that [EMF] exposure can affect both DNA damage and repair processes...and that it can act in concert with chemical agents to potentiate the damaging effects of those agents.'"[167] Nature's pulse can also have a significant effect if we can just clear the electromagnetic smog long enough to sense its reality. Certain behaviors have been associated with the polarized light of the sun as it reflects off of a full moon, increased sun spot activity, auroras and other natural energy sources. Increases in Very Low Frequencies (VLF) can have a significant impact. "More specifically, this atmospheric parameter has been considered a possible trigger for changes in the somatic and emotional well-being of humans, sometimes referred to as weather sensitivity symptoms or meteoropathy. The following review attempts to summarize present knowledge of biological significance

164. Benveniste, Jacques. "Transfer of Biological Activity By Electromagnetic Fields." Fall 1993. Vol. 3, No. 2. EPI1142
165. UPI. "Mysterious Radio Signals May Be Harming Health." *The Columbia Record* (South Carolina). March 27, 1978. EPI867
166. Adey, W. Ross. "Whispering Between Cells: Electromagnetic Fields and Regulatory Mechanisms in Tissue." *Frontier Perspectives*, Vol. 3, No. 3. Fall 1993. EPI313
167. *Microwave News*. "Four Labs Link 50/60 Hz Fields to DNA Breaks; Two Reproduce Effect at Occupational Exposure Levels." November/December 1998. EPI215

of VLF-sferics in humans."[168] These are frequencies utilized in various forms of communications and other military applications. Their potential effect either by lack of operator understanding or the intentional design of the system could have significant impacts on humans, plants and animals.

The body is always compensating for the impacts of energy on the body. If a person thinks about the feel of his body during a power failure – when all of the energy fields of significance are switched off in an instant – it is as if a weight were being lifted from us. The first thing noticed is usually the silence, followed by a release of tension as the body no longer has to attempt to create compensating energy fields for the constant bombardment of modern life and the internal stress it generates.

The Weapon Revolution

A number of new weapons are being developed or are already in operation. The Russians are reported to be ahead in many respects but this is only because the collapse of the old regime has allowed information to flow out of the country from leading scientists. The idea of creating specific brain interference, nervous system complications or heart failure are all targets of the new science of death. "Russia's psychotronic weapons include a psychotronic generator, which produces electromagnetic emanations that can be sent through telephone lines, TV, radio, or even light bulbs; an 'infrasonic sound' generator that destroys all life forms; a 'nervous system generator' known, so far, to paralyze insects; 'ultrasound emanations,' which kill by attacking internal organs without leaving a mark on the skin; and 'noiseless cassettes' featuring voices too low to be heard, which are nevertheless detected by the subconscious."[169]

Cloning from Pigs to Humans?

We are sure there will be a few religious objectors to this new breakthrough in technology. The use of animal tissue, or animal hosts, in growing tissue for use in humans is a cause for great concern. The question of transferability of new disease strains and the repugnance of the use of pigs from a religious perspective, whether Jewish, Hindu or Moslem is certainly a legitimate and again, overriding issue as it relates to the individual person and his/her "right to know." The use of swine was announced as, "Another valid use for cloning technology" which is "currently being researched is the use of pig organs for transplantation into humans. Pig organs are very similar to human organs with the exception of an extra sugar molecule that causes them to be rejected when transplanted into humans. Cloned pigs, genetically engineered to remove this extra sugar molecule,

168. Schienle, R. and Vaitl, D. "Biological effects of Very low Frequency (VLF) Atmospherics in Humans: A review." *Journal of Scientific Exploration*, Vol. 12, No. 3, 1998, pp. 455-468. EPI202
169. *Bulletin of the Atomic Scientists*. "All in the (Russian) mind?" July/August 1998. EPI278

would provide a reliable, steady source of organs for human transplant. The use of cloning technology for these purposes could prove invaluable to the military both in research and in wartime. It is important to exempt these types of technological applications from whatever bans are placed upon human cloning in military applications. However, the cloning of human cells is not the cloning of human beings, and it is the latter application of cloning technology against which international legal and ethical guidelines need to be implemented."170 The military research is already exempt in that they can do their research under cover of national security while the civilized world debates the merits and risks of the science.

The EMF Effect

In 1987, a *Washington Post* articled appeared which drew attention to the rapidly advancing level of electromagnetic field exposures based on increases in communications technologies. "The erection of hundreds of thousands of radio communications facilities around the world and in space during the last three years presents a novel situation: 24-hour-a-day exposure of all life on earth to unprecedented levels of radiation at large numbers of new broadcast frequencies. Almost every country in the world is now faced with how to regulate these new and economically important technologies."171 By the end of 1999 these levels of exposure have continued to rise as unprecedented growth in every related technology advances at even more startling rates. "Unlike some potential environmental hazards, ELF magnetic fields are virtually everywhere, making avoidance difficult. The flow of electric current through power lines creates magnetic fields, which easily penetrate walls of buildings and the body. These low-frequency fields localize near plumbing in houses and under streets, and their strength appears to be related to the types of wiring configurations nearby. They are, for example, found around power stations, welding equipment, subways and movie projectors."172 "Sources of a home's magnetic environment include appliances, overhead power lines, and grounding connections to metallic water pipes. Fields in the home will vary over time, depending on how much current is passing through the electrically conductive sources. Additional contributors to a home's magnetic background may include unusual wiring in the walls, underground power lines, and any nearby high-voltage transmission lines."173

The extremely Low Frequency (ELF) signals are thought to be those which are biologically active and most closely correlate with the Earth's natural rhythms. "Frequency describes the number of complete cycles of a waveform that occur in a second. One complete

170. Donovan, Maj. John L. D.N. Army: "The Implications of Human Cloning on Future Military Forces." http://www.cfcsc.dnd.ca/irc/nh/nh9798/0031.html EPI350
171. Firstenberg, Arthur. "The Politics of Health." *Alive*, Oct. 1999. EPI1168
172. Edwards, Dianne D. "ELF: The Current Controversy." *Science News*, Feb. 14, 1987. EPI756
173. Ralof, Janet. "EMFs Run Aground: Mapping magnetic fields from water and other homely sources." *Science News*, Vol. 144, Aug. 21, 1993. EPI759

cycle per second is 1 Hz, the unit of frequency. In terms of electrical power generation and use, electrical charges are forced to oscillate at 60 Hz in North America and 50Hz elsewhere."[174] These are ELF pulses which pass through virtually everything in their path and through interaction with other fields have an impact on living things. ELFs are energy waves which travel great distances without much change and penetrating, for example, "housing structures. Theoretically, it has been calculated that energy available from ELF phenomena can contribute to neuro-energetic functioning and protein-lipid activity. Correlational and experimental data indicate that ELF fields can influence reaction time, timing behavior, ambulatory behavior, oxygen uptake, endocrine changes, cardiovascular functions, and precipitation-clotting times of colloids."[175] In other words, they do have profound impacts on people.

The health effects of electromagnetic (EM) fields has been a subject of much debate in scientific circles as the research has developed over the years. At the present time, the debate has shifted from science to economics, as vested technology interests continue to downplay the effects of EMFs while science is lining up increasingly on the side of EMFs having significant impacts on people. "In recent years, biomedical researchers have demonstrated that low-frequency EM fields can disrupt the body's natural immune system, modify the production of hormones and help promote the growth of tumors. Some leading researchers now believe that physical events, rather than chemical ones, govern the most basic biological functions – through electrical interactions so elegant and complex that the classical laws of physics may be useless in describing them."[176] The idea that EMFs could increase the effects of chemicals in the body is becoming increasingly clear to researchers. "Electromagnetic fields may interfere with the electrical chit-chat between cells in the body – and cooperate with carcinogens to disrupt normal regulation of cell growth and promote cancer development, says W. Ross Adey of Loma Linda (Calif.) University."[177] Even weak fields once thought to have no effect on health are gaining acceptance as the physical mechanisms are explained by new research. "It is suggested that superconductive Josephson junctions in living systems may provide a physical mechanism with more than enough sensitivity to explain the observed responses of organisms to weak magnetic fields."[178] Some "...researchers are now investigating other, more circuitous mechanisms to explain breast cancer's rise. Though far from conclusive, their findings suggest that an unintended side effect of industrialization is an environment that bathes its inhabitants in a sea of estrogenic agents...Others, such as magnetic fields and certain combustion by-

174. American Industrial Hygiene Association Publications. *Extremely Low Frequency (ELF) Electric and Magnetic Fields.* Nonionizing Radiation Guide Series. EPI635
175. Persinger et al. "Psychophysiological Effects of Extremely Low Frequency Electromagnetic Fields: A Review." *Perceptual and Motor Skills,* 1973, 36, 1131-1159. EPI495
176. Peterson, Cass. "The Zapping of Post-Electrical Man." *Washington Post,* Dec. 6, 1987. EPI632
177. *Science News.* "Cells haywire in electromagnetic field?" Vol. 133, p. 216, 1988. EPI732
178. Cope, F.W. "Biological Sensitivity To Weak Magnetic Fields Due To Biological Superconductive Josephson Junctions." *Physiology, Chemistry & Physics.* Vol. 5, 1973. EPI1188

products, can boost the concentration of estrogens circulating in the blood stream."[179] These interactions are not normally evaluated together but are viewed separately. When seen as functions of both chemical and electrical components a much different picture emerges. In this picture electromagnetic fields are being credited with a larger role in creating changes in the body.

The specific absorption rate (SAR) which the human body is capable of is a major factor in the controversy. These rates are researched by the military in order to determine at what level natural rhythms and energy components of the human body can be overridden, while for civilian science the impacts on health remain central to the discussion. "More recently, concerns have been expressed about cellular telephones and other personal communications services (PCS). There are two main issues. The first relates to the deposition of energy (SAR) in the head. In some cases up to 50% of the device output power may be deposited in the user's head..."[180] This issue is becoming increasingly important as the international debate on cell phones develops, pitting the industry against health scientists. "There is increasing evidence that a number of electromagnetic fields (EMFs) may actually increase the toxicity of chemicals by allowing them more ready access to the sensitive nerve cells of the brain."[181] The increased use of cell phones, placing significant energy concentrations into the brain, will take a toll on the health of users.

In the Aug. 15, 1991, American Journal of Epidemiology, "they report that men whose occupations potentially involved frequent exposure to EMFs had a breast cancer incidence nearly double that of men in other jobs. Men in electrical trades – such as power-line installers and power-plant operators – faced six times the usual risk."[182] "The health effects of EMFs were linked to several cellular mechanisms, such as altered plasma membranes, hormone concentrations, enzyme activities, metabolism, and cellular electrical charge. Studies were reviewed which related EMF exposure to depression."[183] Meanwhile, the "evaluation of the combined effect of electric and magnetic fields for leukemia showed significant elevations of risk for high exposure to both, with a dose-response relation for increasing exposure to electric fields and an inconsistent effect for magnetic fields."[184] "Total mortality and cancer mortality rose slightly with increasing magnetic field exposure. Leukemia mortality, however, was not associated with indices of magnetic field

179. Raloff, Janet. "Ecocancers: Do environmental factors underlie a breast cancer epidemic?" *Science News*, Vol. 144, July 3, 1993. EPI754
180. Stuchly, Prof. Maria A. "Mobile Communication Systems And Biological Effects On Their Users." *Radio Science Bulletin*, No. 275 (Dec. 1995). EPI730
181. Duehring, Cindy. "EMF's Can Increase Chemical Uptake in the Brain." *Medical & Legal Briefs*, Sept./Oct. 1995. EPI627
182. Ezell, Carol. "Power-Line Static: Debates rage over the possible hazards of electromagnetic fields." *Science News*, Vol. 140, Sept. 28, 1991. EPI755
183. Keller-Byrne, J.E. and Akbar-Khanzadeh, F. "Potential Emotional and Cognitive Disorders Associated with Exposure to EMF's. A Review." *AAOHN Journal*, Vol. 45, No. 2, Feb. 1997. EPI135.
184. Miller et al. "Leukemia following Occupational Exposure to 60-HZ Electric and Magnetic Fields among Ontario Electric Utility Workers." *American Journal of Epidemiology*, Vol. 144, No. 2, 1996. EPI633

exposure except for work as an electrician. Brain cancer mortality was modestly elevated in relation to duration of work in exposed jobs and much more strongly associated with magnetic field exposure indices."[185] In 1996 a study, published in the September issue of the journal Epidemiology, "found that women whose jobs exposed them to high levels of electromagnetic radiation had a 43 percent greater chance of being diagnosed with breast cancer than those exposed to minimal radiation: In other words, they were nearly 1 1/2 times as likely to develop breast cancer."[186] The studies on adults are much different than the results when children's health is reviewed because young children have greater risks when immersed in these fields. "Power-frequency electromagnetic fields (EMF's) should be considered a risk factor for childhood cancer, staff at the Environmental Protection Agency (EPA) concluded in 1994. This conclusion appears in a report which was suppressed by EPA's senior managers and never released to the public."[187] But it is not just cancers which are becoming the focus of interest. "Electromagnetic fields, previously implicated in triggering leukemia, brain tumors and breast cancer, may play a far more important role in Alzheimer's disease, a University of Southern California researcher reported Sunday."[188]

What research has also shown is that it is not just the amount of energy which is important but that the frequency, or pulse rate, is even more important. "ELF (extremely low frequency) electromagnetic fields and waves with intensities similar to and slightly larger than those which occur in nature have been shown to be associated with changes in reaction time (Friedman, Becker and Bachman, 1967; Konig, 1962; Reiter, 1964) and verbal behavior (Konig, 1962; Ludwig and Mecke, 1968) in human subjects."[189] "Contemporary neuroscience suggest the existence of fundamental algorithms by which all sensory transduction is translated into an intrinsic brain-specific code. Direct stimulation of these codes within the temporal or limbic cortices by applied electromagnetic patterns may require energy levels which are within the range of both geomagnetic activity and contemporary communication networks."[190] What this implies is, again, that small amounts of energy can harmonize with internal energetic systems in a way which changes them or overrides their natural signals.

The use of radio frequency (RF) energy to influence chemical reactions is proving effective in that the pulse rate is matched to the ELF pulse rate allowing the RF to serve as the carrier of the active

185. Savitz et al. "Magnetic Field Exposure in Relation to Leukemia and Brain Cancer Mortality among Electric Utility Workers." *American Journal of Epidemiology*, Vol. 141, No. 2, 1995. EPI634
186. Saltus, Richard. "Study firms magnetic field, cancer link." *Anchorage Daily News*, Aug. 20, 1996. EPI498
187. *Microwave News*. "Long-Suppressed Draft EPA Report: EMFs Present a Cancer Risk." Vol. XVIII No. 1, January/February 1998. EPI791
188. Maugh II, Thomas H. "Alzheimer's linked to electromagnetic fields." *Los Angeles Times*, Aug. 1, 1994. EPI636
189. Persinger, Michael et al. "Behavioral Changes in Adult Rats Exposed to ELF Magnetic Fields." *International Journal of Biometeor.* 1972, vol. 16, pp. 155-162. EPI1178
190. Persinger, M.A. "On The Possibility Of Directly Accessing Every Human Brain By Electromagnetic Induction Of Fundamental Algorithms." *Perceptual and Motor Skills*, 1995, 80, 791-799. EPI1155

signal. "There is a new type of source technology currently under development in our country and, very likely, other countries as well. This type of directed RF energy is quite different than the narrow-band systems previously described. This type of directed energy is called transient electromagnetic radiation. Instead of generating a train of smooth sine-waves, as the conventional narrow-band systems do, it generates a single spike-like form of energy. This spike-like burst of potential does not have'cycles' or waves and it may be only one or two hundred pico-seconds (psec) in length. 100 psec is the time that it takes light to travel 1.2 inches and often these short time duration pioses are described in 'light-inches'."191

The positive possibilities for the use of this knowledge are also increasing as each new breakthrough demonstrates the need for attention to the field of electromedicine. "Synchronizing cancer treatment with the body's own internal rhythms appears to give ordinary chemotherapy drugs a much stronger punch, perhaps even doubling their power to fight tumors, a major new study shows."192 The results of research in this area is causing scientists who "once scoffed at harnessing light to fight cancer and other diseases, but now they say such 'photodynamic therapy' has potential thanks to potent new drugs that make diseased cells vulnerable to light beams."193 The trend is a major shift in evaluating the effects of the scientific observations so that they can be included in the development of new healing technology. "Doctors in Europe and the United States are planning to harness the imperceptible fields to correct abnormal heart rhythms and to control such conditions as epilepsy – from outside the body. Furthermore, these fields will be used to promote healing of skin, nerves, and tendons. – Studies suggest that these same fields may have deleterious effects."194

On the other side of the issue, the "deleterious effects" of the technology are being applied to the development of silent new killers. It was reported that "The cancer deaths of four Rumanian exiles employed by Radio Free Europe – now under investigation by the FBI – was the work of agents of the Bucharest government, according to a former Rumanian spy chief. Ion M. Pacepa, a close security adviser to Rumanian leader Nicolae Ceaisescu until his defection to the United States in 1978, said in an interview that he believed the RFE officials were killed with a radiation device designed by Romania's DIE intelligence service, with help from the Soviet KGB."195 The significant military advances are discussed at length in various sections of this book.

191. Statement of Mr. David Schriner before the Joint Economic Committee, United States Congress, Feb. 25, 1998. *The Design and Fabrication of a Damage Inflicting RF Weapon by 'Back Yard' Methods.* EPI1187
192. Haney, Daniel Q. "Cancer drugs more effective when linked to body rhythms." *Anchorage Daily News,* May 17, 1994. EPI537
193. Neergaard, Lauran. "Doctors kill cancer with pulses of light." *Anchorage Daily News,* March 4, 1998. EPI499
194. McAuliffe, Kathleen. "The Mind Fields." *Omni,* Feb. 1985. EPI536
195. Gertz, Bill. "Rumanian agents killed exiles with radiation, defector says." *Washington Times,* Dec. 27, 1988. EPI761

Human Photons

In 1997, a scientist rediscovered low level light emissions coming from the human body. These emissions were originally observed by eastern European scientists during the 1930s. At that time, a photographic technique was developed where the human body was placed in a high-frequency electromagnetic field. This stimulated the naturally occurring emissions in a way where they became visible on photographs, which showed small wisps of energy flowing out of the photographed body parts. "This summer Dr. Hyland will give two papers at international conferences outlining his research into the phenomena that biological systems, including the human body generate and emit extremely low intensity radiation in the form of photons (a microscopic packet of light energy), and that these photon emissions are not random but display coherence..."196 The idea that these emissions are "coherent" means that they have unique patterns of oscillation. These patterns indicate what is happening within the body. Researchers we are familiar with in Germany and elsewhere in Europe have been attempting to interpret these patterns for diagnosing disease and disorders in people.

Bio–Terrorists

The focus of public concern regarding bio-terrorists is increasing with each new media report. What is being emphasized is the use of biological or chemical agents in initiating these kinds of attacks. "Perhaps they should reconsider. Berry, who helped run the doomsday simulations that ravaged San Francisco, says 'Security experts are not asking if a biological attack on a civilian population is going to happen, but when...I don't understand why it hasn't happened already.'"197 The reality is that such attacks will occur at some point. "City officials in New York have just completed their first simulated anthrax attack. A bigger drill is planned for sometime during the next few weeks. What happens on these occasions will be used as a model for further simulations in up to 120 cities across the country."198 The other risk assumes that the knowledge of EM stimulation of chemical reactions eventually becomes understood by terrorist organizations. Most likely this information is already available to them. The idea that the chemicals already present in the body could be triggered to have toxic or fatal effects is the basis for the development of these technologies in their military applications.

The debate of the science surrounding electromagnetic fields will continue to build. Economic interests could be seriously harmed when the thresholds of exposure are changed to be in line with the

196. Dept. of Physics, Univ. of Warwick. "Body's Ability to Emit light Arouses New Hopes & Fears on Radiation From Mobile Phones." Press Release. http://www.warwick.ac.uk/news/pr/97 EPI566
197. MacKenzie, Debora. "Bioterrorism Special Report: Bioarmageddon." *New Scientist*, Sept. 19, 1998. http://www.newscientist.com/nsplus/insight/bioterrorism/bioarmageddon.html EPI882
198. Boyce, Nell. "Bioterrorism Special Report: Nowhere to Hide." *New Scientist*, March 21, 1998. http://www.newscientist.com/nsplus/insight/bioterrorism/nowhere.html EPI883

scientific research. In the meantime, the military will continue to create their new weapons, the health of populations will continue to decline and cancer rates will continue to increase.

Chapter 6

Non–Lethal Weapons? The RMA– Revolution in Military Affairs

If we use nonlethal technology to achieve paralysis, eliminate unintentional killing, and erase signs of visible destruction, then perhaps in some situations we can rid the news of sensationalism. Without a riveting story to tell, the media may be silenced.[199]

"In 1787, attendees at the Constitutional Convention first defined the purpose of the United states armed forces. This definition has undergone significant clarification and redefinition over the course of history. What began as the requirement to 'provide for the common defense' has led, most recently, in the National Military Strategy of the United States of America to that of 'fight[ing] and win[ning] our Nation's wars whenever and wherever called upon.'

To most people, that might not seem like such a large leap. There is little question that the writers of the Constitution foresaw that 'Defense' would inevitably lead to fighting wars. But what they may not have envisioned is the ever-growing handful of non-combat actions that the United States armed forces are currently being called upon to undertake on shores far distant from those of the original 13 states.

In recent history, US military might has advanced in what some would argue is a direction diametrically opposed to that of war fighting. This new direction is known as 'military operations other than war' (MOOTW). Admittedly, the division between MOOTW and war becomes difficult to delineate at times; but generally speaking, such operations focus on deterring war and promoting peace, while war encompasses large-scale, sustained combat operations to achieve national objectives or to protect national interests. MOOTW are more

199. Klaaren, Maj. Jonathon W. (USAF) and Mitchell, Maj. Ronald S. "Nonlethal Technology and Airpower: A Winning Combination for Strategic Paralysis." *Air Chronicles.* EPI245

politically sensitive, the military may not be the primary player, and they are most always conducted outside the United States."200

The advance of "non-lethal" weapons is part of the biggest initiative in military systems since the advent of the atomic bomb. These new technologies impact virtually every area of technology in and outside military organizations. The revolution is summarized below:

> "Collectively, most experts believe these innovations reflect an ongoing 'revolution in military affairs' (RMA). The RMA seeks to produce radically more effective – and, as the Friedmans indicate, more humane – militaries by profoundly altering their doctrine, organization, and weaponry through the widespread application of emerging microchip-based technologies, especially advanced computer and communications systems. Many observers believe that the RMA will give the United States a virtually insurmountable military advantage for the foreseeable future."201
>
> The focus is to seek technological solutions to virtually every human situation. American ingenuity has sought to substitute machines for manpower within the U.S., embracing the RMA. Technology has become the cornerstone of America's military planning.
>
> The U.S. itself can no longer afford to maintain many high-tech capabilities separate from the private sector.
>
> Both civilians and military must consider the legal and moral ramifications of using civilian systems for military purposes.
>
> The capabilities of new technology present politicians and soldiers with several options. If information superiority is truly critical, achieving it may require aggressive, draconian measures against international information sources that are not party to the conflict. These measures are of doubtful legal and moral validity, and they have the unintended consequence of antagonizing allies and even bringing the United States into conflict with third parties.
>
> The growing proliferation of the popular new nonlethal technologies presents other problems. Part of the problem comes from the characterization of these systems as "nonlethal, for example. Some of them are potentially deadly and clash with existing treaties which control the use of chemical and biological weapons. The integration of these weapons into the military will create a generation of "console warriors" who wage war without confronting the deadly reality of their actions. Politicians and the military should not assume that combatants will retain the military's traditions that

200. Gray-Briggs, Dr. Abigail and MacIver, Lt. Col. Michael (USAF). "Bombs, Then Bandages." *Airpower Journal,* Summer 1999. EPI229

201. Dunlap, Charles J. Technology: Recomplicating Moral Life for the Nation's Defenders. *Parameters,* Autumn 1999. EPI827

restrain illegal and immoral conduct in battle.

The questions raised are many, but the idea of "console warriors" raises what could be the most serious threat. The concept of fighting wars from computer command posts was exemplified in both the Gulf War and the war in Yugoslavia, where American casualties and direct interaction with hostile troops was minimal. When wars involved direct contact with the enemy the activity of killing was personal and, at the end of the war, all parties understood the impact and could question the morality of their actions. These reflections, after the conflicts have ended, have been important factors in helping people and nations recognize the need for peace and restraint. What happens when men and women are engaged in warfare resembling computer video games or training exercises? The use of these high-tech interfaces, while protecting American lives, serves to desensitize the war fighting participants.

The threat character is also changing to include "economic threats" from powerful economic interests. Developing options for neutralizing these threats seems to belong more to the realm of politics than traditional war fighting. The use of non-lethals is suggested as an alternative tool for militaries engaged in such activity:

> "It is my contention that conflict in the future will be very different from our past experience. Clearly, the Persian Gulf War was a historic event and unlikely to be repeated. The major threat to national security in the future will probably be from economic entities that may or may not have the status of national states.
>
> If future threats are other than from physical force, then a wide range of options will be required. It is likely that some of the threats will be hard to articulate to the American or international populace. Therefore, the dilemma is likely to arise: the demand to be able to respond to a threat, counterbalanced with a reluctance to use lethal force or risk American troops' lives. For instance, how far do we let an off-shore computer incursion go before it is labeled a 'threat.'
>
> What if they were backed by a foreign government or powerful multinational conglomerate? Non-lethal weapons will provide many options that will raise such questions."202

"Perhaps more importantly, analysis of the ethical basis of RMA is needed. Military strategists often overlook the fact that the employment of force occurs within and is structured by an elaborate normative framework. This has a historical foundation based on just war theory, the Judeo-Christian ethical tradition, and international law as well as a superstructure constantly modified by specific military and political developments. In the 20th century, total war, strategic bombing, nuclear weapons, limited war, and revolutionary people's

201. Alexander, J.B. "Potential Non-Lethal Policy Issues." Los Alamos National Laboratory, LA-UR 92-3206. EPI249

war forced adaptation of the normative framework Americans use when employing force. The RMA will require a new assessment. We must decide whether innovative military capabilities are, in fact, acceptable and desirable. That can only happen through open debate. The military must be a vital participant, but not the sole one. But as the institution most intensely aware of both the opportunities and dangers offered by emerging technology and concepts, the military must – as our notion of civil-military relations evolves to meet changing conditions – serve as a catalyst of this debate."203 Although this call for debate was made in the earlier 1990s it has yet to reach the majority of the public, which has already been assessed billions of dollars in taxes for its development.

Nonlethal and Nondestructive?

"Serious interest in 'non-lethality' as a technology and as a distinct class of weapons is recent. One study, 'Nonlethal and Nondestructive Combat in Cities Overseas,' proved to be a seminal assessment of potential non-lethal concepts. The study assessed numerous potential applications and non-lethal technologies for operations in urban areas. This early evaluation became the template for current technology research and development. Today's assortment of emerging non-lethal technologies grew from these concepts following the termination of the cold war. In a search for relevance, the national labs turned from nuclear warfare technology to less conventional research areas as 'non-lethality.' As a result, non-lethal concepts are a product of a 'technology-push,' and therefore, lacked traditional, well-defined war fighting requirements, established doctrine, and initial support.

Non-lethal weapons are intended to have one or both of the following characteristics: 1) they have relatively reversible effects on personnel or material, 2) they affect objects differently within their area of influence.

Implicit in this definition are several important points that are relevant to the discussion. The first is the concept of non-lethal intent. Non-lethal weapons, when properly employed, should significantly reduce lethal effects. However, there is no guarantee of 'zero' fatalities or permanent injuries. Certainly, even the most benign weapons technologies may create lethal effects under some conditions. It is the intent that separates this class of weapons from conventional munitions. Unintended lethal effects must be considered, and may modify, employment strategies and tactics."204

In a hearing before the Foreign Affairs Subcommittee of the

203. Metz, Steven and Kievit, James. "The Revolution in Military Affairs and Conflict Short of War." Strategic Studies Institute, U.S. Army War College, July 25, 1994. EPI516
204. Siniscalchi, Col. Joseph (USAF). "Non-Lethal Technologies: Implications For Military Strategy." Air War College, Air University, April 1997. AU/AWC/RWP177/97-04. EPI235

European Parliament,205 the issue of these new technologies was discussed. I was one of those called to testify along with a number of other people. One of the most interesting speakers was from the International Red Cross in Geneva, Switzerland, who gave an excellent presentation on "non-lethals." One of the points which he made involved the definition of "non-lethal." Part of the definition involved the idea that such weapons would result in a less than 25% kill factor for those exposed to them. He explained the fallacy in this by noting that land mines would even fit this definition because they did not kill over 25% of their victims. He explained that lasers which could permanently blind a person could also fit the definition. He also gave the example of "sticky foam" being used on an adversary and that this might not kill the person unless it landed on the victim's face and caused a slow and agonizing death by suffocation. The main point made was that non-lethals could indeed be lethal. Many of the panelists concluded that the term non-lethal was not accurate in describing these new systems and seemed more like a ploy to gain acceptance for the new technology.

Another relevant point made in the hearing was the frequency of use of these weapons in non-combat situations or policing actions. Comparisons between Bosnia and Northern Ireland were made. It was pointed out that in conflicts where rubber bullets and other non-lethal systems were available they tended to be used with greater frequency because the troops using them believed that they would not kill. Others in conflict situations using weapons clearly designed for killing used much greater restraint. In several years "peace keepers," armed with modern weapons, had not fired a shot in Bosnia whereas in Northern Ireland there were often injuries and deaths from the use of "non-lethals."

What to do with the "Peace Dividend?"

According to some, there is no peace dividend worth saving. The hawkish statements of some leaders shows their desire to turn the savings into a quiet arms race nobody knows they are in. "These already-paid-for technologies – according to a can-do letter written last year by Newt Gingrich, who has followed his futurist mentor, Toffler, onto the nonlethals bandwagon – 'are our real peace dividend.' Nonlethal weapons, Gingrich added, will 'preserve the defense industrial base, stimulate jobs in high-technology industry, and provide needed new options to local police and law enforcement authorities.'"206 Congressman Gingrich was an advocate for these technologies for years, having written the foreword for *Low Intensity Conflict and Modern Technology* as far back as 1986. That particular document set the stage for many of the new and emerging technologies. During Congressman Gingrich's watch, the merging of

205. February 6, 1998, Brussels, Belgium. European Parliament's Foreign Affairs Subcommittee on Security and Disarmament. EPI715
206. Shorto, Russell. "Armageddon: Killing Them Softly." *GQ*, March 1995. EPI390

the Defense and Justice Departments' efforts in the development and use of these new weapons took place.

While officials still show a fascination for the gadgetry of war, the creation of more sinister technology continues to advance. "Quantico Marine Base, home of the Pentagon's Joint Non-Lethal Weapons Program, hosted the first annual Non-Lethal Technology and Academic Research Symposium. In attendance were U.S. senators, marines, professors, defense vendors, and police officers. Defense vendors displayed high-tech non-lethal weapons, including sound wave and beam devices designed to temporarily disable suspects. Another interesting addition was the $220,000 SARGE, or Surveillance and Reconnaissance Ground Equipment, which is an all-terrain robot with infrared cameras that fires bean bags and smoke rounds at targets."207

The Future of War

The Council on Foreign Relations (CFR) made some interesting observations which are quickly being integrated into the military strategy framework. Part of this new thinking includes the ability to make war without declarations or detection. The ability to wage war without detection should be cause for great concern on the part of our allies as well as adversaries. Who will determine when, or even if, the public would be alerted to the conflict or its implications? The CFR makes the following points:

> "U.S. restraint will not prevent development of all non-lethal weapons by others. Russia, the United Kingdom, France, Italy, and Israel are said to have made significant efforts to develop non-lethal capabilities. Some non-lethal weapons can be assembled from components commercially available to terrorists as well as to governments. Research and development of non-lethal technologies will contribute to knowledge of defenses and antidotes. Some research and perhaps deployment should be undertaken in secret, both to attempt to limit proliferation and to retain the benefits of surprise.
>
> The Nairobi Convention, to which the United States is a signatory, prohibits the broadcast of electronic signals into a sovereign state without its consent in peacetime. Of course the contemporary world provides many situations between full peace and all-out war. The concept of a 'declaration of hostilities' or of a 'failed state' may be appropriate in such circumstances, not only with regard to the use of electronic signals but to the use of enhanced sanctions and non-lethal weapons as well.

In the longer run a policy of this nature could affect

207. Vogel, Steve. "Trained Not To Kill." *Washington Post*, May 6, 1999. Source: NLECTC *Law Enforcement & Technology News Summary*, May 13, 1999. EPI971

decisions by rogue states regarding costly and/or treaty-violating acquisition of some types of weapons of mass destruction or advanced conventional capability because of their knowledge that advanced non-lethal capabilities may provide the means of effective retaliation without causing large civilian casualties, thus making such U.S. action credible as a deterrent. Conversely, a long twilight war against terrorist groups and terrorist-supporting regimes may require a level of secrecy to preserve the effectiveness of non-lethal technologies and create uncertainty as to the agent, foreign or domestic, of disruptive events, and the degree of ultimate potential for destabilization or for support of domestic opposition."208

Hiding Behind the Curtain

Most ethical scientists find weapon developments which violate the basic internal mechanisms of mind and body highly distasteful. In order to develop these new systems health professionals have to either become openly involved or used by the military establishment which exploits their life sciences research for their own ends. A classified conference was sponsored by the military in order to disclose the possibilities to policing organizations and others:

"About 400 scientists who are developing nonlethal technologies – such as radiofrequency (RF) radiation, electromagnetic pulse (EMP), extremely low frequency (ELF) fields, lasers and chemicals – exchanged ideas at a classified meeting hosted by the John Hopkins University Applied Physics Lab in Laurel, MD, November 16-17.

Dr. Clay Easterly of Oak Ridge National Lab in Oak Ridge, TN, led a session on the use of ELF/EMFs. 'My major point was that there seem to be some biological sensitivities or responses [to ELF fields] that could in the future be useful for nonlethal technology,' Easterly told Microwave News. But he emphasized that information in the open literature can be applicable: 'There seem to be some phenomena not associated with thermal effects that could be useful.'

Easterly said that, while the military is primarily interested in the use of non-ionizing radiation to disable enemy electronics, his presentation dealt with the possibility of developing measures that would affect people.

Dr. George Baker of the Defense Nuclear Agency in Washington titled his paper 'RF Weapons: A Very Attractive Nonlethal Option.'

208. Council on Foreign Relations. *Non-Lethal Technologies: Military Options and Implications.* 1995. EPI764

Dr. Edward Teller and Attorney General Janet Reno were scheduled as keynote speakers."209

Special Operations Forces in 2025

Special operations will take completely on new dimensions in the next century. Forecasts in the open literature point to the year 2025 as the target for profound change while the pace of progress dictates something quite different. Many of the suggestions in futuristic writing on military technologies reflects a failure to grasp the true speed of technological and scientific change. The future projections are in many cases already here now or on the short path to reality.

> "Dim Mak (or Dim Hsueh) is a once forbidden technique in Chinese kung fu. The literal translation is 'The Poison Hand' (or 'Touch of Death'). Dim Mak's technique teaches to strike a vital point, with a certain force, at a certain time, and kill.
>
> The mastery of this art requires long hours of hard training with patience, perseverance, and study. It masterfully focuses a precise strike, accounting for both position and direction, with a variable degree of power, depending on the point of impact, at a target. It also requires near-perfect knowledge of the enemy system, and is highly dependent on both the weather and the time of day for a successful strike. The Dim Mak strike provides for many levels of lethality, from paralysis to death in several hundred days.
>
> The attributes of Dim Mak are mirrored in those of special operations forces in 2025. These forces will be highly dedicated, motivated, specially trained, and uniquely equipped...In 2025, the SOF precision operation's capability will demand a continuous stand-ready posture on a global watch and, that a moment's notice, the ability, to mobilize, deploy, locate, identify, and engage specific targets. Using varying levels of effect or lethality, SOF can then withdraw and redeploy without a trace.
>
> **Assumptions:** Nonstate actors with the power to threaten US interests will exist in 2025. These actors may include the multinational corporations, terrorist organizations, drug cartels, criminal organizations, and possibly energy or resource coalitions. Nonstate actors will be less sensitive to political influence and economic pressure will have very little effect on their organization or operation. State and nonstate actors challenging US interests may emerge with an expanded technological edge over the US. These actors may appear

209. *Microwave News.* "Military on Nonlethal Weapons: 'A Very Attractive Option.'" November/December 1993. EPI1233.

slowly and cautiously or may come on the global scene unexpectedly.

The US must be able to attack selected targets which are not vulnerable to precision-guided munitions or conventional explosives. These targets may need servicing with tunable destructive weapons which limit or eliminate collateral damage. These high-value targets could be people, facilities, or electronic databases. Enemy targets, valuable to the US, will be protected by passive and active means. Deep underground bunkers and mobile targets will present the greatest challenge to US targeteers.

Capabilities Required and Concept of Operations: Special Operations Forces (SOF) in 2025, like today, will focus on high-risk, highly specialized, high-consequence-of-failure missions; and will require nearly 100% guarantee of success, 'Zero Tolerance/Zero Error.' Political sensitivity is so significant that only a tailored organization with special skills, training, and equipment can accomplish these missions and assure success.

In 2025, weapons of mass destruction (WMD) will include nuclear weapons, poor man's nukes (biological and chemical weapons), and a new 'deadly' WMD – 'Information Bombs' (IB).

Enabling Capabilities and Supporting Technologies: Under the charge of the US Army Communications Electronic Command (CECOM), the 21st Century Land Warrior Program is developing and field testing an individual soldier computer/radio kit. This kit is expected to weigh approximately two pounds and strap onto the soldier of the twenty-first century providing voice, data, and imagery to each soldier and throughout the chain of command. This system is expected to greatly enhance the overall effectiveness of Army combatant units. Though two pounds of equipment does not sound cumbersome by today's standard, the packaging of this equipment will not meet precision-operations team requirements. Team communications equipment will need to be light, mobile, and unrecognizable as a communication system. Miniaturization should allow communication and supporting equipment to be embedded into mission apparel or uniforms. Interfacing with the equipment could be through implanted ear and throat pieces. Contact lenses could be the display screens of tomorrow affording normal fields of view. These lenses could display necessary visual data for mission tasking and additionally act as sensors for data collection or enhancing as night vision goggles do today. Controls for such equipment must also be unrecognizable and activated through gestures, voice, touch, or even thought control.

Data Fusion: ARPA is working on a 'neuron-based biosensor that can feature nerve cells growing directly on microchips capable of sensing toxic substances.'

Mobility: Technological advances also offer promise in powerplant and propulsion systems. National Aeronautics and Space Administration (NASA), working in conjunction with aircraft manufacturers, has invested heavily in aircraft engine technology. One effort having a direct military application is a Mach 4 civil transport with reduced nitrogen oxides emissions, and quieter engines. An additional area offering promise is the use of magnetic-based rotation of ionized air as a substitute for physical turbine blades.

Low Earth Orbiter: Additional hypersonic platforms in research at this time are rapid response/global reach aircraft system and space launch/support system. The rapid response/global reach aircraft system is projected to fly at speeds greater than Mach 8 with global reach.

HVT Engagement (Targeting Items): However, future deceptive technologies such as chameleon camouflage and deceptive holographic imaging would assist SOF in nondetectability.

Additional options for pinpoint designation would be to optimize nanotechnology and develop a ROBOBUG, a fly on the wall, or some form of nanotech emitter to proceed or be placed on a target's desired mean point of impact (DMPI) awaiting signal capture by an air system with an adequate weapons payload.

HVT Engagement (targeting people): The Hollywood movie, Runaway™, with Tom Selleck provides an interesting form of projectile for development. It is fired from a hand-held weapon, resembling a gun, but houses homing missiles with an individual DNA signature applied. Fired within the sensor range of the target, the missile goes active.

Star Trek the Next Generation™ , episode 157, called 'The Vengeance Factor,' showed an interesting form of targeting people which may [be] a SOF precision operations tool for 2025. The story involves a planet with a history of clan wars, and one clan developed a bio-virus that would only affect a certain clan of people. The developing clan was not directly affected, but could carry the virus, nearly undetected within their own bloodstream. By merely a touch to anyone of the enemy clan, can be death. The possibility of a SOF precision operations team, being able to infiltrate into an enemy target area, apply a predetermined or tunable level of lethality to enemy personnel simply by touch, would minimize the need for additional support equipment and weapons – thus, allowing the forces to blend into the cultural environment.

HVA Recovery (Targeting the System Process): A special operations team should have the capability to walk into a roomful of individuals, and within a split-second neutralize all the bandits, sort all the bogeys (presenting appropriate decision making data to all precision team members), and exclude all the friendlies. Primary use for this capability would be in hostage rescue.

To provide this level of coverage requires advances in two areas, first the sensory/display area and then the fire control/weapons system. The sensory array could be tuned for target ID via DNA sensing, or possibly a form of pheromone sensing like that of pre-covert target marking, or as simple as those individuals with weapons are bad and all others require further forms of interrogate. These sensory inputs could then be filtered and combined with other team-gathered information, near instantaneously, and displayed within the visor of an ultimate warrior targeting helmet or a modified tactical information display helmet. Of course, the sensory/ targeting system must operate in all light conditions and weather environments. The targeting data then is instantly fed to a hand-held slaved weapons system which will appropriately target the captors.

Ether Targeting (Ether): The ether targeting environment also drives needs for peculiar skills and equipment. Specifically, adversaries will certainly avail themselves of high-fidelity sniffers and sensors to detect net invasion. By 2025, electrons will be identifiable as DNA strands allowing individuals to detect, identify, and target particular transmissions for manipulation. Unlike most warfare, cyberwar and commercial war open Pandora's box – defining truth."210

"Defining truth," is the real issue behind infowar and the development of new technologies which merge in a manner which can lead to easy manipulation of visual images, voice recordings and even direct interference with mental functions. The ability to change the very essence of reality becomes possible when those in control have the ability to synthesize any thought, emotion, or the more obvious – the media's reports.

210. Cerniglia, Col. James A. *The Dim Mak Response of Special Operations Forces to the World of 2025.* Air Force 2025, Aug. 1996. EPI670

Terrorists

Fear has become the central theme in the drive for more sophisticated kill systems. Terrorists and narcotraffickers remain high on the list of villains we must guard against, according to military planners and research laboratories.

On the civilian side the fear is being converted to cash. "Michael Stedman, president of the Wellesley, Mass., ethical intelligence consulting firm Business Intelligence Network Systems, predicts now that the United States has adopted counterterrorism as a major policy concern, armored vehicle companies will enjoy a boost in sales. One such company is O'Gara-Hess & Eisenhardt Armoring Co. (OGH&E) of Fairfield, Ohio. Owner Bill O'Gara says that in the early 1990s, the company sold about two cars a year to corporate customers, and now sells several hundred a year. OGH&E offers vehicles with varying degrees of protective armoring and other security features. In addition, the firm supplies the U.S. Army and Air Force with 600 up-armored Humvees."211

The fear factor is also played up in weapons of mass destruction (WMD) scenarios. "A significant concern for the U.S. is the psychological fear of a WMD attack (e.g., April 24 package in [Washington] D.C.) The anxiety generated by such fear may pose far more difficult problems than the physical threat itself. The public must be made aware of the many limitations of the WMD threat and that there are ways to respond effectively, which their local authorities are developing."212 "As a venue for infliction of pain, terrorists have long favored mass transit. Their focus has shifted, however, from international airliners to domestic rail and bus systems. Thus, combating terrorism has become especially problematic due to increased involvement from disparate agencies, including law enforcement, emergency services, structural engineers, and heavy equipment operators. Fortunately, numerous cities nationwide have begun preincident planning and extensive interagency training to circumvent this disability."213 Meanwhile, "Attorney General Janet Reno trumpeted a new FBI office that will offer local police, fire and rescue workers one-stop shopping for federal training and equipment to respond to chemical, biological or nuclear attacks by terrorists."214 And, "On April 30, 1998, the Attorney General delegated authority to the Assistant Attorney General for the Office of Justice Programs (OJP) to administer grants to assist state and local public safety personnel in acquiring the specialized equipment and training

211. Stedman, Michael. "U.S. Counter-terrorism Policy Boosts Prospects for Armored Cars." *Security, Technology & Design,* April 1999. Vol. 26, No. 4. Source: NLECTC *Law Enforcement & Technology News Summary,* May 13, 1999. EPI975
212. *Defense Issues.* "Defending America Against New Breed of Terror." Vol. 12, No. 31. http://www.defenselink.mil/speeches/1997/s19970428-holmes.html EPI360
213. Nelson, Kurt R. "Mass Transit: Target of Terror." *FBI Law Enforcement Bulletin,* Jan. 1999. Vol. 68, No. 1. Source: NLECTC *Law Enforcement & Technology News Summary,* Feb. 18, 1999. EPI1028
214. Sniffen, Michael J. "FBI Opens New Anti-Terrorist Office." *Newsday.com,* Oct. 16. 1998. http://www.newsday.com/ap/rnmpwh1v.htm EPI1146

necessary to safely respond to and manage domestic terrorist activities, especially those dealing with chemical and biological agents, and nuclear, radiological and explosive devices. On May 8, 1998, the Assistant Attorney General, OJP, announced the establishment of the OJP Office for State and Local Domestic Preparedness Support. (OSLDPS)."[215]

Solutions for those eventual attacks are being developed. "Government chemists at Sandia National Laboratories in Albuquerque, N.M., have created chemical and disease neutralizing foam using toothpaste and hair care products ingredients. Reports by chemists indicate that the foam can render harmless such chemical and biological agents as nerve and mustard gases, viruses and bacteria, including anthrax. Inventors Dr. Maher E. Tadros and Dr. Mark Tucker say the foam could be effective in biological or chemical attacks within a building. The laboratory said it is working with officials in New York and Washington, D.C., to discuss placing foam dispensers in subway systems. According to Tadros, an ideal application would be to combine foam dispensers with a sprinkler system."[216]

The European Concern Rises
The Technology of Political Control

The technology of political control and its impacts has gained attention in the European community as they come to grips with its implications for democracies and individuals. A significant report was released which expresses these concerns in clear terms:

> "Thus, the purpose of this report is to explore the most recent developments in the technology of political control and the major consequences associated with their integration into processes and strategies of policing and internal control. The report ends each section with a series of policy options which might facilitate more democratic, open and efficient regulatory control, including specific areas where further research is needed to make such regulatory controls effective.
>
> A brief look at the historical development of this concept is instructive. Twenty years ago, the British Society for Social Responsibility of Scientists (BSSRS) warned about the dangers of a new technology of political control. BSSRS defined this technology as 'a new type of weaponry'...'It is the product of the application of science and technology to the problem of neutralizing the state's internal enemies. It is mainly directed at civilian populations, and is aimed as much at hearts and minds as at bodies.' For these scientists, 'This new weaponry ranges from means of monitoring internal dissent to devices for controlling demonstrations; from new techniques of

215. U.S. Dept. of Justice, OJP Office for State and Local Preparedness Support. "What's New at OJP/OSLDPS." http://www. ojp.usdoj.gov/osldps/ EPI221
216. Browne, Malcolm W. "Chemists Create Foam to Fight Nerve Gasses." *New York Times*, March 16, 1999. Source: NLECTC *Law Enforcement & Technology News Summary*, March 18, 1999. EPI999

interrogation to methods of prisoner control. The intended and actual effects of these new technological aids are both broader and more complex than the more lethal weaponry they complement.'

The Role & Function of Political Control Technologies: What is emerging in certain quarters is a chilling picture of ongoing innovation in the science and technology of social and political control, including: semi-intelligent zone-denial systems using neural networks which can identify and potentially punish unsanctioned behavior; the advent of global telecommunications surveillance systems using voice recognition and other biometric techniques to facilitate human tracking; data-veillance systems which can match computer held data to visual recognition systems or identify friendship maps simply by analyzing the telephone and e-mail links between who calls whom; new sub-lethal incapacitating weapons used both for prison and riot control as well as in sub-state conflict operations other than war; new target acquisition aids, lethal weapons and expanding dum-dum like ammunition which although banned by the Geneva conventions for use against other state's soldiers, is finding increasing popularity amongst SWAT and special forces teams; discreet order vehicles designed to look like ambulances on prime time television but which can deploy a formidable array of weaponry to provide a show of force in countries like Indonesia or Turkey, or spray harassing chemicals or dye onto protesters. Such marking appears to be kid glove in its restraint but tags all protesters so that the snatch squads can arrest them later, out of the prying lenses of CNN.

The technology of political control produces a continuum of flexible options which stretch from modern law enforcement to advanced state suppression. It is multi-functional and has led to a rapid extension of the scope, efficiency and growth of policing power, creating policing revolutions both with Europe, the US and the rest of the world. The key difference being the level of democratic accountability in the manner in which the technology is applied. Yet because of a process of technological and decision drift these instruments of control, once deployed quickly become 'normalized.' Their secondary and unanticipated effects often lead to a paramilitarisation of the police – often because the companies which produce them service both markets.

Innovations In Crowd Control Weapons: Likewise there is a need to consider halting the use of peppergas in Europe until independent evaluation of its biomedical effects is undertaken. Special Agent Ward, the FBI officer who cleared OC in the USA was found to have taken a $57,000 kickback to give it the OK. Other US military scientists warned of dangerous side effects including neurotoxicity and a recent estimate by the International Association of Chief Police Officers suggested at least 113 peppergas linked fatalities in the US – predominately from positional asphyxia. Amnesty International has said that the use of pepper spray by Californian police against peaceful environmental activists, is 'cruel, inhuman and degrading treatment of such deliberateness and severity that it is tantamount to torture.' (Police deputies pulled back protestors heads, opened their eyes and 'swabbed' the burning liquid directly on to their eyeballs).

In the early Nineties, much to the disbelief of serious researchers, a new doctrine emerged in the US – non-lethal warfare. Its advocates were predominately science fiction writers such as (Toffler, A., & Toffler, H., 1994) and (Morris, J., & Morris, C. 1990, 1994), who found a willing ear in the nuclear weapons laboratories of Los Alamos, Oak Ridge and Lawrence Livermore. The cynics were quick to point out that non-lethal warfare was a contradiction in terms and that this was really a 'rice-bowls' initiative, dreamt up to protect jobs in beleaguered weapons laboratories facing the challenge of life without cold war.

This naive doctrine found a champion in Col. John Alexander (who made his name in the rather more lethal Phoenix assassination programs of the Vietnam War) and subsequently picked up by the US Defense and Justice Departments. After the controversial and overly public beating of Rodney King (who was subdued by 'an electroshock 'taser' before being attacked); the excessive firepower deployed by all sides in the Waco debacle (where the police used chemical agents which failed to end the siege); and the humiliations of the US military missions in Somalia – America was in search of a magic bullet which would somehow allow the powers of good to prevail without being hurt. Yet US doctrine in practice was not that simple, it was not to replace lethal weapons with 'non-lethal' alternatives but to augment the use of deadly force, in both war and 'operations other than war', where the main targets include civilians. A dubious Pandora's box of new weapons has emerged, designed to appear rather than be safe. Because of the 'CNN factor' they need to be media friendly, more a case of invisible weapons than war without blood. America now has an integrated product team consisting of the US Marines, US Air Force, US Special Operations Command, US Army, US Navy, DOT, DOJ, DOE, Joint Staff, and CINCS Office of Sec.Def. Bridgeheads for this technology are already emerging since one of the roles of this team is to liase with friendly foreign governments.

Last year the interim report advised that the Commission should be requested to report on the existence of formal liaison arrangements with the US, for introducing advanced non-lethal weapons into the EU. The urgency of this advice was highlighted in November 1997 for example, when a special conference on the 'Future of Non-Lethal Weapons', was held in London. A flavor of what was on offer was provided by Ms. Hildi Libby, a systems manager of the US Army's Non-lethal Material Program.

Ms. Libby described the M203 Anti-personnel blunt trauma crowd dispersal grenade, which hurtles a large number of small 'stinging' rubber balls at rioters. The US team also promoted acoustic wave weapons that used 'mechanical pressure wave generation' to 'provide the war fighter with a weapon capable of delivering incapacitating effects, from lethal to non-lethal'; the non-lethal Claymore mine – a crowd control version of the more lethal M18A1; ground vehicle stoppers; the M139 Volcano mine which projects a net (that can cover a football sized field) laced with either razor blades or other 'immobilization enhancers' – adhesive or sting; canister launched area denial systems; sticky foam; vortex ring guns – to apply vortex ring gas impulses with flash, concussion and the option of quickly changing between lethal and non-lethal operations; and the underbarrel tactical payload delivery system – essentially an M16 which shoots either bullets, disabling chemicals, kinetic munitions or marker dye.

One of the unanticipated consequences of these weapons is that they offer a flexible response which can potentially undermine non-violent direct action. Used to inflict instant gratuitous punishment, their flexibility means that if official violence does tempt demonstrators to fight back, the weapons are often just a switch away from street level executions.

New Prison Control Systems: Apart from mechanical restraint, prison authorities have access to pharmacological approaches for immobilizing inmates, colloquially known as 'the liquid cosh.' These vary from psychotropic drugs such as anti-depressants, sedatives and powerful hypnotics. Drugs like Largactil or Seranace offer a chemical strait-jacket and their usage is becoming increasingly controversial as prison populations rise and larger numbers of inmates are 'treated'. In the USA, the trend is for punishment to become therapy: 'behavior modification' – Pavlovian reward and punishment routines using drugs like anectine, producing fear or pain, to recondition behavior. The possibilities of testing new social control drugs are extensive, whilst controls are few. Prisons form the new laboratories developing the next generation of drugs for social reprogramming, whilst military and university laboratories provide scores of new psychoactive drugs each year.

Interrogation, Torture Techniques & Technologies: Helen Bamber, Director of the British Medical Foundation for the Treatment of the Victims of Torture, has described electro-shock batons as 'the most universal modern tool of the torturers' (Gregory, 1995). Recent surveys of torture victims have confirmed that after systematic beating, electroshock is one of the most common factors (London, 1993); Rasmusson, 1990)."[217]

The root issues pointed out by our allies in Europe touch on the ethical issues which are being raised in light of the capability of the United States and others. The United States "has been contemplating how to minimize 'collateral damage' in future operations. Technologies could include: radio frequency and micro-waves, lasers, supercaustics, polymers, smoke, and electromagnetic pulse generators – to name a few. Lasers could be deployed and used to deny access to sensitive areas. Largely unspoken is the fact that some of the research underway could lead the US close to violating international conventions to which it is a party."[218] These conventions are not clear in terms of some of the newest technologies although they are clear with respect to the use of chemicals or biologicals agents which are illegal. The weapons being developed which use electromagnetic fields or variations on this theme are not well addressed in treaty agreements.

Black Budgets

"Today, every day, close to $100 million flows through underground pipelines from the Treasury to the Pentagon to fuel the national-security machinery of the United States. The black budget – 'black' in the sense of being unseen, covert, hidden from light – is the secret treasury of the nation's military and intelligence agencies. It is appropriated and spent with the barest public debate or scrutiny."[219] These are the budgets which fund much of the new technology. Without much oversight billions of dollars are spent in the creation and development of the nation's most sophisticated systems, even the new "non-lethal" weapons. "But little is known publicly about the extent of the US government's projects in nonlethal weaponry. Virtually all are classified as 'special access,' a stricter security level than 'top secret.' Only a handful of people are allowed knowledge of

217. An Omega Foundation Summary & Options Report For The European Parliament. *An Appraisal Of The Technologies Of Political Control.* Sept. 1998. EPI128.
218. Starr, Barbara. "Non-lethal weapon puzzle for US Army." *International Defense Review*, April 1993. EPI860
219. Weiner, Tim. "The Pentagon's Secret Stash." *Mother Jones*, March/April 1992. EPI451

even the existence of special-access programs, let alone the details of what they involve."220

RF Weapons

The United States Air Force has been interested in radio frequency (RF) weapons ever since it was first noticed that certain radio frequency energy could have significant effects on humans and hardware alike. "Public discussion of RF/MW weapons has focused on disrupting technology. But a recent article in the *Airpower Journal* revealed for the first time that the military is developing high-powered microwave weapons for use against human beings...RF/MW and EMF–based weapons are also being studied for civilian law enforcement."221 The direction of the research begins to take more open form during the 1980s. The Air Force points out several areas of interest in developing RF weapons as follows:

"Radiofrequency (RFR) Radiation"

> **Introduction:** Biotechnology requirements in the next three decades must consider significant advances in electronic (electromagnetic radiation) warfare, since both offensive and defensive systems will add significant radiation stress to humans in a wide range of military operations. We can expect increases in available on-board power; development of sophisticated methodologies for detecting, tracking, identifying and attacking; and ultimately the development of systems to inflict intense pulses of electromagnetic energy on an adversary.
>
> As the technological race continues, knowledge of mechanisms of action of RFR with living systems and the assessment of pulse RFR effects will demonstrate the vulnerability of humans to complex pulsed electromagnetic radiation fields in combination with other stresses...
>
> **Assessment and Development of Pulsed Radiofrequency Radiation Effects**
>
> a. Objectives
>
> (1) Develop techniques to deposit radiofrequency radiation (RFR) at selected organ sites.
>
> (2) Develop mathematical models and physical measurement capabilities (microdosimetry) to track the real-time RFR energy distribution within organ sites as a function of physiological responses such as diffusion and blood flow.
>
> (3) Establish thresholds and other response rates

220. Kiernan, Vincent. "War over weapons that can't kill." *New Scientist*, Dec. 11, 1993. EPI673
221. *Microwave News*. "RF Weapons: Disabling People and Electronics." January/February 1996. EPI688

for selected biological effects as a function of RFR wave parameters (shape, width, repetition rate, resource groups and intensity).
(4) Develop laboratory tools to simulate likely real-time RFR encounters in Air Force operations (from VLF to millimeter wave frequencies).

RFR Forced Disruptive Phenomena

a. Objectives

(1) Define the ability of RFR to interrupt, degrade or direct human central nervous system functioning.
(2) Define the ability of RFR to interrupt or degrade physiological functions such as cardiac output and respiration.
(3) Define the ability of RFR to interact with chemical and other physical agents, and to assess their combined impact on humans.

A rapidly scanning RFR system could provide an effective stun or kill capability over a large area. System effectiveness will be a function of waveform, field intensity, pulse widths, repetition frequency and carrier frequency. The system can be developed using tissue and whole animal experimental studies, coupled with mechanisms and waveform effects research.

Microresonance and receptor site mechanisms research will suggest specific frequencies which may interfere with or enhance drug or chemical agent effects. Confirmatory experiments in animals will be necessary. Using relatively low level RFR, it may be possible to sensitize large military groups to extremely dispersed amounts of biological or chemical agents to which the unirradiated population would be immune."222

The use of radio frequency energy as a carrier for a silent death has reached varying degrees of completion. It is now possible to disrupt the entire living system with weapons growing out of this research. The heating and more dramatic effects were first discovered and applied to the first generation of these new instruments. "A thermal gun would have the effect of heating the body to 105 to 107 degrees F, thereby incapacitating any threat, based on the fact that even a slight fever can affect the ability of a person to perform even simple tasks. This approach is built on four decades of research relating radio frequency exposure to body heating. A seizure gun would use electromagnetic energy to induce epileptic-like seizures in persons within a range of a particular electromagnetic field. The magnetophosphene gun is designed around a biophysical mechanism which evokes a visual response and is thought to be centered in the

222. Southwest Research Institute. *Final Report On Biotechnology Research Requirements For Aeronautical Systems Through The Year 2000.* Prepared for: The Air Force Office of Scientific Research. July 30, 1982. EPI707

retina, known as magnetophosphenes. This effect is experienced when a person receives a blow and sees 'stars.' This same effect can be produced with electromagnetic energy."223 As far back as the early 1990s this new tool was under development. "Low frequency infrasound systems were considered for use in Somalia but rejected, as were radio frequency systems. The latter focus a beam of radio frequency energy on the targeted individual. This causes a rise in body temperature to between 105 to 107 degrees Fahrenheit, producing fever-like disabling symptoms... Oak Ridge National Laboratory is developing a thermal gun of this type..."224

"Bioeffects research now being conducted by the Radiofrequency Radiation Branch examines effects at the subcellular, cellular, and whole organism levels. The research is conducted through the Tri-Service Electromagnetic Radiation Panel, which is chartered through the Deputy Undersecretary of Defense for Environmental Security. In order to examine carcinogenicity potential, some studies expose small laboratory animals to RFR over virtually their entire life span. Other research focuses on basic mechanisms of RFR bioeffects. Also emphasized are studies on the effects of millimeter wave frequency and high power microwave radiation on ocular and nervous system function. Some new directed energy weapons systems use short, intense pulses of microwave energy to incapacitate opponent electronic systems. A major research effort is focused on determining the biological effects of these novel pulses in order to establish protection criteria necessary before these systems can be tested and fielded. Bioeffects issues are critical to the success of new non-lethal weapons. Because of our core bioeffects expertise, we have become a major test facility for the bioeffects of non-lethal weapons."225 The new "Technologies could include: radio frequency and microwaves, lasers, supercaustics, polymers, smoke, and electromagnetic pulse generators, to name a few...The US Army has even looked into infrasound – very low frequency sound – as a riot- or crowd-control agent. Infrasound generators could be turned against humans, causing disorientation, nausea and vomiting."226

The new systems have already been built and are available. Even "backyard inventors" are creating these new systems with a handful of off-the-shelf parts and easily obtainable materials. "Fancy building your own Klingon disrupter? An ex-US navy engineer has done just that for the bargain basement price of $500. The gadget fiend has built a 'gun,' using readily available hardware, that can disable almost any piece of electronic equipment from 20 feet away."227 This same system if tuned to the right frequency could

223. Oak Ridge National Laboratory. Physiological Responses Applicable to development of Less-Than-lethal Weapons. EPI674
224. Richardson, Doug. "Non-lethal options." *Defence & Security Review.* http://www.atalink.co.uk/DSR/CLIENT/richweap.htm EPI505
225. Air Force Research Laboratory, Brooks AFB. Radio Frequency Radiation Bioeffects Research at the United States Air Force Research Laboratory. http://www.brooks.af.mil/AFRL/HED/hedr/hedr.html EPI187
226. Starr, Barbara. "Non-Lethal Weapon Puzzle For US Army." *International Defense Review*, April 1993. EPI57
227. Sherriff, Lucy. "Build your own Klingon disrupter." *The Register*, Sept. 9, 1999. EPI1186

also be used against a person by inducing a heart attack or creating other effects. "Portable microwave weapons being field-tested by the U.S. Special Forces can quietly cut enemy communications but also can cook internal organs. 'I don't know that nonlethality is all that humane,' concludes Myron L. Wolbarsht, a Duke University ophthalmologist and expert on laser weapons."228 These advances just begin with hand held devices. "A 1996 Air Force Scientific Advisory Board report on future weapons, for instance, includes a classified section on a radio frequency or 'RF Gunship.' Other military documents confirm that radio-frequency antipersonnel weapons programs are underway."229

One of the other areas where RF is being exploited is in creating artificial electromagnetic pulses (EMPs). These energy surges override and cripple sophisticated and simple electronic circuits. These technologies are being developed under dual-use programs for both military and police use:

> "Jaycor has recently extended the pulse-power testing technology developed under Department of Defense programs for electromagnetic pulse and high-power microwave simulation to civilian applications with substantive success. Jaycor has developed a technology demonstration system for law enforcement, anti-terrorist operations, and military operations other than war (OOTW) to safely stop fleeing vehicles. The system is a potential answer to the prevention of the tragic endings to numerous high-speed chases that occur every year.
>
> Jaycor has a variety of nonlethal weapons in development for both military and law enforcement applications. One of those devices, dubbed Sticky Shocker™ for its ability to both stick to a human target and electrically stun the person, is nearing completion of engineering development. This project is being sponsored by the Defense Advanced Research Projects Agency and the National Institute of Justice through the Joint Program Steering Group.
>
> Jaycor is using its expertise in electromagnetics to develop innovative and cost-effective methods for protecting new and existing systems from hostile exposures to intense radio frequency (RF) radiation. Advanced computer codes and models, which are verified using Jaycor's high-power microwave laboratories, are used to characterize the system's response to RF radiation. These response models are integrated into computer programs to support design engineers. The program leads users through a step-by-step RF protection design process."230

228. Ricks, Thomas E. "Nonlethal Arms: New Class of Weapons Could Incapacitate Foe Yet Limit Casualties." *Wall Street Journal*, Jan 4, 1993. EPI570
229. Pasternak, Douglas. "Wonder Weapons." *U.S. News & World Report*, July 7, 1997. EPI642
230. Jaycor. Less-Than-lethal Technologies, Products. http://www.jaycor.com/eme/nlp.htm EPI712

Jaycor is one of the companies actively developing these technologies for the Justice Department.

> "The National Institute of Justice (NIJ) and the Department of Defense have been developing new non-lethal weapons, including laser flashlights, nets, and projectiles. LE Systems developed the LaserDazzler in a project sponsored by the NIJ and the Defense Advanced Research Project Agency. Laser flashlights like the LaserDissuader and the LaserDazzler disorient a subject without causing lasting damage to the eyes. These devices look just like a regular flashlight, which also offers officers the advantage of surprise. The LaserDissuader uses an adjustable red 650 nanometer laser diode, supported by a complex electronics package, and top-of-the-line optics, which can be operated in continuous or flicker mode. In addition, the LaserDazzler flashes a series of random green bursts of light of up to 50 meters even in daylight, to distract a subject. LE Systems is looking for ways to reduce the dazzler's size, weight, and cost, while also searching for a means to commercially market the product. Other new nonlethal weapons include ring airfoil and electric stun projectiles. The Sticky Shocker, for example, clings to the target and administers pulses near 50 kV every few microseconds at a rate of 10 to 15 pulses per second. The maker, Jaycor, has also created a wireless stun gun with a range of 25 feet, without resorting to cables. Finally, the NIJ has provided funds to Delta Defense to create a pepper spray projectile, with the intention of having a 100-foot launch range able to penetrate a household windowpane of glass."[231]

As with all new weapons, counter-measures also need to be developed. Defensive systems are being created to protect the developers of this technology from the fruits of their labors when the enemy chooses to test their new systems on us.

The New Weapons Mix

"Department of Defense policy defines nonlethal weapons as 'weapons systems that are explicitly designed and primarily employed so as to incapacitate personnel or materiel, while minimizing fatalities, permanent injury to personnel, and undesired damage to property and the environment.'"[232] Sounds good. "Imagine a world where land

231. Siuru, Bill. "Developments for the Military and Law Enforcement Now Apply to Corrections." *Correction Technology & Management*, March/April 1999. Vol. 3, No. 2. Source: NLECTC *Law Enforcement & Technology News Summary*, May 13, 1999. EPI976
232. Coppernoll, Lt. Col. Margaret-Anne. "The Nonlethal Weapons Debate." http://www.nwc.navy.mil/press/review/1999/spring/art5-sp9.htm EPI248

mines don't blow up but give off an eerie sound that makes intruders feel sick. Or a war where attackers don't use missiles to stop tanks but microwaves to shut down engines."233

The Air Force is only one organization which is pursuing the technology but they seem to publish the best material for those of us interested in this area of research. In one recent document a great deal is revealed:

> "How would we employ this nonlethal technology? We should use it early in a conflict and in such a way that targeted leaders are unaware of its application. The objective of this strategy would be to disrupt leadership to such an extent that it would reconsider going to war. Innovative weapons and approaches for conducting these types of operations offer opportunities to apply the military instrument of power and stop a potential outbreak of war. By using technology to get into the enemy's networks, we could use electronic bullets from a remote site to destroy specific components of the regime's command and control equipment. Nonlethal weapons for attacking electricity already exist in the U.S. arsenal. Also at our disposal are microbes or chemicals that alter petroleum products, rendering them useless. One can effectively disrupt most of a nation's transportation system through nonlethal means. Airpower could drop microbes or chemical agents on roads and airports to ruin them or to damage the rubber tires of vehicles that use the roads. We could drop different agents or caustics on rail lines to deteriorate the lines or to prevent train cars from generating the friction they need to move. We could also affect the economic infrastructure by infiltrating the state's electronic financial network and causing general economic chaos among the government and its people.
>
> In addition to cost considerations, several other factors justify the incorporation of nonlethal weapons into our military arsenal: public opinion, the media, and dual use technology. Public opinion shapes the decisions of America's leadership regarding armed conflict. Because nonlethal warfare limits bloodshed, it will be endorsed by the American public as a positive approach for conducting future wars. In an age of instant communication, capabilities available to the media have an increasingly important impact on military operations. The media serves as a conduit of information – not only to the American public, but also to the rest of the world. We need to eliminate the notoriety associated with war. If we use nonlethal technology to achieve paralysis, eliminate unintentional killing, and erase signs of visible destruction, then perhaps in some situations we can rid the news of sensationalism. Without a riveting story to tell, the media may

233. Raphael, Michael. "Future Weapons may avert deaths." *Anchorage Daily News.* EPI425

be silenced. One last advantage of nonlethal warfare is its applicability to the civilian sector. Developing these weapons with a dual use in mind will greatly assist the efforts of our law enforcement communities. Currently, little is available to law enforcement short of deadly force. A means of safely subduing a suspect without using deadly force would be a significant addition to the war on crime. Such uses of nonlethal weapons are endless. Drug interdiction, border patrols, antiterrorism and riot control are good examples.

Most of the near term work with nonlethal weapons will continue to be geared to antimateriel uses. However, current treaties must be renegotiated to take into account other nonlethal technologies. Certain chemical and biological uses of nonlethal technology may be acceptable, given the nonlethal aspects of their use. Although international agreements currently proscribe the use of chemical or biological warfare in water and food supplies, these agreements came at a time when offensive chemical and biological warfare sought to kill the enemy. New forms of chemicals and microbes would not kill; instead they would merely have a temporary effect on the population and conceivably could save lives by averting combat. Such weapons most likely would be in chemical or biological form. Chemicals placed in the water could indirectly affect agriculture and population by discoloring the water to make it appear undrinkable, slowing crop growth, or even temporarily altering the mental states of potential enemies. Clearly, we must address the incapacitation of humans and the moral dilemma that surrounds this emotional issue.

Another controversial issue is the use of mind altering drugs to influence the population of enemy states. According to Dr. Stuart Yudofsky of Baylor University, psychopharmacology (the science of drugs that affect the mind) is on 'the brink of revolution.' Previously, psychopharmacology had concentrated on the development of drugs that modify the brain chemistry of mentally ill patients, which led to the development of drugs such as Prozac during the late 1980s. Presently, scientists are studying 'normal' brains and determining which chemicals cause certain personality traits. Imminent breakthroughs in this area will lead to the possibility of 'made-to-order, off-the-shelf personalities.' Additionally, these new drugs are supposed to have no serious side effects and no addictive properties. Potentially, psychopharmacology has great application for nonlethal warfare and should be followed closely to ensure that its offensive and defensive potentials are well understood. Although the United States may choose not to pursue mind altering drugs as a weapon, other states may hold a different view. For that reason, it is imperative that we understand this capability. In short, technologies of the future will be able to incapacitate humans. If this option is the most efficient way to obtain strategic objectives, should we limit its use?"[234]

I marvel at the idea of military personnel contemplating the ethical consideration of "drugging" the enemy. Some might recall the problems the Army had when they cooperated with the CIA in testing chemicals on US military personnel in the 1960s. The violations of international agreements are clear when it comes to drugging people. Such activity is more commonly called Chemical Warfare. The most interesting facts in the text deal with the fear the military has of media coverage of its active engagements and the need to wage war without visible impacts. Herein is the risk – if such invisible wars can be waged, who will decide when they are fought and who will control their outcomes?

The public relations efforts surrounding non-lethal systems is slowly weaving its way into the mainstream of acceptance on the basis of its greater humanity. "Achieving military objectives while conducting special operations or while providing humanitarian assistance calls for using another metric for measuring success – obtaining minimum casualties with little or no collateral damage. Nonlethal weapons such as high-power microwaves, 'sticky foam,' or sonic weapons, among others, show great promise for achieving military objectives. The use of nonlethal weapons that have the capability to neutralize, stun, disable, disorient, or confuse the enemy without having lasting effects is therefore highly desirable."[235] The Air Force provided the following partial list of these new systems:

"Acoustic generators – Anti-materiel biologicals
Anti-materiel chemicals
Anti-personnel optical/dazzling munition
Calmative agents
Chemical immobilizers
Chemical mobility impairments
Counter-sensor optical
High-power microwave generator
Holographic projections
Information system perturbation
Neural inhibitors
Non-nuclear electromagnetic pulse
Wireless stun technology"[236]

234. Klaaren, Maj. Jonathon W. (USAF) and Mitchell, Maj. Ronald S. "Nonlethal Technology and Airpower: A Winning Combination for Strategic Paralysis." *Air Chronicles.* EPI245
235. Institute for National Strategic Studies. "The present Defense Research Strategy." http://www.ndu.edu/ndu/inss/books/dodsnt/ch4.html EPI218
236. *Inside the Air Force.* "Pentagon To Set Priorities In Non-Lethal Technologies, Weapons." Vol. 5, No. 15, April 15, 1994. EPI189

The Merging of the Justice Department and Military Technology

On July 21, 1994, Dr. Christopher Lamb, Director of Policy Planning, issued a draft Department of Defense directive which would establish a policy for non-lethal weapons. The policy was intended to take effect January 1, 1995, and formally connected the military's non-lethal research to civilian law enforcement agencies.

The government's plan to use pulsed electromagnetic and radio frequency systems as a nonlethal technology for domestic Justice Department use rings the alarm for some observers. Nevertheless, the plan for integrating these systems has moved forward. Coupling these uses with expanded military missions is even more disturbing. This combined mission raises additional constitutional questions for Americans regarding the power of the federal government.237

In interviews with members of the Defense Department the development of this policy was confirmed.238 In those February 1995, discussions, it was discovered that these policies were internal to agencies and were not subject to any public review process.

In its draft form, the policy gives highest priority to development of those technologies most likely to get dual use, i.e. law enforcement and military applications. According to this document, non-lethal weapons are to be used on the government's domestic "adversaries." The definition of "adversary" has been significantly enlarged in the policy:

> "The term 'adversary' is used above in its broadest sense, including those who are not declared enemies but who are engaged in activities we wish to stop. This policy does not preclude legally authorized domestic use of the nonlethal weapons by United States military forces in support of law enforcement."239

This allows use of the military in actions against the citizens of the country that they are supposed to protect. This policy statement begs the question; who are the enemies that are engaged in activities they wish to stop, what are those activities, and who will make the decisions to stop those activities?

An important aspect of non-lethal weapon systems is that the name non-lethal is intentionally misleading. The Policy adds, "It is important that the public understand that just as lethal weapons do not achieve perfect lethality, neither will 'non-lethal' weapons always be capable of precluding fatalities and undesired collateral damage."240 In other words, you might still destroy property and kill people with the use of these new weapons.

237. Department of Defense Directive, Policy for Non-Lethal Weapons, Office of the Assistant Secretary of Defense, Draft. July 21, 1994. EPI1234
238. Interviews in late February 1995 by Nick Begich.
239. Department of Defense Directive, Policy for Non-Lethal Weapons, Office of the Assistant Secretary of Defense, Draft. July 21, 1994. EPI1234
240. Ibid.

Fear in the Ranks

In "Radical Destabilizing Effects of New Technologies" written by Thomas K. Adams for the U.S. Army War College's publication, *Parameters* (Autumn 1998) it is pointed out that important developments in three areas are ongoing. Information systems, biotechnology and nanotechnology are mutually reinforcing in their development and are changing the very nature of knowledge disbursement. The advances in these areas, according to the article, are transferring enormous power and potential to the general public.

Technology is advancing in a way which is creating a diffusion of power best exemplified by the Internet. The Internet places huge research resources at the fingertips of anyone willing to ask a question and pursue a line of thinking. The results can be incredible. While for many individuals it represents an opportunity to expand and advance knowledge, for military planners the idea that knowledge – allowing access to powerful technologies can not be restricted – creates a great deal of fear. The article expressed concern that the availability of the technology could shift power in a way which could create greater breakups of composite states such as the former Soviet Union and increase the possibility in China and other parts of the world.

The future, to a great extent, is already here. What remains of this predicted future to occur has probably already been designed or will be in the next few years. Already the privacy of individuals is compromised by every purchase we make where the information is digitalized. From the list of goods purchased at a grocery store with a scanner and charged to a debit or credit card, to all telephone calls and other forms of communication – all are transparent to those who have access to the systems. In the future, given the pace of miniaturization and information processing, it will soon be possible to monitor all forms of communication, create miniature surveillance equipment at a cost where the monitoring of inner cities, then whole cities and regions will be possible.

In March 1998, a paper, "Non-Lethal Technologies: Implications for Military Strategy," was authored by U.S. Air Force Colonel Joseph Siniscalchi and published by the Center for Strategy and Technology, Air War College, Maxwell Air Force Base, Alabama. The paper suggests that a "focus on Global Management" is the direction military thought is taking, because the overriding unifying force of the great powers of the United States, Europe, Japan, China and Russia is now a shared and interdependent economic system driven by expansion and growth. The lack of competing ideologies, with the exception of China, removes the primary threats to global security and replaces them with new ones.

These new military threats are groups or "non-state actors" motivated by religious causes, nationalism, ethnic rivalries and narco-interests. Dealing with these groups in the territorial boundaries of other countries limits military intervention, or at least it was assumed so until the United States attacked a suspected terrorist organization inside Afghanistan. The distance from adversaries is also increasing, primarily because of the accuracy and range of conventional arms and because of the proliferation of weapons of mass destruction. The military insists that because of the changes in the nature of the conflicts that there is greater need to bring forward the newest weapons with the hope that these new systems will minimize noncombatant casualties, reduce property destruction and increase control in areas judged to be a threat.

The proliferation of first and second generation non-lethal weapons will occur quickly because the technologies and equipment are not unique and are widely available to those with the knowledge to use and assemble them. These same advances make possible the use of these new technologies against governments, with the increases in electronic sophistication making developed countries' systems even more vulnerable to attack.

An additional risk with the use of these new non-lethal systems is the risk of conflict escalation. If a country is unable to counterattack in kind it will likely resort to conventional war fighting methods, terrorism or weapons of mass destruction.

The idea that non-lethal weapons could be used in conflicts with the emerging threats of "non-state" actors presupposes that all of the conflict participants are operating with the same "logic or rational thought basis" or that they make decisions based on similar value-sets. This is an inaccurate assumption given the history of conflicts involving these players. The fact is, they do not react in predictable ways and to expect them to be coerced by new systems is probably unrealistic and may serve only to increase the combatants' resolve.

The newer technologies offer militaries and states the possibility of non-visible combat. The idea that a country's communications, power generation and transfer systems, and all forms of electronic data processing can be shut down is mind boggling. Yet this is exactly what the various referenced documents suggest with the use of electromagnetic pulse (EMP) weapons. Adding these capabilities to economic sanctions would increase the immediacy of the effect of such sanctions and eliminate all access to supply. As an example:

> "...disrupting television, radio, and commercial communications can isolate a state's leadership, or denying electrical production can grind an economy to a halt."241

> "...The following are examples of non-lethal technologies that are employed to augment sanctions. To enforce sanctions, EMP munitions, delivered via cruise or air launched missiles, can disable suspect shipping within a designated

241. March 1998, "Non-Lethal Technologies: Implications for Military Strategy," by U.S. Air Force Colonel Joseph Siniscalchi, Center for Strategy and Technology, Air War College, Maxwell Air Force Base, Alabama. EPI235

restricted area. EMP sea mines may be employed in the restricted area to deter any maritime traffic. The port activities can be disrupted via air-launched EMP weapons to disable electronic components of infra-structure equipment and the electronic ignition of transportation vehicles at selected port areas."242

Not Just for the Military Police Like Them Too!

The following was taken from the United States Department of Justice's report on the increased use of new technology. As these words are written, the authors note that many of these earlier reports have resulted in the development and deployment of these systems by 1999.

"U.S. Department of Justice, National Institute of Justice, Office of Communication and Research Utilization. Report on the Attorney General's Conference on Less Than Lethal Weapons" by Sherri Sweetman, March 1987

Basically, two distinct categories of less than lethal weapons would be useful in crowd control: one to control major groups of people, and another to deal with individual instigators.

Participants also discussed the use of various wave lengths and forms of administration of electromagnetic energy as a non-lethal weapon. A substantial amount of preliminary research has been conducted in this area. Flashing or stroboscopic light has been found to produce a disorienting effect (termed photic driving or photic stimulation) at frequencies close to the alpha rhythm of the brain (12.5 cycles per second in most people). Stroboscopic light at exactly that frequency will induce seizures in approximately 1.5 percent of the population. One conference delegate reported testing 100 subjects, using flashing light near but not at the alpha rhythm frequency. Discomfort or disorientation was produced with an intensity of light down to 4 cycles per second, and the effect was still produced when the light was introduced from the side or through closed eyelids. A sharp leading edge to the waveform was found to be more effective than a round waveform.

The effect of stroboscopic light has been studied by a number of groups. In military applications, for instance, pulsed strobes in open terrain were found to cause disorientation, stumbling, and inability to concentrate. The disorienting effect produced by light flashing at an

242. Ibid.

appropriate frequency is not limited to nighttime. Sunlight filtering through helicopter rotors has also been reported to produce nausea or seizures. Reflected light and closed eyelids do not negate the effect. Lights flashing on airplanes at night may also produce disorientation. The fact that the brain can be severely affected by optic stimulation of a specific type offers clear possibilities for the development of less than lethal weapons – in particular those designed for crowd control (where it should be possible to protect law enforcement personnel from the effects of the light by means such as special protective glasses).

It is also quite likely that certain human physiological systems can be affected by exposure to various specific frequencies of electromagnetic radiation. One conference participant noted that scientific knowledge of human physiology is progressing to the point where it may soon be possible to target specific physiologic systems with specific frequencies of electromagnetic radiation to produce much more subtle and fine-tuned effects than those produced by photic driving. There is some evidence (and a good deal of supposition) that sustained, extremely low frequency (ELF) radiation can produce nausea or disorientation. One researcher has subjected animals to ELF electromagnetic radiation through electrode implants, and feels that similar results could be produced from afar, without electrodes. One participant suggested that ideally, one might like to develop the ability to design these electromagnetic fields for specialized use, for instance to produce sleep or confusion. It is known that sleep can be induced by electrodes in the brain, and Russian scientists claim to be able to produce sleep from afar (electrosleep).

Other frequencies may have significant impacts as well. It has been reported that a man who stepped in front of a microwave communications transmitter felt various disorienting effects. A participant suggested that in contrast to the long time periods that might be necessary (1/2 hour to 4 hours) to produce disorienting effects using ELF, other frequencies could potentially stun a person within 100 milliseconds. Needless to say, very careful and extended testing would be essential and the potential for irreversible physiological damage may be high. The damage may be far more subtle than that caused by a gun and, as a result, more difficult to detect, control and restrain.

It is easy to say that less than lethal weapons must be tested during the development process, but actually performing those tests is more difficult, because any weapons developed must be tested on animals, and eventually on human volunteers. A medical steering committee will have to establish acceptable limits of safety, since it is unlikely that any weapon can be 100 percent effective. The medical steering committee, together with Department of Justice personnel, including the FBI, must draw up a testing protocol – possibly with the help of advisory personnel from the National Institutes of Health and the Food and Drug Administration – and closely supervise the testing to ensure that it is as safe and as humane as possible.

Even if new weapons themselves win public acceptance, the process of testing those weapons on animal and human volunteers will require honest and skillful explanation to the public. The nature of the weapons to be tested and the necessity for the tests must not be a secret of the kind whose "leak" would result in an exaggerated expose and associated public outcry.

Much of the information necessary to construct these data bases exists already, but needs to be collected and organized in a usable form. A major problem is access to information. The military has undoubtedly conducted research and testing pertinent to the development of less than lethal weapons, but much of such work is classified. A similar problem may exist in accessing a corporation's proprietary information. Since the collection and dissemination of information was viewed as a key element in the successful development and refinement of less than lethal weapons, conference participants urged that efforts be made to resolve these access problems.

One suggestion was that development of less than lethal weapons be tied to defense contracts, since those contractors are typically large, strong, and technologically sophisticated.

Research for the Department of Defense and for the National Aeronautics and Space Administration has encouraged the development of thousands of companies that focus on technological research and engineering design.

A related suggestion emphasized the need to involve the medical community from the start, and perhaps to establish an independent panel of medical doctors to testify as to the safety of any weapons developed or in development."243

The predicted public outcry has begun. However, the gradual introduction of each technology softens the public's resolve to oppose these new technologies. By introducing these a few at a time the public does not see how the totality of the technology will impact them personally. The impacts of these new technologies on privacy issues and other matters of law only become apparent when the full scope of these initiatives is considered in the context of the revolution in military affairs. A more complete picture of the interest of the Department of Justice is gained by reviewing their solicitation for

243. Sweetman, Sherri. *Report on the Attorney General's Conference on Less than lethal Weapons.* National Institute of Justice, March 1987. EPI762

proposals which will introduce the technology to law enforcement organizations as follows:

> This solicitation seeks proposals for the development, testing, evaluation, adoption and implementation of new and innovative technologies and techniques to support and enhance law enforcement, courts, and/or corrections operations, particularly at the State and local level, and with an eye toward successfully developing new commercial products for their near-term use.
>
> There are four major project category areas, which may contain overlaps, to guide the formulation of proposals. Proposals may address one or more of these category areas:
>
> 1) General Technology Thrust Areas for Law Enforcement, Courts and Corrections.
> 2) Special Law Enforcement, Courts and Corrections Technology Development and Demonstration Projects.
> 3) Behavioral and Organizational Impact of Technological Advances.
> 4) Creative Technology Solutions to Law Enforcement, Courts, and Corrections problems.
>
> This solicitation is not intended to fund the purchase of existing systems, but rather it is intended to support the development of new technologies or the innovative adaptation of existing technologies to deal effectively with law enforcement, courts, and/or corrections problems.
>
> The National Institute of Justice (NIJ), the research and development agency of the U.S. Department of Justice (DOJ), is responsible under the Crime Act for providing funding for technology development projects to enhance law enforcement, courts, and corrections technologies; and the demonstration and evaluation of innovative approaches to the development, application, and utilization of leading edge technologies.
>
> **Focus Areas**
>
> Less-Than-Lethal (LTL) Area. Technologies are sought that will provide new or significantly improved options to conventional use-of-force methods sanctioned by law enforcement or corrections agencies to counter violent behavior or effect a resisted arrest. Scenarios where LTL technologies, devices, and/or techniques might be appropriately employed include, but are not limited to: (1) confrontation with subjects who assault officers or refuse to comply with lawful orders, and who may be armed with a non-firearm weapon; (2) encounters with individuals on foot

advancing toward or fleeing from an officer when ordered to stop; (3) riots, civil disturbances, or crowd control; (4) encounters with fleeing vehicles (vehicle stopping techniques); (5) arrestees, prisoners, or individuals who become violent or uncooperative; and (6) barricaded individuals or groups – with or without hostages.

Situational Awareness and Crime Prevention Area. This area addresses various applications that require unique or innovative sensor, detection, monitoring, surveillance, communications, and processing techniques and products. Emphasis should be placed on developing concepts that are easy to use, require low power (primarily for man-portable or unattended applications), and inexpensive to purchase and operate. Applications and need areas include, but are not limited to, the following: (1) through-wall sensors; (2) surveillance and monitoring (perimeter/physical security; asset protection; officer or offender tracking by Global Positioning System or other technologies; prison staff/inmate tracking; domestic surveillance or stalking alarms; court surveillance; probation/parole personnel monitoring; etc.); (3) sniper detection/location; (4) explosives/drug detection; (5) covert wires or sensors; (6) interagency communications interoperability; (7) continuous communication coverage in urban or rural environments; (8) secure communications; (9) image/audio processing and enhancement (including noise and false alarm reduction); (10) video/data capture, transmission (including video conferencing), and/or compression (including noise and false alarm reduction); (11) language translation; (12) speaker identification; (13) techniques to determine 'original' digital recordings for evidential purposes; (14) data exchange remediation; (15) 911 system to handle/locate cellular callers and more efficiently handle information flow from caller to responders; (16) caller/officer locator transmitter; (17) concealed weapons detection; and (18) information management in prison and court systems.

Forensic Sciences Area. Forensic sciences technologies are used to identify and develop evidence to solve criminal cases. A major goal of this focus area is to develop DNA technology that can reduce costs and processing time for DNA testing and that can enable DNA testing to be performed at the crime scene. Technologies that can address this goal include, but are not limited to, mass spectrometry, robotics, and micro-chips. Other Forensic Science technology areas of interest include, but are not limited to: DNA/serology, firearms, trace evidence, odontology, questioned documents, fingerprints, pathology, entomology, and toxicology. Of these, technologies which identify or develop the scientific bases of forensics are of interest, especially odontology, questioned documents, and trace evidence.

Special Law Enforcement, Courts, and Corrections Technology Development and Demonstration Projects

These projects include efforts in:

(a) Information Technology and database Integration Assessment and Adoption;
(b) Concealed Weapons detection;
(c) Vehicle Stopping Technology;
(d) Non-invasive drug testing;
(e) Crime mapping;
(f) Integrated Smart Gun/Laser Systems;
(g) Electronic Monitoring; and
(h) Improved Judicial Processing, Court proceedings, and Corrections Monitoring"[244]

These huge government financed labs have discovered new markets in domestic police organizations. The development of these initiatives focuses some of the best minds involved in weapons research. "The big research labs and strategic think tanks in the past couple of years have been busy selling the police and the public at large on the concept of non-lethal weapons. When reporters are invited to write about this stuff, they often watch demonstrations of sticky foam, a gooey substance used to immobilize rioting prisoners, and are supposed to be impressed at the humanity of this instrument of peace enforcement. High-powered microwave weapons, meanwhile, in the public relations context, have been presented as instruments used to disable ground and air transportation vehicles by zapping their electronics."[245]

Cooperation continues to increase between the military and civilian law enforcement. Urban drills and exercises are taking place to test the ability of the military in these situations. "Urban Warrior, now in its third day, is the culmination of a series of experiments designed to test new concepts and technologies for fighting throughout the streets, sewers and buildings of the world's urban areas. The experiment is also providing the corps with the opportunity to practice coordinating with local fire and police officials during simulated natural disasters and terrorist attacks involving chemical weapons."[246]

244. National Institute of Justice. Solicitation for Law Enforcement, Courts and Corrections Technology Development, Implementation and Evaluation, August 1996. EPI136
245. Cassidy, Peter. "Guess Who's the Enemy." *The Progressive*, Jan. 1996. EPI765
246. Verton, Daniel. "Urban warfare tech may alter Corps." *Federal Computer Week*, March 15, 1999. EPI1116

Deadly Force and Fleeing Felons

The logic applied to the deployment of these new technologies is being motivated in part by court rulings and the need to apprehend rather than kill fleeing suspects. "In March 1985, the U.S. Supreme Court ruled by a 6-3 vote that more than half the States' laws and many law enforcement agencies' regulations on police use of deadly force were unconstitutionally permissive. In Tennessee vs Garner, an apparently unarmed 15-year old was fatally wounded by police as he fled the scene of a burglary. Through this decision, the Supreme Court imposed a national minimum standard of force for the first time.

The decision invalidated laws in nearly half the states that allowed using 'deadly force' to prevent the escape of someone suspected of a felony. However, the court limited its ruling by providing that if a suspect is armed and poses 'a significant threat of death or physical injury to the officer or others,' police use of deadly force is not prohibited.

The Court held that '...the use of deadly force to apprehend an apparently unarmed, non-violent fleeing felon is an unreasonable search and seizure under the fourth amendment.'

Tennessee v. Garner sharply limited situations in which lethal force could be used by police and added significantly to the interest in developing less-than-lethal devices, particularly for use against fleeing felons."247

Slower than a Speeding Bullet

"Dr. Rusi Taleyarkhan and other researchers at the Oak Ridge National Laboratory have developed a prototype rifle that uses an aluminum-based gunpowder replacement that can control the speed of a bullet, which is based on theories of what happens when molten aluminum comes in contact with water. Therefore, a rifle could be available in the next year that has an electronic firing system guided by a target-finding laser with the capability to control the speed of the bullets fired. The guns are of interest to the Energy Department and the Pentagon, since both agencies need nonlethal weapons to protect their nuclear installations."248 "It sounds like something from a science fiction tale – a rifle that can be adjusted so its user fires bullets at varying speeds, such as 'stun,' 'disable,' or 'destroy.'...But researchers say they already have developed an aluminum-based gunpowder replacement that can control the speed of a bullet."249 This in another variation on the nonlethal weapon theme which in many cases may make sense. However even a "slow" bullet can kill if it hits the eye or some other entry point on the body. The question

247. Hayeslip, David et al. NIJ Inlative on Less-Than-Lethal Weapons. National Institute of Justice. March 1993. EPI312
248. Mansfield, Duncan. "Speeding Bullet." AP, July 19, 1999. Source: NLECTC *Law Enforcement & Technology News Summary*, July 22, 1999. EPI921
249. Mansfield, Duncan. "Scientists develop propellant that can vary speed of bullets." *Anchorage Daily News*, July 20, 1999, p. A-4. EPI160

which is raised is what history shows – if a weapon is perceived as nonlethal it is more apt to be used.

Acoustic Weapons

The use of sound as a weapon has been considered over the years. One of the problems has been in focusing sound waves into a small energetic package which can be used to deliver an impact on a person or object. This is not the only area of interest when it comes to sound and its use in warfare and law enforcement. The United States Air Force once more lays out the direction of this technology and its implications:

> "**Acoustics.** There has been considerable interest in using acoustics for potential non-lethal weapons. The acoustical weapons generate a low frequency sound (below 50Hz) that can disorient or cause nausea in personnel. The distress is reported to be temporary and stops when the acoustic source is stopped. At high power settings, these weapons may have an anti-material capability if 'tuned' to the appropriate frequencies.
>
> **Directed Energy – High Power Microwave (HPM).** High powered microwaves are normally considered an anti-material weapon, but they may have significant anti-personnel capabilities as well. Some directed energy weapons, such as microwaves, are able to produce a variety of effects on humans to include increasing levels of pain, incapacitation, and disorientation. Research is ongoing. If the range and power of a future capability is sufficient, a high-powered microwave weapon may be used for area denial or as a force protection capability."[250]
>
> "**High-intensity Sound.** High-intensity sound sets the ear drum in motion. These vibrations cause the inner ear to initiate nerve impulses that the brain registers as sound. The inner ear regulates the spatial orientation of the body. If the ear is subjected to high-intensity sound, the individual may experience imbalance. Low-frequency, high-intensity sound may cause other organs to resonate, causing a number of physiological results, including death.
>
> The British use high-intensity sound as a means of riot control in Northern Ireland. The Curdler is a device that emits a high 'shrieking noise at irregular intervals.' The sound is emitted at lower levels than the pain threshold.
>
> **Infrasound.** This is a powerful ultralow frequency (ULF)

250. Siniscalchi, Col. Joseph (USAF). "Non-Lethal Technologies: Implications For Military Strategy." Air War College, Air University, April 1997. AU/AWC/RWP177/97-04. EPI235

sonic weapon that can penetrate buildings and vehicles and can be directional and tunable. As a weapon, infrasound, lowfrequency sound, entails the same concerns as high-intensity sound. After being exposed to highintensity infrasound, a subject suffers from disorientation and reduced ability to perform sensorymotor tasks. At elevated levels, experimental animals cease breathing temporarily. The principles and findings regarding highintensity sound would apply to infrasound. The suffering would be no greater than that experienced by conventional weapons. The suffering must be proportionate to the military objectives. The sound must be applied so that damage to noncombatants is incidental in light of the military objective.

Sonic Bullets. These are packets of sonic energy that are propelled toward the target. The Russians apparently have a portable device that can propel a 10-Hertz (Hz) sonic packet the size of a baseball hundreds of yards. When employed against humans, the energy can be selected to result in nonlethal or lethal damage. The sonic bullet uses direct sonic energy. If the energy can be controlled so that it is used only against lawful combatants, the concerns surrounding acoustical weapons may be reduced or eliminated.

Voice Synthesis. This is the ability to clone a person's voice and broadcast a synthesized message to a selected audience. The propaganda value of this technique in our highly media dependent world would be enormous. We currently have the ability to control the broadcasts of foreign radio and television stations by using orbiting platforms packed with electronic gear.

In considering whether it is legal to clone a person's voice in order to gain a military advantage, it is important to determine whose voice is being cloned. In most cases, it would be realistic to expect that the voice cloned would be that of a political leader or a military officer. The cloned voice might give orders to the enemy combatant that might prove detrimental to the combatant. The combatant would most likely be under an obligation to follow those orders. That obligation, however, is owed to his own chain of command and is not under the law of armed conflict. Treacherous acts, those which abuse an obligation to be truthful under the law of armed conflict, are illegal. But if there is no obligation to be truthful under the law of armed conflict, then the misinformation amounts to a lawful ruse.

Conventional Warfare. Nonlethal weapons can also be used in conventional conflicts. Electromagnetic pulse (EMP) weapons can be used to disable grounded aircraft or vehicles rendering them useless on a temporary or even permanent basis. These weapons can also be used to down airborne aircraft although this would hardly be considered nonlethal One key to effective warfighting doctrine is to attack an enemy's critical nodes of command and communications as well as other infrastructures. While smart weapons can attack specific complexes and bunkers, nonlethal weapons offer the opportunity to disable entire nodes on a much grander scale. For example, the remote injection of a computer virus into an enemy's command and control system could be devastating. Likewise, certain biological agents that are designed to attack silicon or other computer components could effectively destroy computerized warfighting equipment. Super caustics can be sprayed on roads to deteriorate tank tracks and truck tires. Antitraction compounds can render mountain roads impassable, and embrittlement compounds could be sprayed on virtually any mechanical device – rendering them ineffective over a period of time. Combustion alteration technology agents could be used to shut down an entire harbor or airfield. Of course, practical matters such as method of delivery, persistence, concentration, and efficiency of these agents versus more lethal weapons must be considered."[251]

The use of these new technologies raises its own set of questions, particularly concerning the manipulation of media. The questions of national security again rub up against basic American values of the freedom of the press and communications generally. Granted that in wartime the rules are suspended, but the questions persist. Could it be asserted, for instance, that in peacetime the objectives of the government could be better served by silent interference with perceived enemies?

Some of the observations made by trying to get rid of birds on runways resulted in the development of, once again, dual use systems – in this case, against birds and humans. "Researchers are attempting to reduce aircraft-bird strikes by warning the birds away with an audible modulation on a conventional radar. The technique might even have far-reaching applications as a nonlethal weapon for the military and police as a crowd-control weapon."[252] The interesting thing about this little device is that it operated as a signal modulated (pulsed) on a conventional radar. In other words the radar signal still performed its primary function as a radar while carrying the extra acoustic signal. Any number of carriers could be used in this way to mask, behind a safe technology, a hidden weapon system which would only be detectable to the most sophisticated adversary.

251. Cook, III, Maj. Joseph W. et al. "Nonlethal Weapons: Technologies, Legalities, and Potential Policies." *Air Chronicles.* http://www.airpower.maxwell.af.mil/airchronicles/apj/mcgowan.html EPI246

252. Nordwall, Bruce. "Radar Warns Birds Of Impending Aircraft." *Aviation Week & Space Technology,* March 10, 1997. EPI1105

The effects of acoustics were not just observed by the Air Force. They were pursued by the United States Army as well. "The US Army has even looked into the use of infrasound – very low frequency sound – as a riot or crowd control agent. Infrasound generators could be turned against humans, causing disorientation, nausea and vomiting. The effects disappear when the sound is turned off."253

The interest of the military did not go unnoticed but began to raise questions from others. "In order to provide reliable information for future assessments of how acoustic-based weapons should be addressed by international laws of warfare and of human rights, I have done an exhaustive literature survey and conducted theoretical analyses. My questions included: What kind of sound sources could be used? How far does strong sound propagate? What are the effects on humans? Is there a danger of permanent damage?"254 The limits of the possibilities have yet to be defined as new breakthroughs in this area continue to be made.

There are still surprises unfolding in this area, which will emerge as one of the leading technological discovery areas of the next century. The applications of acoustics, from the following reports, may even extend well beyond the weapons arena, touching on transportation or other technologies involving moving objects. "Tim Lucas says he made a radical discovery while working at the Los Alamos National Laboratory in New Mexico that enables him to create more energy through sound waves than was ever thought possible. 'It's not an incremental improvement in an existing technology,' Lucas says, 'it's suddenly doing something which before was completely impossible.'"255...a team of scientists at Intersonics, Inc. of Northbrook, Illinois, has recently accomplished what may be the first prestidigitation, lifting a one-ounce steel ball bearing using nothing but sound waves.256

Stun Guns
Frozen in Their Tracks

"Unlike cattle prods, which can immobilize and cause localized pain, stun weapons are designed to inflict immediate and severe pain that can temporarily incapacitate a person. Victims are tortured, often repeatedly, with shocks applied to the armpits, neck, face, chest, abdomen, inner sides of the legs, soles of the feet, inside of the mouth and ears, on the genitals, and inside the vagina and rectum."257 "'Torturers seem to be discovering that electroshock

253. Starr, Barbara. "Non-Lethal weapon puzzle for US Army." *International Defense Review*, April 1993, pp. 319-320. EPI309
254. Altmann, Jurgen. "Acoustic Weapons? Sources, Propagation and Effects of Strong Sound." Acoustical Society of America, ASA/EAA/DAGA '99 Meeting: Lay language Papers. http://www.acoustics.org/altmann.html EPI809
255. Hill, Jim. "Invention may do for sound what laser did for light." *CNN Interactive*, Dec. 2, 1997. http:www.cnn.com/TECH/9712/sound.wavr.energy/ EPI271
256. *Omni*. "Acoustic Levitation." EPI185
257. Welsh, James. "Electroshock torture and the spread of stun technology." *The Lancet*, Vol. 349, April 26, 1997. EPI786

stun weapons are ideal for their evil purposes – cheap, easy to conceal and hard to trace,' says Brian Wood, who tracks the weapons internationally for Amnesty. Time's own investigation found... disturbing evidence that stun guns from the U.S., Asia and Europe wind up in countries whose governments practice torture."[258] Why should this have been any surprise at all is beyond us. When stun, or electroshock weapons, are already widely used in the United States by police and civilians, their migration into other regions should not be a surprise.

The stun belt is another such weapon raising the hackles of human rights activists and others. "Some defense lawyers and civil rights activists criticize the belts, arguing that the belt's shock is too powerful and could either be abused or malfunction. Sheriff's officials, including instructors who receive the shock to qualify for training, concede that it is indeed painful, but that the belts are necessary for the very small percentage of inmates who cannot otherwise be controlled."[259] Sounds good to some, but what about the "rule of law" and who will have the authority to selectively apply it? "The use of what the Bureau of Prisons refers to as 'custody control belts' in the United States has attracted the attention of Amnesty International. Not only does the group view the devices as the upholding of cruel, inhuman, or degrading treatment or punishment, it says the United States isn't even following international standards concerning the use of new weapons. 'The U.S. authorities have failed to live up to this standard as electroshock weapons have proliferated around the country's law enforcement agencies...without rigorous independent testing, evaluation and monitoring,' according to a recent Amnesty report. Some 20 states are said to have authorized the use of 'stun belts,' the Amnesty term for the devices that can deliver a 50,000-volt shock that can last for eight seconds, and cause severe pain and incapacitate the wearer."[260] "The stun belt looks to be a weapon which will almost certainly result in cruel, inhuman, or degrading treatment, a violation of international law...The use of the stun belt in U.S. prisons will inevitably encourage prison authorities – including those in torturing states – to do likewise..."[261] Again, the United States sets an example by its own actions and then condemns others who use these new technologies in perhaps even more cruel and invasive ways.

In one recent version of the electroshock weapons, the device is fired at a distance, without physical contact with the victim, when rendering a paralyzing blow. "A non-lethal weapon for temporarily immobilizing a target subject by means of muscular tetanization in which the tetanization is produced by conducting a precisely-modulated electrical current through the target. The transmission of

258. Waller, Douglas. "Weapons of Torture." *Time*, April 6, 1998, pp. 52-53. EPI196
259. Shaver, Catherine. "New Tool in Courts: Stun Belts." *Washington Post*, Dec. 29, 1998. Source: NLECTC *Law Enforcement & technology News Summary*, Dec. 31, 1998. EPI1046
260. Walsh, Edward. "Stun Belts Shock Human Rights Group." *Washington Post*, June 15, 1999. Source: NLECTC *Law Enforcement & Technology News Summary*, June 17, 1999. EPI943
261. Cusac, Anne-Marie. "Life In Prison: Stunning Technology." *The Progressive*, Vol. 60, No. 7, July 1996. EPI574

this current to the distant target is via two channels of electrically conducted air."262

These weapons should be severely restricted or eliminated. The potential for, and actual abuse, of these weapons should be apparent to all.

Holography

Holography has been referenced, to a very limited degree, and is only vaguely mentioned in discussions of nonlethal weapons. The concept is that an image can be holographically projected so that it appears as a real thing or ghostlike apparition. The image can be endowed with speech which seems to emanate from the image itself. This type of technology could be used for creating panic and fear in unsophisticated combatants who are unaware of the technology. The following references point to the viability of this technology:

> "HyperSonic Sound is a new technology in sound production that employs ultrasonic tones in a new patented and patents-pending process to produce sounds directly in the air. The laser-like HSS ultrasonic beam can project audible sound to virtually any listening environment creating many new sound applications previously impossible with existing speaker technology."263

> "Future funding may be allocated to radio frequency devices for use against people and material; non-blinding lasers to confuse personnel, night vision or EO equipment; holography as a tool for psychological warfare; and calmative agents"264

> "Potential weapons identified in the memo include 'dazzling radiator' munitions for causing temporary blindness, 'traction inhibitors' for making roads impassable to military vehicles, 'neural inhibitors,' stun guns and even 'holographic projections' for creating illusions on the battlefield."265

The trick in future wars might be in determining what is real, imagined or projected. Who knows what novel applications might be made with this technology; perhaps a UFO invasion, the appearance of God or some person known by the local population.

Sticky Foam and More

The use of sticky foam is hyped in almost every press report on nonlethal weapons as a "good example" of the technology. "The Marine Corps Systems Command and the U.S. Army are together developing non-lethal rigid foam. The Marine Corps is adapting an

262. US Patent #5,675,103, Oct. 7, 1997. Non-Lethal Tetanizing Weapon. Inventor: Herr, Jan Eric. EPI332
263. American Technology Corporation. Chairman's Message. May 17, 1999. http://www.smallcapresearch.com/news.htm EPI811
264. Starr, Barbara. "USA defines policy on non-lethal weapons." *Jane's Defence Weekly,* March 6, 1996. EPI684
265. Lancaster, John. "Pentagon, Justice Dept. Set Plans for Sharing Nonlethal Technology." *Washington Post,* March 23, 1994. EPI696

off-the-shelf product for military use to seal doors and windows, disable moving vehicles of all kinds, and secure razorwire. The goal is for the material to be rigid within 10 to 15 seconds, and fully set within 15 minutes and to increase the life of such products to a minimum of 42 months."266 What is not considered with the use of foam are the consequences to the victim if the foam covers his face. Another problem is that of an armed suspect. In a late addition to the sticky foam concept, the gun dilemma was addressed by electrifying the foam to paralyze the victim. The problems should be apparent here as well – when the muscles of the victim contract and fire the gun he then falls into the foam, suffocating in convulsions. Somehow, even the most "safe" technologies leave a person with the impression that perhaps a bullet might be the preferred option of the individual on the receiving end of this innovation.

"Less lethal refers to the application of force that is not designed to cause death or serious injury, and experts say it is most useful in unique situations, such as suicidal threats and conflicts where deranged individuals are armed with edged weapons. In the past, such sensitive situations often forced officers to shoot the suicidal person, inflicting severe wounds and frequently – death. Not so with less lethal weapons, officers now have a wide range of options at their disposal. Pepper sprays (OC) and the Taser, which is an electronic device that shoots small darts discharging 50,000 volts into the suspect's body, enjoy the most popularity among law enforcement officers. ...Less lethal projectiles come in an assortment of classifications: specialty impact munitions like MK Ballistics are loaded with steel pellets designed to cut through automobiles and windshields; a fragmenting munition can be fired into the engine compartment to slash hoses, cut wires, and otherwise destroy a vehicle's mechanical components; while baton and stinger rounds are particularly useful in discouraging rioters."267

These are but a few of the new mix of innovation being added to law enforcement and military arsenals. Many of these technologies have a place in these situations but the ethics of weapons development and use must be in conformity to law. Moreover, the concept of revisiting treaties seems to be a topic of discussion every time a new breakthrough is made. These Treaties may need to be revisited, however, the initiative should be driven by humanitarian interests and not by militaries which are inconvenienced by public scrutiny and debate.

266. *Law Enforcement Technology*. "Locking Down the Enemy With Foam." May 1999, Vol. 26, No. 5. Source: NLECTC *Law Enforcement & Technology News Summary*, June 3, 1999. EPI963
267. Flynn, Michael. "The Future Holds More Less Lethal." *Law Enforcement Technology*, Oct. 1998, Vol. 25, No. 10. Source: NLECTC *Law Enforcement & Technology News Summary*, Oct. 22, 1998. EPI1088

Chapter 7

Mind Control
The Ultimate
"Brave New World"

"It would also appear possible to create high fidelity speech in the human body, raising the possibility of covert suggestion and psychological direction...Thus, it may be possible to 'talk' to selected adversaries in a fashion that would be most disturbing to them."[268]

The idea that the brain can be made to function at a more efficient and directed level has been the subject of research by scientists, mystics, health practitioners and others for as long as mankind has contemplated such matters. In the last decade, advances in the science of the brain have begun to yield significant results. The results of the research are startling, challenging and, if misused, will be frightening. The certainty to be expected from the research is that it will continue to proceed.

The idea that people can be impacted by external signal generators which create, for example, pulsed electromagnetic fields, pulsed light and pulsed sound signals is not new. The following information demonstrates some of the possibilities and gives hints of the potentials of the technology. On the positive side, researchers in the field of light and sound are making huge progress in a number of areas, including working with learning disabilities, attention deficit disorders, stroke recovery, accelerated learning, drug/alcohol addiction and enhanced human performance. The research has shown that certain brain states can be influenced in a way which causes changes within the brain itself. These changes allow individuals the possibility of influencing specific conditions in the mind and body otherwise thought beyond our direct control.

268. *New World Vistas: Air and Space Power for the 21st Century - Ancillary Volume*; Scientific Advisory Board (Air Force), Washington, D.C.; Document #19960618040; 1996; pp. 89-90. EPI402

The military and others interested in such things have also focused a large amount of research into this area for the purpose of enhancing the performance of soldiers while degrading the performance of adversaries.

What is known is that great strides in the area of behavior control are now possible with systems developed and under development by most sophisticated countries on the planet. These new technologies represent a much different approach to warfare which our government is describing as part of the Revolution in Military Affairs. While these new technologies offer much for military planners they offer even more to citizens generally. Their potential use in military applications and "peacekeeping" creates the need for open debate of this new realm of intelligence gathering, manipulation and warfare. The most basic ethical questions regarding use of these technologies have not been adequately addressed.

At the same time that defense and intelligence gathering capabilities are being sought, independent researchers are fully engaged in seeking positive uses for the technology. The potentials of the technology, like all technology, are great as both a destructive or constructive force for change. The idea of enhancing physical and mental performance while bypassing what heretofore was a long and arduous road to achieve the same results is exciting. Maintaining the research in the open literature and insuring that constructive uses are encouraged is critical.

I began looking into technologies for stimulating brain performance about fifteen years ago. At the time, there were limited tools available compared to what is now possible. Now it is possible to obtain light and sound, electrocranial and biofeedback tools for use in this exploration. Moreover, there are audio materials also available for use with most of these tools. These audio materials can be used for learning languages, behavior modification or enhanced performance. The biofeedback side of the new technology is being used to train people to reach specific desired brain states for optimum performance.

The use of light and sound devices for stimulating brain activity which is conducive to accelerated learning and relaxation is a growing area of interest to many people. Moreover, the use of these tools in conjunction with biofeedback has been the subject of quickly evolving research. The combined technologies of brain state inducement and biofeedback offer exciting possibilities. It has been found with the combination that a person, in a matter of several weeks, can learn to purposefully modify his/her brain activity in a way which would have taken a Zen master twenty years to accomplish. It has been shown that some children with attention deficit disorders can be taught to regulate their brain activities so that they can learn efficiently without chemicals. It has been demonstrated that recovering stroke victims can more rapidly recover when working with brain-biofeedback practitioners and these new tools.

The research is also teaching us a good deal about our suggestibility in terms of influences which impact our behavior. The underlying message that comes with the new technology is the necessity of providing safeguards against misuse. Additionally, recognition of the everyday stimulation we all get and the effect of these information inputs on our learning processes becomes more clear. Human suggestibility, particularly when in a fatigued condition, has been exploited by terrorists, cults and others in pursuit of their own aims. The passive suggestibility of radio and television as we weave in and out of the semi-sleep states is for the most part not even recognized. The passive learning situations become even more relevant when we consider how we "receive the news" in our daily lives. The ability to influence thinking, behavior and performance is indeed a two-edged sword.

The 1980s and 90s were focused on building up the physical body. The next century will see a focus on building the mind and optimizing mental performance. The idea of merging the new technologies into education is interesting and also calls into question who will decide what is learned. In the interim, the possibilities are incredible for those interested in such pursuits. The control of our mental function is no different than the control of the muscles in our bodies. Learning to control or coordinate the activity of our minds will propel our bodies through a much more productive and fuller life. The new tools may offer just such opportunities.

On the other side of the issue is the potential for misuse and exploitation of the science. Military planners, law enforcement officials and others are now seeking the covert use of these technologies for controlling the ultimate "information processor" – The Human Being.

MK-ULTRA

"Dr. Gottleib, born August 3, 1918, was the CIA's real-life 'Dr. Strangelove' – a brilliant bio-chemist who designed and headed MK-ULTRA, the agency's most far-reaching drug and mind-control program at the height of the Cold War. Though the super-secret MK-ULTRA was ended in 1964, a streamlined version called MK-SEARCH was continued – with Gottleib in charge – until 1972."[269] During this period substantial interest in mind control was stimulated by Soviet use of microwaves. In 1988, "thirty-five years after security officers first noticed that the Soviets were bombarding the U.S. embassy in Moscow with microwave radiation, the U.S. government still has not determined conclusively – or is unwilling to reveal – the purpose behind the beams."[270] The government did know what was happening. The Soviets had developed methods for disrupting the purposeful thought of humans and was using their knowledge to impact diplomats in the United States embassy in Moscow.

269. Foster, Sarah. "Cold War legend dies at 80: Famed as CIA's real-life 'Dr. Strangelove.'" *Worldnetdaily*, March 9, 1999. EPI279
270. Reppert, Barton. "The Zapping of an Embassy: 35 Years Later, The Mystery Lingers." AP, May 22, 1988. EPI1112

In 1994 a report concerning the MKULTRA program was issued containing the following information:

> "In the 1950s and 60s, the CIA engaged in an extensive program of human experimentation, using drugs, psychological, and other means, in search of techniques to control human behavior for counterintelligence and covert action purposes.
>
> In 1973, the CIA purposefully destroyed most of the MKULTRA files concerning its research and testing on human behavior. In 1977, the agency uncovered additional MKULTRA files in the budget and fiscal records that were not indexed under the name MKULTRA. These documents detailed over 150 subprojects that the CIA funded in this area, but no evidence was uncovered at that time concerning the use of radiation.
>
> The CIA did investigate the use and effect of microwaves on human beings in response to a Soviet practice of beaming microwaves on the U.S. embassy. The agency determined that this was outside the scope of the Advisory Committee's purview.
>
> ...The Church Committee found some records, but also noted that the practice of MKULTRA at that time was 'to maintain no records of the planning and approval of test programs.'...MKULTRA itself was technically closed out in 1964, but some of its work was transferred to the Office of Research and Development (ORD) within the DS&T under the name MKSEARCH and continued into the 1970s.
>
> The CIA worked closely with the Army in conducting the LSD experiments. This connection with the Army is significant because MKULTRA began at the same time that Secretary of Defense Wilson issued his 1953 directive to the military services on ethical guidelines for human experiments.
>
> Throughout the course of MKULTRA, the CIA sponsored numerous experiments on unwitting humans. After the death of one such individual (Frank Olson, an army scientist, was given LSD in 1953 and committed suicide a week later), an internal CIA investigation warned about the dangers of such experimentation. The CIA persisted in this practice for at least the next ten years. After the 1963 IG report recommended termination of unwitting testing, Deputy Director for Plans Richard Helms (Who later became Director of Central Intelligence) continued to advocate covert testing on the ground that 'positive operational capability to use drugs is diminishing, owing to a lack of realistic testing. With increasing knowledge of state of the art, we are less capable of staying up with the Soviet advances in this field.'...Helms attributed the cessation of the unwitting testing to the high risk

> of embarrassment to the Agency as well as the 'moral problem.' He noted that no better covert situation had been devised than that which had been used, and that 'we have no answer to the moral issue.'"271

They did have the answers to the moral questions on human experimentation but chose to ignore them, destroy the records, hide the truth and still continue in their efforts. Nothing has changed as each participating organization, using national security laws, avoids disclosure and accountability. The records which were destroyed contained the evidence necessary to perhaps send some participants to jail for society's version of behavior modification. Once again – there was no accountability and no recognition of the rights of the individuals damaged by these experiments.

Mind Wars

"For the first time in some 500 years, a scientific revolution has begun that will fundamentally change the world as much as the Renaissance and Enlightenment did. A handful of extraordinary new advances in science are taking humans quickly and deeply into areas that will have profound implications for the future."272 One of these areas is control of the human mind. The issues surrounding behavior modification, mind control and information warfare become crystal clear as the facts unfold. The following is taken from a current military document which clarifies their position in the emerging area of research, taking a much different direction than the one described above:

The Information Revolution and the Future Air Force

Colonel John A. Warden, III, USAF

> We're currently experiencing, on an unprecedented global basis, three simultaneous revolutions, any one of which would be more than enough to shock and confound us. The first revolution, a geopolitical revolution, sees a single dominant power in the world for the first time since the fall of Rome. The opportunities that are inherent in this situation are extraordinary, as are the pitfalls. Unfortunately, there is no one around that has first hand experience in how to deal with that kind of single power dominant world.
>
> The second revolution, and there's a lot of discussion about this so far, is the information revolution. As other people have mentioned, it is following inexorably in tandem behind Moore's law of computing power. Attendant to it,

271. Advisory Committee Staff, Committee on Human Radiation Experiments. Methodological Review of Agency Data Collection Efforts: Initial Report on the Central Intelligence Agency Document Search. June 27, 1994. (HG) EPI579

272. Petersen, John L. *The Road To 2015: Profiles of the Future.* Waite Group Press™, 1994. ISBN 1-878739-85-9 EPI849

though, is not the creation of new ideas and technologies, but also an exponential growth in the velocity of information dissemination, and for us, that is of extraordinary importance. A key part of this information revolution has an awesome impact on competition. The business that introduced a new product ten years ago could count on probably five years before it had to look seriously at potential competitors based overseas. Today, you're lucky if you can count on five months or even five weeks before you are facing the overseas competitor. In today's world, success simply demands rapid introduction of successively new products or military systems. Success now goes to the organization which exploits information almost instantly, while failure is the near certain fate of the organization which tries to husband or hide ideas. Real simple – use it or you're going to lose it.

The third revolution, which is a little bit more complex, is the military/technological revolution, or in some places called the revolution in military affairs. I'm convinced that this is the first military technological revolution ever because we now have, for the first time, a conceptually different way to wage war. We can wage war in parallel now. In the past, communications and weapons technology, especially weapons accuracy, have constrained us to waging serial war. This changes almost everything.

Biological Process Control: As we look forward to the future, it seems likely that this nation will be involved in multiple conflicts where our military forces increasingly will be placed in situations where the application of full force capabilities of our military might cannot be applied. We will be involved intimately with hostile populations in situations where the application of non-lethal force will be the tactical or political preference. It appears likely that there are a number of physical agents that might actively, but largely benignly, interact or interfere with biological processes in an adversary in a manner that will provide our armed forces the tools to control these adversaries without extensive loss of life or property. These physical agents could include acoustic fields, optical fields, electromagnetic fields, and combinations thereof. This paper will address only the prospect of physical regulation of biological processes using electromagnetic fields.

Prior to the mid-21st century, there will be a virtual explosion of knowledge in the field of neuroscience. We will have achieved a clear understanding of how the human brain works, how it really controls the various functions of the body, and how it can be manipulated (both positively and negatively). One can envision the development of electromagnetic energy sources, the output of which can be pulsed, shaped, and focused, that can couple with the human body in a fashion that will allow one to prevent voluntary muscular movements, control emotions (and thus actions), produce sleep, transmit suggestions, interfere with both short-term and long-term memory, produce an experience set, and delete an experience set. This will open the door for the development of some novel capabilities that can be used in armed conflict, in terrorist/hostage situations, and in training. New weapons that offer the opportunity of control of an adversary without resorting to a lethal situation or to collateral casualties can be developed around this concept. This would offer significant improvements in the capabilities of our special operation forces. Initial experimentation should be focused on the interaction of electromagnetic energy and the neuromuscular junctions involved in voluntary muscle control. Theories need to be developed, modeled, and tested in experimental preparations. Early testing using in vitro cell cultures of neural networks could provide the focus for more definitive intact animal testing. If successful, one could envision a weapon that would render an opponent incapable of taking any meaningful action involving any higher motor skills, (e.g. using weapons, operating tracking systems). The prospect of a weapon to accomplish this when targeted against an individual target is reasonable; the prospect of a weapon effective against a massed force would seem to be more remote. Use of such a device in an enclosed area against multiple targets (hostage situation) may be more difficult than an individual target system, but probably feasible.

It would also appear to be possible to create high fidelity speech in the human body, raising the possibility of covert suggestion and psychological direction. When a high power microwave pulse in the gigahertz range strikes the human body, a very small temperature perturbation occurs. This is associated with a sudden expansion of the slightly heated tissue. This expansion is fast enough to produce an acoustic wave. If a pulse stream is used, it should be possible to create an internal acoustic field in the 5-15 kilohertz range, which is audible. Thus, it may be possible to 'talk' to selected adversaries in a fashion that would be most disturbing to them.

In comparison to the discussion in the paragraphs above, the concept of imprinting an experience set is highly speculative, but nonetheless, highly exciting. Modern electromagnetic scattering theory raises the prospect that ultrashort pulse scattering through the human brain can result in reflected signals that can be used to construct a reliable estimate of the degree of central nervous system arousal. The concept behind this 'remote EEG' is to scatter off of action potentials or ensembles of action potentials in major central nervous system tracts. Assuming we will understand how our skills are imprinted and recalled, it might be possible to take this concept one step further and duplicate the experience set in another individual. The prospect of providing a 'been there – done that' knowledge base could provide a revolutionary change in our approach to specialized training. How this can be done or even if it can be done are significant unknowns. The impact of success would boggle the mind!"273

The above report was a forecast for the year 2020. However the reality is that these technologies already exist and there are a number of patents in the open literature which clearly show the possibilities. This research is not new but goes back to the 1950s. "A new class of weapons, based on electromagnetic fields, has been added to the muscles of the military organism. The C^3I doctrine is still growing and expanding. It would appear that the military may yet be able to completely control the minds of the civilian population."274 The targeting of civilian populations by the military is a significant departure from its history. In the past the military has used persuasion through real information rather than using deliberate deception and mind manipulation to win populations over. "A decoy and deception concept presently being considered is to remotely create the perception of noise in the heads of personnel by exposing them to low power, pulsed microwaves. When people are illuminated with properly modulated low power microwaves the sensation is reported as a buzzing, clicking, or hissing which seems to originate (regardless of the person's position in the field) within or just behind the head. The phenomena occurs at average power densities as low as microwatts per square centimeter with carrier frequencies from .4 to 3.0 GHz. By proper choice of pulse characteristics, intelligible speech may be created. Before this technique may be extended and used for military applications, an understanding of the basic principles must be developed. Such an understanding is not only required to optimize the use of the concept for camouflage, decoy and deception operations but is required to properly assess safety factors of such microwave exposure."275 Actual testing of certain systems has proven "that movements, sensations, emotions, desires, ideas, and a variety of psychological phenomena may be induced, inhibited, or modified by electrical stimulation of specific areas of the brain. These facts have changed the classical philosophical concept that the mind was beyond experimental reach."276

The first widespread interest in the subject of mind control hit the mainstream of military think tanks after the Korean War when

---

273. USAF Scientific Advisory Board. *New World Vistas: Air And Space Power For The 21st Century - Ancillary Volume.* 1996. EPI402
274. U.S.EPA. *Summary and Results of the April 26-27, 1993 Radiofrequency Radiation Conference, Volume 2: Papers.* 402-R-95-011, March 1995. EPI728
275. Oscar, K.J. *Effects of low power microwaves on the local cerebral blood flow of conscious rats.* Army Mobility Equipment Command. June 1, 1980. EPI1195
276. Delgado, Jose M.R. *Physical Control of the Mind: Toward a Psychocivilized Society.* Harper & Row, Publishers, New York, 1969. EPI850

returning prisoners of war exhibited significant behavioral changes. In 1956 the following was written into the United States Congressional Record:

> "Reports of the treatment of American prisoners of war in Korea have given rise to several popular misconceptions, of which the most widely publicized is 'brainwashing.' The term itself has caught the public imagination and is used, very loosely, to describe any act committed against an individual by the Communists. Actual 'brainwashing' is a prolonged psychological process, designed to erase an individual's past beliefs and concepts and to substitute new ones. It is a highly coercive practice which is irreconcilable with universally accepted medical ethics. In the process of 'brainwashing,' the efforts of many are directed against an individual. To be successful, it requires, among other things, that the individual be completely isolated from normal associations and environment."[277]

The ethical considerations have not changed, but the military's position on the ethics has changed as they have gained significant capabilities in these areas. "Psychological warfare is becoming increasingly important for U.S. forces as they engage in peacekeeping operations. 'In the psychological operations area, we're always looking to build on our existing technologies, so much of this is evolutionary,' Holmes said. 'It is critically important that we stay ahead of the technology curve."[278] The temptation to dabble in this area has now overcome the ethical considerations.

A recent Russian military article offered a slightly different slant to the problem, declaring that "humanity stands on the brink of a psychotronic war" with the mind and body as the focus.

These weapons aim to control or alter the psyche, or to attack the various sensory and data-processing systems of the human organism. In both cases, the goal is to confuse or destroy the signals that normally keep the body in equilibrium.

According to a Department of Defense Directive, information warfare is defined as "an information operation conducted during time of crisis or conflict to achieve or promote specific objectives over a specific adversary or adversaries." An information operation is defined in the same directive as "actions taken to affect adversary information and information systems." These "information systems" lie at the heart of the modernization effort of the US armed forces and manifest themselves as hardware, software, communications capabilities, and highly trained individuals

Information warfare has tended to ignore the role of the human body as an information – or data-processor, in this quest for

---

277. U.S. Senate. *Communist Interrogation, Indoctrination and Exploitation Of American Military, and Civilian Prisoners.* Committee on Government Operations, Subcommittee On Investigations. 84th Congress, 2nd Session. Dec. 31, 1956. EPI1131
278. Cooper, Pat. "U.S. Enhances Mind Games." *Defense News*, April 17-23, 1995. EPI1154

dominance except in those cases where an individual's logic or rational thought may be upset via disinformation or deception...Yet, the body is capable not only of being deceived, manipulated, or misinformed but also shut down or destroyed – just as any other data-processing system. The "data" the body receives from external sources – such as electromagnetic, vortex, or acoustic energy waves – or creates through its own electrical or chemical stimuli can be manipulated or changed just as the data (information) in any hardware system can be altered. If the ultimate target of information warfare is the information-dependent process, "whether human or automated," then the definition implies that human data-processing of internal and external signals can clearly be considered an aspect of information warfare.279

On a much grander scale, the use of mind control was contemplated as far back as 1969 by a former science advisor to President Johnson. "Gordon J.F. Macdonald, a geophysicist specializing in problems of warfare, has written that accurately timed, artificially excited strokes, 'could lead to a pattern of oscillations that produce relatively high power levels over certain regions of the earth...In this way, one could develop a system that would seriously impair the brain performance of very large populations in selected regions over an extended period...'"280 This capability exists today through the use of systems which can stimulate the ionosphere to return a pulsed (modulated) signal which at the right frequency can override normal brain functions. By overriding the natural pulsations of the brain chemical reactions are triggered which alter the emotional state of targeted populations.

Subliminal Messages and Commercial Uses

One of the areas where this new technology is being used is in systems to dissuade shoplifters using sound below the range of hearing. "Japanese shopkeepers are playing CDs with subliminal messages to curb the impulses of the growing band of shoplifters. The Mind Control CDs have sound-tracks of popular music or ocean waves, with encoded voices in seven languages...warning that anyone caught stealing will be reported to the police."281 A number of patents have been developed to influence behavior in this way. The following summations are taken from some of these patents dealing with both audio and video programing only this time we are the program:

> "An auditory subliminal programming system includes a subliminal message encoder that generates fixed frequency security tones and combines them with a subliminal message signal to produce an encoded subliminal message signal which

279. Thomas, Timothy L. "The Mind Has No Firewall." *Parameters*, Vol. XXVIII, No. 1, Spring 1998. EPI525
280. Brzezinski, Zbigniew. *Between Two Ages: America's Role in the Technetronic Era.* Viking Press, New York. 1970. EPI787
281. McGill, Peter. "'Mind Control Music' Stops Shoplifters." *The Sydney Morning Herald*, Feb. 4, 1995. EPI95

is recorded on audio tape or the like. A corresponding subliminal decoder/mixer is connected as part of a user's conventional stereo system and receives as inputs an audio program selected by the user and the encoded subliminal message."[282]

"Ambient audio signals from the customer shopping area within a store are sensed and fed to a signal processing circuit that produces a control signal which varies with variations in the amplitude of the sensed audio signals. A control circuit adjusts the amplitude of an auditory subliminal anti-shoplifting message to increase with increasing amplitudes of sensed audio signals and decrease with decreasing amplitudes of sensed audio signals. This amplitude controlled subliminal message may be mixed with background music and transmitted to the shopping area."[283]

"Data to be displayed is combined with a composite video signal. The data is stored in memory in digital form. Each byte of data is read out in sequential fashion to determine: the recurrence display rate of the data according to the frame sync pulses of the video signal; the location of the data within the video image according to the line sync pulses of the video signal; and the location of the data display within the video image according to the position information."[284]

"This invention is a combination of a subliminal message generator that is 100% user programmable for use with a television receiver. The subliminal message generator periodically displays user specified messages for the normal television signal for specific period of time. This permits an individual to employ a combination of subliminal and supraliminal therapy while watching television."[285]

The above may seem a bit complicated, however, they can be summarized. These patents are designed to provide a way to hide messages in video or audio formats masking any suggestions that the programmer wishes to convey. These kinds of messages bypass the conscious mind and are acted upon by the person hearing them – they are not sorted out by the active mind. Although these technologies are being developed for personal use and as security measures, consider the possibilities for abuse by commercial interests

282. US Patent #4,777,529, Oct. 11, 1988. Auditory Subliminal Programming System. Inventors: Schultz et al. Assignee: Richard M. Schultz and Associates, Inc. EPI265
283. US Patent # 4,395,600, July 26, 1983. Auditory Subliminal Message System and Method. Inventors: Lundy et al. EPI264
284. US Patent # 5,134,484, July 28, 1992. Superimposing Method and Apparatus Useful for Subliminal Messages. Inventor: Willson, Joseph. Assignee: MindsEye Educational Systems Inc. EPI290.
285. US Patent #5,270,800, Dec. 14, 1993. Subliminal Message Generator. Inventor: Sweet, Robert L. EPI288

where the messages might be "buy, buy, buy," "drink more, don't worry" or some other self-serving script. Should these systems be regulated? By whom and under what conditions?

New Standards for What is a Memory?

"Nevada is currently the only state to allow witness testimony of a person who has undergone hypnosis. As of October 1, 1997, courts hearing both civil and criminal cases can take a hypnotically refreshed testimony, as long as the witness, if a minor, has had the informed consent of parent or guardian, and the person performing the hypnosis is any of the following: a health care provider, a clinical social worker licensed in accordance with 641B of Nevada Revised statue, or a disinterested investigator."286

This issue will surely become more complex as technology advances in terms of evidence. When the day arrives when it is possible to completely change or alter memory as was suggested earlier by military officers, what then? How will we separate the real from the unreal? What will be the impact on the burden of proof in courts as it relates to "reasonable doubt"? Again the emergence of the technology has to first be recognized as real before laws can be constructed and systems established for controlling misuse. Think how long it has taken the courts to even recognize hypnotherapy as valid science. We are hopeful that we will not have to wait so long for legislative bodies to take the initiative to address these issues.

Auditory Effect?

The questions which this section raises are profound. Is it possible to transmit a signal to the brain of a person, from a distance, which deposits specific sounds, voice or other information which is understandable? Is it possible to transfer sound in a way where only the targeted person can hear the "voice in the head" and no one else hears a thing? Is it possible to shift a person's emotions using remote electromagnetic tools? The answer to each of these questions is a resounding – Yes! The state of the science has passed even the most optimistic predictions and the capabilities are here now.

Military literature suggests that this is possible. A series of experiments, patents and independent research confirm that this technology exists today. While giving testimony to the European Parliament in 1998, I demonstrated one such device to the astonishment of those in attendance. This particular device required physical contact in order to work and was nearly forty years old. This area is one of the most important because it points to the ultimate weapons of political control – the ability to place information directly into the human brain, bypassing all normal filtering mechanisms.

286. Hall, E. Gene. "Watch Carefully Now: Solving Crime in the 21st Century." *Police*, June 1999. Vol. 23, No. 6. Source: NLECTC *Law Enforcement & Technology News Summary*, June 17, 1999. EPI944

The Department of Defense put forward the following information in 1995 which would be used for direct communications with military personnel. "Communicating Via the Microwave Auditory Effect; Awarding Agency: Department of Defense; SBIR Contract Number: F41624-95-C-9007." The description of this technology was written as follows:

> **"Title: Communicating Via the Microwave Auditory Effect**
>
> Description: An innovative and revolutionary technology is described that offers a means of low-probability-of-intercept Radio Frequency (RF) communications. The feasibility of the concept has been established using both a low intensity laboratory system and a high power RF transmitter. Numerous military applications exist in areas of search and rescue, security and special operations."[287]

The feasibility was not only demonstrated in the laboratory but also in the field using a radio frequency carrier. In the case of the Gulf War, we had always suspected that the reason that the Iraqis gave up in mass was not because of the heavy bombardments but because they were being hit with new "non-lethal" systems which created fear and perhaps even worse. Our research uncovered reports which now confirm our suspicions as fact.

"What the 'Voice of the Gulf' began broadcasting, along with prayers from the Koran and testimonials from well-treated Iraqi prisoners, was precise information on the units to be bombed each day, along with a new, silent psychological technique which induced thoughts of great fear in each soldier's mind..."[288] This makes a great deal of sense today given what has become increasingly known about mind control weapons. "According to statements made by captured and deserting Iraqi soldiers, however, the most devastating and demoralizing programming was the first known military use of the new, high tech, type of subliminal messages referred to as ultra-high-frequency 'Silent Sounds' or 'Silent Subliminals.'"[289] The use of these new techniques, we believe, went well beyond the injection of fear and may have involved more powerful signal generators which caused the other symptoms which the world observed including head pain, bleeding from the nose, disorientation and nausea – all possible with so-called non-thethal weapons. The questions which now remain: Are they still using the techniques like an electronic concentration camp in order to control the population? Is this part of the way in which modern governments will suppress rogue nations?

287. Dept. of Defense (awarding Agency). Communicating Via the Microwave Auditory Effect. SBIR Contract Number: F41624-95-C-9007. EPI277
288. ITV News Bureau, Ltd. "A Psy-Ops Bonanza On The Desert." 1991. http://www.mindspring.com/~silent/xx/daisy.htm EPI568
289. ITV News Bureau, Ltd. "High Tech Psychological Warfare Arrives In The Middle East." 1991. http://www.mindspring.com/~silent/xx/news.htm EPI567

The development of the technology followed a very traceable history which began in the early 1960s at the height of the cold war. In 1961 Dr. Allen Frey wrote, "Our data to date indicate that the human auditory system can respond to electromagnetic energy in at least a portion of the radio frequency (RF) spectrum. Further, this response is instantaneous and occurs at low power densities, densities which are well below that necessary for biological damage. For example, the effect has been induced with power densities 1/60 of the standard maximum safe level for continuous exposure."290 This observation had incredible ramifications because it meant that within certain ranges RF could create a sound within the brain of a person at energy concentration levels considered too small to be significant.

Later that year a patent was issued to Dr. Puharich which stated in part, "The present invention is directed to means for auxiliary hearing communication, useful for improving hearing, for example, and relates more specifically to novel and improved arrangements for auxiliary hearing communications by effecting the transmission of sound signals through the dental structure and facial nervous system of the user."291 This crude device produced a signal which could be heard in the brain by inducing a vibration which was transferred through the bone into the inner ear where it was then carried to the brain via the nervous system. Puharich continued researching along this line, gaining an additional patent in 1965.292. Both of these inventions required physical contact with the head of the subject.

By 1962 Dr. Allan Frey had advanced his work and was able to create sound at a distance from the subject using a pulsed (modulated) radio transmitter. "Using extremely low average power densities of electromagnetic energy, the perception of sounds was induced in normal and deaf humans. The effect was induced several hundred feet from the antenna the instant the transmitter was turned on, and is a function of carrier frequency and modulation."293 What was occurring in this research were the first attempts to "tune" into the brain of a human in the same manner as "tuning" into a radio station. The same energy was being used, it was just at a different frequency with a slight vibration (modulation) on the carrier wave which delivered the signal.

In 1968 G. Patrick Flanagan was issued a patent for a device which also required physical contact with the skin of the subject. "This invention relates to electromagnetic excitation of the nervous system of a mammal and pertains more particularly to a method and apparatus for exciting the nervous system of a person with electromagnetic waves that are capable of causing that person to

290. Frey, Allan H. "Auditory System Response to Radio Frequency Energy." *Aerospace Medicine,* Dec. 1961. Vol. 32, pp. 1140-1142. EPI370
291. US Patent #2,995,633, Aug. 8, 1961. Means for Aiding Hearing. Inventors: Puharich et al. EPI256
292. US Patent #3,170,993, Feb. 23, 1965. Means For Aiding Hearing By Electrical Stimulation Of The Facial Nerve System. Inventors: Puharich et al. EPI1119
293. Frey, Allan H. "Human Auditory System Response To Modulated Electromagnetic Energy." *Journal of Applied Physiology,* 17(4): 689-692. 1962 EPI544

become conscious of information conveyed by the electromagnetic waves."294 This invention was much different than what others had created by that time because this device actually sent a clear audible signal through the nervous system to the brain. The device could be placed anywhere on the body and a clear voice or music would appear in the head of the subject. This was a most unbelievable device which had actually been invented in the late 1950s. It had taken years to convince patent examiners that it worked. The initial patent was only granted after the dramatic demonstration of the device on a deaf employee of the US Patent Office. In 1972 a second patent was issued to G. Patrick Flanagan after being suppressed by the military since 1968. This device was much more efficient in that it converted a speech waveform into "a constant amplitude square wave in which the transitions between the amplitude extremes are spaced so as to carry the speech information."295 What this did is establish the code of modulation or timing sequences necessary for efficient transfers into the nervous system where the signals could be sent to the brain and decoded as sound in the same way that normal sound is decoded. The result was a clear and understandable sound.

The military interest was present since the first inventions were patented, but in 1971 a system was designed which would allow troops to communicate through a radio transmitter which would render the enemy deaf and disoriented while allowing "friendly" combatants to communicate at the same time. The device was described as follows: "Broadly, this disclosure is directed to a system for producing aural and psychological disturbances and partial deafness of the enemy during combat situation. Essentially, a high directional beam is radiated from a plurality of distinct transducers and is modulated by a noise, code, or speech beat signal. The invention may utilize various forms and may include movable radiators mounted on a vehicle and oriented to converge at a desired point, independently positioned vehicles with a common frequency modulator, or means employed to modulate the acoustical beam with respect to a fixed frequency. During combat, friendly forces would be equipped with a reference generator to provide aural demodulation of the projected signal, thereby yielding an intelligible beat signal while enemy personnel would be rendered partially deaf by the projected signal as well as being unable to perceive any intelligence transmitted in the form of a modulated beat signal."296 What this says simply is that at-a-distance personal communication could be achieved by one's own forces while denying it to another and disabling adversaries at the same time.

In 1974, using a microwave, it was noted that the signal was changed (transduced) by the receiver into an acoustic signal. This was

294. US Patent #3,393,279, July 16, 1968. Nervous System Excitation Device. Inventor: Flanagan, Gillis Patrick. Assignee: Listening Incorporated. EPI261

295. US Patent #3,647,970, March 7, 1972. Method and System of Simplifying Speech Waveforms. Inventor: Flanagan, Gillis P. EPI259

296. US Patent #3,566,347, Feb. 23, 1971. Psycho-Acoustic Projector. Inventor: Flanders, Andrew E. Assignee: General Dynamics Corporation. EPI260

the signal that was "heard" inside or just behind the head. The report stated: "...it was noticed that the apparent locus of the 'sound' moved from the observer's head to the absorber. That is, the absorber acted as a transducer from microwave energy to an acoustic signal. This observation, to the best of our knowledge, has not been described in the literature and may serve as a mechanism mediating the 'hearing' of pulsed microwave signals."[297]

By 1989 the science took another leap forward with the combination of the modulated signal on a microwave carrier. This provided a much more efficient delivery of the sound. It was reported that, "Sound is induced in the head of a person by radiating the head with microwaves in the range of 100 megahertz to 10,000 megahertz that are modulated with a particular waveform. The waveform consists of frequency modulated bursts. Each burst is made up of ten to twenty uniformly spaced pulses grouped tightly together. The burst width is between 500 nanoseconds and 100 microseconds. The pulse width is in the range of 10 nanoseconds to 1 microsecond. The bursts are frequency modulated by the audio input to create the sensation of hearing in the person whose head is irradiated."[298] Two patents were filed that year which addressed this breakthrough. The first "invention relates to devices for aiding of hearing in mammals. The invention is based upon perception of sounds which is experienced in the brain when the brain is subjected to certain microwave radiation signals."[299] And the second confirmed the earlier observations by stating that "Sound is induced in the head of a person by radiating the head with microwaves in the range of 100 megahertz to 10,000 megahertz that are modulated with a particular waveform. The waveform consists of frequency modulated bursts. Each burst is made up of ten to twenty uniformly spaced pulses grouped tightly together."[300]

In 1992 another patent described: "A silent communications system in which nonaural carriers, in the very low or very high audio frequency range or in the adjacent ultrasonic frequency spectrum, are amplitude or frequency modulated with the desired intelligence and propagated acoustically or vibrationally, for inducement into the brain, typically through the use of loudspeakers, earphones or piezoelectric transducers."[301] This device had limited practicality in that it required that the person be in contact or close proximity to the sending device. When examined together, each of these patents are seen to be discrete steps toward a new weapon system.

In 1995 it was reported that in the early research, clear sound signals had been sent and received. It is difficult now to determine what level of military or other research was being advanced in these

297. Sharp et al. "Generation of Acoustic Signals by Pulsed Microwave Energy." *IEEE Transactions On Microwave Theory And Techniques,* May 1974. EPI817
298. US Patent #4,877,027, Oct. 31, 1989. Hearing System. Inventor: Brunkan, Wayne. EPI1124
299. US Patent #4,858,612, Aug. 22, 1989. Hearing Device. Inventor: Stocklin, William L. EPI270
300. US Patent #4,877,027, Oct. 31, 1989. Hearing System. Inventor: Brunkan, Wayne B. EPI262
301. US Patent #5,159,703, Oct. 27, 1992. Silent Subliminal Presentation System. Inventor: Lowry, Oliver M. EPI285

areas. History was clear from Congressional Reports that this entire area was of great interest to the intelligence communities. "Drs. Allan Frey and Joseph Sharp conducted related research. Sharp himself took part in these experiments and reported that he heard and understood words transmitted in pulse-microwave analogs of the speakers sound vibrations. Commenting on these studies, Dr. Robert Becker, twice nominated for the Nobel Peace Prize, observed that such a device has obvious applications in covert operations designed to drive a target crazy with voices, or deliver undetectable instructions to a potential assassin."302

Later, in 1996 "A wireless communication system undetectable by radio frequency methods for converting audio signals, including human voice, to electronic signals in the ultrasonic frequency range, transmitting the ultrasonic signal by means of acoustic pressure waves across a carrier medium, including gases, liquids, or solids, and reconverting the ultrasonic acoustical pressure waves back to the original audio signal."303 was developed. Although this was meant to be used with both receiving and sending hardware, what was determined were the modulation methods for transferring the signal.

The real work was yet to be made public in the form of patents. However, the military claims in the area were starting to surface. What was known from experience was that patents were being held back by the government and confiscated by the military. When this intellectual property was seized the inventors were given a choice – work for the government or you can not continue research, or even talk about, your invention under a national security order. Those who did not cooperate could have their work effectively shut down.

Brain to Computer Connections

Major steps are being made to connect biology to information technology. "Researchers said they took a key first step toward creating electronic microchips that use living brain cells. The researchers said they had learned how to place embryonic brain cells in desired spots on silicon or glass chips and then induce the brain cells to grow along desired paths."304 In addition, "Scientists have succeeded for the first time in establishing a colony of human brain cells that divide and grow in laboratory dishes, an achievement with profound implications for understanding and treating a wide range of neurological disorders from epilepsy to Alzheimer's disease."305 The other possibility is that both brain cells and computer hardware could be built in the laboratories creating, perhaps, the first biologically augmented computers.

302. Scientists for Global Responsibility. "Non-Lethal Defence: The New Age Mental War Zone." Issue 10, 1995. EPI810
303. US Patent #5,539,705, July 23, 1996. Ultrasonic Speech Translator and Communication System. Inventor: Akerman, M. Alfred et al. Assignee: Martin Marietta Energy Systems. EPI293
304. Bishop, Jerry. "Nervy Scientists Move Toward Union Of Living Brain Cells With Microchips." *Wall Street Journal*, Feb. 1, 1994, p. B3. EPI49
305. Specter, Michael. "Scientists make brain cells grow." *Anchorage Daily News*, May 4, 1990. EPI527

What's on Your Mind?

A significant initiative was started for use in creating counter drug measures: the "Brain Imaging Technology Initiative. This initiative establishes NIDA regional neuroimaging centers and represents an interagency cooperative endeavor funded by CTAC, Department of Energy (DOE), and NIDA to develop new scientific tools (new radiotracers and technologies) for understanding the mechanisms of addiction and for the evaluation of new pharmacological treatments."306 Through neuroimaging, not only could the stated objective be achieved, but through imaging, a person's emotional states could be mapped, chemical influences determined and perhaps even specific thoughts could be read.

In 1975: "Developments in ways to measure the extremely weak magnetic fields emanating from organs such as the heart, brain and lungs are leading to important new methods for diagnosing abnormal conditions."307

In 1995 a system for capturing and decoding brain signals was developed which includes a transducer for stimulating a person, EEG transducers for recording brainwave signals from the person, a computer for controlling and synchronizing stimuli presented to the person and at the same time recording brainwave signals, and either interpreting signals using a model for conceptual, perceptual and emotional thought to correspond to the EEG signals of thought of the person or comparing the signals to normal brain signals from a normal population to diagnose and locate the origin of brain dysfunctional underlying perception, conception and emotion.308 In other words, the device reads your mind by comparing your brain activity to other people's.

In 1996 came this Orwellian development: "a method for remotely determining information relating to a person's emotional state, as waveform energy having a predetermined frequency and a predetermined intensity is generated and wirelessly transmitted towards a remotely located subject. Waveform energy emitted from the subject is detected and automatically analyzed to derive information relating to the individual's emotional state. Physiological or physical parameters of blood pressure, pulse rate, pupil size, respiration rate and perspiration level are measured and compared with reference values to provide information utilizable in evaluating interviewee's responses or possibly criminal intent in security sensitive areas."309 This technology could be used for determining what a person might do, given his totally discernible interior emotions. This technology walks through any behavior wall a person can erect and goes straight to the brain to see what might be on a person's mind.

306. ONDCP,CTAC. Counterdrug Research and Development Blueprint Update: http://www.whitehousedrugpolicy.gov/scimed/blueprint99/execsumm.htm EPI305
307. Cohen, David. "Magnetic Fields of the Human Body." *Physics Today*, Aug. 1975. EPI1179
308. US Patent #5,392,788, Feb. 28, 1995. Method And Device For Interpreting Concepts And Conceptual Thought From Brainwave Data And For Assisting For Diagnosis Of Brainwave Disfunction. Inventor: Hudspeth, William J. EPI1129
309. US Patent #5,507,291, April 16, 1996. Method And An Associated Apparatus For Remotely Determining Information As To Person's Emotional State. Inventors: Stirbl et al. EPI1130

Inducing behavior rather than just reading a person's emotional state is the subject of one scientist's work in Canada. "Scientists are trying to recreate alien abductions in the laboratory... The experiment, to be run by Professor Michael Persinger, a neuroscientist at Laurentian University, of Sudbury, Ontario, consists of a converted motorcycle helmet with solenoids on its sides that set up magnetic fields across a subject's head."310 This experiment was carried out and was the subject of a Canadian Broadcasting System exposé on mind control. The segment ran on a program called "Undercurrents" in February 1999. One of the authors of this book also appeared in that program along with several others interested in this field.

Dr. Persinger for over 20 years "has been working on a theory that connects not only UFO's and earthquakes, but also powerful electromagnetic fields and an explanation of paranormal beliefs in terms of unusual brain activity. He has also found that stimulating another area, the temporal lobes, can produce all sorts of mystical experiences, out-of-body sensations and other apparently paranormal phenomena."311 The work of this doctor suggests that these experiences may be the results of activity in the brain and not the actual experiences of the individuals. He has had some measure of success in recreating many of these experiences in his subjects. Dr. Persinger is also known for his work in studying the effects of ELF on memory and brain function.312

In 1991 a method for changing brain waves to a desired frequency was patented.313 A 1975 patent discussed a similar technology, a device and method for "sensing brain waves at a position remote from a subject whereby electromagnetic signals of different frequencies are simultaneously transmitted to the brain of the subject in which the signals interfere with one another to yield a waveform which is modulated by the subject's brain waves. The interference waveform which is representative of the brain wave activity is retransmitted by the brain to a receiver where it is demodulated and amplified. The demodulated waveform is then displayed for visual viewing and then routed to a computer for further processing and analysis. The demodulated waveform also can be used to produce a compensating signal which is transmitted back to the brain to effect a desired change in electrical activity therein."314 In simple terms, the brain's activity is mapped in order to read a person's emotional state, conceptual abilities or intellectual patterns. A second signal can be generated and sent back into the brain which overrides the natural signal, causing the brain's energy patterns to

310. Watts, Susan. "Alien kidnaps may just be mind zaps." *Sydney Morning Herald*, Nov. 19, 1994. EPI816

311. Opall, Barbara. "U.S. Explores Russian Mind-Control Technology." *Defense News*, Jan. 11-17, 1993. EPI818

312. Persinger, M. et al. "Partial Amnesia For a Narrative Following Application of Theta Frequency EM Fields." *Journal of Bioelectricity*, Vol. 4(2), pp. 481-494 (1985). EPI372

313. US Patent #5,036,858, Aug. 6, 1991. Method And Apparatus For Changing Brain Wave Frequency. Inventors: Carter et al. EPI1127

314. US Patent #3,951,134, April 20. 1976. Apparatus And Method For Remotely Monitoring And Altering Brainwaves. Inventor: Malech, Robert G. Assignee: Dorne & Margolin Inc. EPI1122

shift. This is called "brain entrainment" which causes the shift in consciousness. There are many uses, of a positive nature, for this kind of technology, as was mentioned at the front of this section, the important factor being who controls the technology and for what purpose. In the leading scientific journal, *Nature,* the following encapsulating statement appeared:

> "But neuroscience also poses potential risks, he said, arguing that advances in cerebral imaging make the scope for invasion of privacy immense...it will become commonplace and capable of being used at a distance, he predicted. That will open the way for abuses such as invasion of personal liberty, control of behavior and brainwashing."[315]

Dancing to the Tune of an Unknown Drummer

In "a dramatic demonstration of mind reading, neuroscientists have created videos of what a cat sees by using electrodes implanted in the animal's brain. 'Trying to understand how the brain codes information leads to the possibility of replacing parts of the nervous system with an artificial device,' he said."[316] The scientist commenting on this technology saw the future possibility of brain activity mapping being used in creating electronic components to replace damaged parts of the system. The use of mind mapping had other possibilities as well. Similar research was pursued by Dr. José Delgado at one of the country's leading research institutions in controlling the behavior of humans and animals. Actual testing of certain systems proved "that movements, sensations, emotions, desires, ideas, and a variety of psychological phenomena may be induced, inhibited, or modified by electrical stimulation of specific areas of the brain."[317] By 1985, Dr. Delgado was able to create these effects using only a radio signal sent to the brain remotely, using energy concentrations of less than 1/50th of what the Earth naturally produces. This discovery implied that frequency, waveform and pulse rate (modulation) were the important factors rather than the amount of energy being used. In considering this it makes sense because the human body does not require high electromagnetic power concentration to regulate its normal functioning – the key was in finding the "tuning" mechanisms for locating the right "receiving station" in the brain.

By 1998, publicly released information was being discussed as a result of information openly flowing out of Russia. A meeting was held to assess the threat: the "main purpose of the March meetings was described in the Psychotechnologies memo as to 'determine whether psycho-correction technologies represent a present or future threat to U.S. national security in situations where inaudible

---

315. *Nature.* "Advances in Neuroscience 'May Threaten human rights.'" Vol 391. Jan. 22, 1998. EPI116
316. Kahney, Leander. "A Cat's Eye Marvel." *Wired News*, Oct. 7, 1999. http://www.wired.com/news/technology/story/22116.html EPI832
317. Delgado, Jose M.R. *Physical Control of the Mind: Toward a Psychocivilized Society.* Harper & Row, Publishers. New York, 1969. EPI850

commands might be used to alter behavior.'"[318] The threat assessment was likely to begin to condition Americans for the pubic acknowledgement of one of the government's long held secrets – the human mind and body could be controlled remotely without a trace of evidence being left behind. In another quote, one of the leading researchers in this area began to announce his findings: "But psychological-warfare experts on all sides still dream that they will one day control the enemy's mind. And in a tiny, dungeon like lab in the basement of Moscow's ominously named Institute of Psycho-Correction, Smirnov and other Russian psychiatrists are already working on schizophrenics, drug addicts and cancer patients."[319] The results of this research have been investigated and demonstrated to members of the intelligence community in the United States and have even been demonstrated on a Canadian television documentary by Dr. Smirnov.

"Fantasies are thought processes involving internal monologues and imaginative sequences which can motivate healthy people to constructive behavior; likewise, they can inspire unbalanced individuals to destructive or dangerous behavior. One conclusion from that research was that fantasy played a major role among violent criminals. Researchers learned that criminals often daydreamed their fantasies, and then practiced elements of those fantasies before committing their crime. FBI agents determined that violent criminals often exhibit telltale signs as children and as adults. Hence, disturbed employees or students may demonstrate signs of violent fantasies to close observers. Troubled individuals may be obsessively interested in music with violent lyrics, or may have a drug or alcohol problem. When these signs reveal themselves, they should be reported to a threat management team, which can then neutralize the threat, either by therapy, if rehabilitation is possible, or by firing the employee. Workplace and school violence is usually preceded by warning signs."[320] This issue is also an interesting one as the next chapter will show. The ability to determine a "predisposition" for a behavior does not mean that a person will make the "choice" to act on the feelings and internal thoughts. Every person on the planet can remember times when his thoughts were dangerous, immoral or otherwise un-acceptable, falling below the standards set by societal and cultural "norms." Yet, we could have these thoughts in the privacy of our own mind. The trend in the application of mind control technology now would make our most private internal thoughts, as we wrestle with the temptations and choices of everyday life, subject to scrutiny by government and employers. Who will define the rules for psycho-correction? Who will decide what is ethical and right in this area as it develops over the next decade?

318. *Defense Electronics.* "DoD, Intel Agencies Look at Russian Mind Control Technology, Claims." July 1993. EPI538
319. Elliot, Dorinda and Barry, John. "A Subliminal Dr. Strangelove." *Newsweek,* Aug. 21, 1994. EPI542
320. Depue, Roger L. and Depue, Joanne M. "To Dream, Perchance to Kill." *Security Management,* July 1999, Vol. 43, No. 6. Source: NLECTC *Law Enforcement & Technology News Summary,* July 8, 1999. EPI932

Control of the Mind and Body

The predominant brain wave frequencies indicate the kind of activity taking place in the brain. There are four basic groups of brain wave frequencies which are associated with most mental activity. The first, beta waves, (13-35 Hertz or pulses per second) are associated with normal activity. The high end of this range is associated with stress or agitated states which can impair thinking and reasoning skills. The second group, alpha waves (8-12 Hertz), can indicate relaxation. Alpha frequencies are ideal for learning and focused mental functioning. The third, theta waves (4-7 Hertz), indicate mental imagery, access to memories and internal mental focus. This state is often associated with young children, behavioral modification and sleep/dream states. The last, ultra slow, delta waves (.5-3 Hertz), are found when a person is in deep sleep. The general rule is that the brain's predominant wave frequency will be lowest, in terms of pulses per second, when relaxed, and highest when people are most alert or agitated.321

External stimulation of the brain by electromagnetic means can cause the brain to be entrained or locked into phase with an external signal generator.322 Predominant brain waves can be driven or pushed into new frequency patterns by external stimulation. In other words, the external signal driver or impulse generator entrains the brain, overriding the normal frequencies, and causing changes in the brain waves; which then cause changes in brain chemistry; which then cause changes in brain outputs in the form of thoughts, emotions or physical condition. As you are driven, so you arrive – brain manipulation can be either beneficial or detrimental to the individual being impacted depending on the level of knowledge or the intentions of the person controlling the technology.

In combination with specific wave forms, the various frequencies trigger precise chemical responses in the brain. The release of these neurochemicals causes specific reactions in the brain which result in feelings of fear, lust, depression, love, etc. All of these, and the full range of emotional/intellectual responses, are caused by very specific combinations of these brain chemicals which are released by frequency-specific electrical impulses. "Precise mixtures of these brain juices can produce extraordinarily specific mental states, such as fear of the dark, or intense concentration."323 The work in this area is advancing at a very rapid rate with new discoveries being made regularly. Unlocking the knowledge of these specific frequencies will yield significant breakthroughs in understanding human health. Radio frequency radiation, acting as a carrier for extremely low frequencies (ELF), can be used to wirelessly entrain brain waves.

321. Mega Brain, New Tools and Techniques for Brain Growth and Mind Expansion, by Michael Hutchison, 1986. EPI1235
322. US Patent #5,356,368, Oct. 18, 1994. Method and Apparatus for Inducing Desired States of Consciousness. Inventor: Monroe, Robert. Assignee: Interstate Industries, Inc. EPI286
323. Mega Brain, New Tools and Techniques for Brain Growth and Mind Expansion, by Michael Hutchison, 1986, pg 114. EPI1235

The control of mind and body using various forms of electromagnetic energy, including radio signals, light pulsations, sound and by other methods has resulted in several inventions and innovations. The positive health effects and uses have been pursued by private researchers around the world. In 1973 an "apparatus for the treatment of neuropsychic and somatic disorders wherein light–, sound–, VHF electromagnetic field– and heat– sources, respectively, are simultaneously applied by means of a control unit to the patient's central nervous system with a predetermined repetition rate. The light radiation and sound radiation sources are made so as to exert an adequate and monotonous influence of the light-and-sound-radiation on the patient's visual analyzers and auditory analyzers, respectively."[324] This results in the brain following the external simulating source in triggering brain pattern changes which effect the brain immediately and directly. In a simple application patented in 1977 one "invention provides a device, for improving upon the aforesaid application in assisting the induction of natural sleep. As stated above, this invention is concerned specifically with an improvement that will permit the creation of several waveforms such that an analgesic noise device can approximate soothing sounds of nature, i.e., waves, rain, wind."[325] These kinds of devices are available everywhere and are noted for their calming effects in helping people relax and sleep.

In 1980 another patent was issued which disclosed "a method and apparatus for producing a noise-like signal for inducing a hypnotic or anesthetic effect in a human being. The invention also has application in crowd control and consciousness level training (biofeedback). The invention may also be used in creating special musical effects."[326] This device would have a profound effect in controlling individuals to a point otherwise only achievable through the application of hypnotherapy or drugs. A couple years later another device was engineered to create these types of effects, again using very subtle energy resulting in: "Brain wave patterns associated with relaxed and meditative states in a subject are gradually induced without deleterious chemical or neurological side effects."[327]

Various systems were perfected and patents issued for

324. US Patent #3,773,049, Nov. 20, 1973. Apparatus for the Treatment of Neuropsychic and Somatic Diseases with Heat, Light, Sound and VHF Electromagnetic Radiation. Inventors: Rabichev et al. EPI257
325. US Patent #4,034,741, July 12, 1977. Noise Transmitter and Generator. Inventors: Adams et al. Assignee: Solitron Devices Inc. EPI267
326. US Patent #4,191,175, March 4, 1980. Method and Apparatus for Repetitively Producing a Noise-Like Audible Signal. Inventor: Nagle, William L. EPI269
327. US Patent # 4,335,710, June 22, 1982. Device For the Induction of Specific Brain Wave Patterns. Inventor: Williamson, John D. Assignee: Omnitronics Research Corporation. EPI292

controlling brain activity.[328,329,330,331,332,333,334,335] These inventions generated a whole array of breakthroughs for controlling a person's emotional state, concentration, and pain levels, and creating other effects as well. In 1990, the results of a study strongly indicated "that specific types of subjective experiences can be enhanced when extremely low frequency magnetic fields of less than 1 milligauss are generated through the brain at the level of the temporal lobes. Vestibular feelings (vibrations, floating), depersonalization (feeling detached, sense of a presence) and imaginings (vivid images from childhood) were more frequent within the field exposed groups than the sham-field exposed group."[336]

In another 1996 "new age" invention, quartz crystals are used to create stress relief by slowing brain activity. "Physiological stress in a human subject is treated by generating a weak electromagnetic field about a quartz crystal. The crystal is stimulated by applying electrical pulses of pulse widths between 0.1 and 50 microseconds each at a pulse repetition rate of between 0.5k and 10k pulses per second to a conductor positioned adjacent to the quartz crystal thereby generating a weak magnetic field. A subject is positioned within the weak magnetic field for a period of time sufficient to reduce stress."[337] It is interesting that "new age" thinkers have played on "crystal magic" as a way to get in "tune" with oneself and relax and here is a quartz crystal being included as a component of this invention. Again, the crossovers between fiction and science continue to appear.

Consciousness training is also a big theme in cults, religious organizations and others pursuing the "new age." Science has now gained a greater understanding of how the mind and brain work so that what used to take years, or even decades, to achieve, can now be mastered in weeks, days or even minutes. For instance, in 1996 a method and apparatus for the use in achieving alpha and theta brainwave states and effecting positive emotional states in

328. US Patent #4,717,343, Jan. 5, 1988. Method of Changing a Person's Behavior. Inventor: Densky, Alan. EPI284
329. US Patent #4,834,701, May 30, 1989. Apparatus for Inducing Frequency Reduction in Brainwave. Inventor: Masaki, Kazumi. Assignee: Hayashibara, Ken. EPI266
330. US Patent #4,889,526, Dec 26, 1989. Non-Invasive Method and Apparatus for Modulating Brain Signals Through an External Magnetic Or Electric Field To Reduce Pain. Inventors, Rauscher, Elizabeth A. and Van Bise, William L. Assignee: Megatech Laboratories, Inc. EPI268
331. US Patent #4,227,516, Oct. 14, 1990. Apparatus for Electrophysiological Stimulation. Inventors: Meland et al. EPI283
332. US Patent #4,883,067, Nov. 28, 1989. Method and Apparatus for Translating the EEG into Music To Induce and Control Various Psychological and Physiological States and to Control a Musical Instrument. Inventors: Knispel et al. Assignee: Neurosonics, Inc. EPI282
333. US Patent # 5,123,899, June 23, 1992. Method and System for Altering Consciousness. Inventor: Gall, James. EPI289.
334. Patent # 5,352,181, Oct. 4, 1994. Method and Recording For Producing Sounds and Messages To Achieve Alpha and Theta Brainwave States and Positive Emotional States in Humans. Inventor: Davis, Mark E. EPI291
335. US Patent #5,289,438, Feb. 22, 1994. Method and System For Altering Consciousness. Gall, James. EPI333
336. Ruttan, Leslie A. et al. "Enhancement of Temporal Lobe-Related Experiences During Brief Exposures To Milligauss Intensity Extremely Low Frequency Magnetic Fields." *Journal of Bioelectricity*, 9(1), 33-54 (1990). EPI311
337. US Patent #5,562,597, Oct. 8, 1996. Method and Apparatus for Reducing Physiological Stress. Inventor: Van Dick, Robert C. EPI294

humans,"338 was developed. Two years later another patent was issued which could create desired consciousness states; in the training of an individual to replicate such states of consciousness without further audio stimulation; and in the transferring of such states from one human being to another through the imposition of one person's EEG, superimposed on desired stereo signals, on another individual, by inducement of a binaural beat phenomenon."339 Thought transference? This is interesting in that it speaks to the ideas alluded to earlier by the military in changing the memory of a person by imposing computer manipulated signals which would integrate with the normal memory of a person. The possibility of abuse are obvious and the opportunity for personal advancement is also great. Consider the possibility of gaining education by the transfer of data directly into the human brain by these new methods rather than the standard methods of learning. A serious consideration in developing these types of memory transfer systems will be the fact that they bypass normal intellectual filters – they are deposited into the brain as fact, without question or careful consideration. What happens when the new information conflicts with existing information? Would it be possible to include hidden information meant to unduly influence things like religious beliefs, politics or consumption of goods and services?

The possibilities are immense and the ethical and moral questions surrounding these matters equally large. We can no longer avoid the debate, in fact the debate is lagging far behind the scientific advancements. In the interim, there are some simple things we could all do to enhance our own, or our children's, learning capacity by applying simple and available knowledge. For instance: "researchers at the Center for the Neurobiology of Learning and Memory at University of California, Irvine, have determined that 10 minutes of listening to a Mozart piano sonata raised the measurable IQ of college students by up to nine points."340 This is a simple thing of great use to anyone seeking self improvement.

Weapons of the Mind

> "The results of many studies that have been published in the last few years indicate that specific biological effects can be achieved by controlling the various parameters of the electromagnetic (EM) field. A few of the more important EM factors that can be manipulated are frequency, wave shape, rate of pulse onset, pulse duration, pulse amplitude, repetition rate, secondary modulation, and symmetry and asymmetry of the pulse. Many of the clinical effects of electromagnetic

338. US Patent # 5,586,967, Dec. 24, 1996. Method and Recording or Producing Sounds and Messages To Achieve Alpha and Theta Brainwave States in Humans. Inventor: Davis, Mark E. EPI296

339. US Patent #5,213,562, May 25, 1993. Method of Inducing Mental, Emotional and Physical States of Consciousness, Including Specific Mental Activity in Human Beings. Inventor: Monroe, Robert. Assignee: Interstate Industries, Inc. EPI287

340. Hotz, Robert Lee. "Listening to Mozart a real - but temporary - IQ builder, study says." *Anchorage Daily News*, Oct. 15, 1993. EPI529

radiation were first noticed using direct current applied directly to the skin. Later the same effects were obtained by applying external fields. Electromagnetic radiation has been reported in the literature to induce or enhance the following effects:

1. Stimulation of bone regeneration in fractures.
2. Healing of normal fractures.
3. Treatment of congenital pseudarthosis.
4. Healing of wounds.
5. Electroanesthesia.
6. Electroconvulsive therapy.
7. Behavior modification in animals.
8. Altered electroencephalograms in animals and humans.
9. Altered brain morphology in animals.
10. Effects of acupuncture.
11. Treatment of drug addiction.
12. Electrostimulation for relief of pain.

These are but a few of the many biological effects and uses that have been reported over the past decade. They are not exhaustive and do not include many of the effects reported in the Soviet and East European literature.

As with most human endeavors, these applications of electromagnetic radiation have the potential for being a double-edged sword. They can produce significant benefits, yet at the same time can be exploited and used in a controlled manner for military or covert applications. This paper focuses on the potential uses of electromagnetic radiation in future low-intensity conflicts.

POTENTIAL MILITARY APPLICATIONS OF EMR

The exploitation of this technology for military uses is still in its infancy and only recently has been recognized by the United States as a feasible option. A 1982 Air Force review of biotechnology had this to say:

'Currently available data allow the projection that specially generated radiofrequency radiation (RFR) fields may pose powerful and revolutionary antipersonnel military threats. Electroshock therapy indicates the ability of induced electric current to completely interrupt mental functioning for short periods of time, to obtain cognition for longer periods and to restructure emotional response over prolonged intervals.

Experience with electroshock therapy, RFR experiments and the increasing understanding of the brain as an electrically mediated organ suggested the serious probability that impressed electromagnetic fields can be disruptive to purposeful behavior and may be capable of directing and or interrogating such behavior. Further, the passage of approximately 100 milliamperes through the myocardium can lead to cardiac standstill and death, again pointing to a speed-of-light weapons effect.

A rapidly scanning RFR system could provide an effective stun or kill capability over a large area. System effectiveness will be a function of wave form, field intensity, pulse widths, repetition frequency, and carrier frequency. The system can be developed using tissue and whole animal experimental studies, coupled with mechanisms and waveform effects research.

Using relatively low-level RFR, it may be possible to sensitize large military groups to extremely dispersed amounts of biological or chemical agents to which the unirradiated population would be immune.

The potential applications of artificial electromagnetic fields are wide ranging and can be used in many military or quasi-military operations.

Some of these potential uses include dealing with terrorist groups, crowd control, controlling breeches of security at military installations, and antipersonnel techniques in tactical warfare...One last area where electromagnetic radiation may prove to be of some value is in enhancing abilities of individuals for anomalous phenomena."[341]

Quite the paper for 1984. Stimulating anomalous phenomena was another interesting point revealed in this paper. What could this mean? In one press report, the interest of the CIA was disclosed when it was announced that for "20 years, the United States has secretly used psychics in attempts to help drug enforcement agencies, track down Libyan leader Moammar Gadhafi and find plutonium in North Korea, the CIA and others confirm. The ESP spying operations – code named 'Stargate' – were unreliable, but three psychics continued to work out of Ft. Meade, at least into July, researchers who evaluated the program for the CIA said Tuesday."[342] It is also worth pointing out that this report coincided with the public disclosure by military personnel of this project. The story was revealed in a book called *Psychic Warrior.* John Alexander, working out of Los Alamos, and a major proponent of this area of research, wrote that "there are weapons systems that operate on the power of the mind and whose lethal capacity has already been demonstrated...The psychotronic weapon would be silent, difficult to detect, and would require only a human operator as a power source."[343]

341. Dean, Lt. Col. David. *Low-Intensity Conflict and Modern Technology.* Air University Press, June 1986. EPI709
342. Cole, Richard. "ESP spies, 'Stargate' are psychic reality." *Saint Paul Pioneer Press*, Nov. 30, 1995. EPI491
343. Aftergood, Steven. "The "Soft Kill" Fallacy." *Bulletin of the Atomic Scientists*, September/October 1994. EPI281

An "RF weapon currently under development is the high powered very low frequency (VLF) modulator. Working in the 20-35 KHz spectrum, the frequency emits from a 1-2 meter antenna dish to form into a type of acoustic bullet. The weapon is especially convenient because the power level is easily adjustable. At its low setting, the acoustic bullet causes physical discomfort – enough to deter most approaching threats. Incrementally increasing the power nets an effect of nausea, vomiting and abdominal pains. The highest settings can cause a person's bones to literally explode internally. Aimed at the head, the resonating skull bones have caused people to hear 'voices.' Researched by the Russian military more extensively than by the U.S., the Russians actually offered the use of such a weapon to the FBI in the Branch Davidian standoff to make them think that 'God' was talking to them. Concerned with the unpredictability of what the voices might actually say to the followers, the FBI declined the offer. Another RF weapon that was ready for use back in 1978 was developed under the guise of Operation PIQUE. Developed by the CIA, the plan was to bounce high powered radio signals off the ionosphere to affect the mental functions of people in selected areas, including Eastern European nuclear installations."344

The use of the ionosphere in the CIA's experiments reminds one of the possibilities now available with systems such as HAARP, which was developed fifteen years later. What is clear in all of this is that these systems have been developed and hidden from public view. The practice continues to this day.

"The next area of non-lethal weapons is primarily used against machinery...these devices can either cause the machinery to stop functioning or to render it vulnerable to further, more lethal attacks. In addition to this effect, man has become very dependent upon the use of machines and is often rendered helpless in a situation when they become dysfunctional. Therefore, it is only appropriate that they are covered here. The primary anti-machinery arsenal includes the microwave weapon, the non-nuclear electromagnetic pulse, and the laser weapon.

U.S. Special Operations Command has in its arsenal the portable microwave weapon. The capability of such a weapon is varied in that it can not only disrupt enemy communications, but can also superheat internal organs. Of course, directing this type of weapon towards personnel eliminates it from the non-lethal classification. Developed in the Los Alamos National Laboratory, the weapon forms its signal similar to the RF weapons discussed above in that it directs the energy into a high-powered pulse and destroys transistors and other electrical equipment...On an even smaller scale, a portable EMP weapon could be carried by ground forces to destroy the electrical components in an armored vehicle or tank. This capability is being

344. Suhajda, Joseph M. "Non-Lethal Weapons for Military Operations Other Than War." http://www.usafa.af.mil/wing/34edg/airman/suhajd~1.htm EPI348

developed with police forces to emit a pulse that would stop a car almost immediately."345

These systems offer both promise and risks as we move into the next century. What will be the public reaction to these systems? We suggest that the reaction will cause a significant change in the uses and further development of these technologies. Additionally, we suspect that monitoring systems would be developed which would allow for the detection of these technologies in order to control abuse.

Beamers?

Sometimes referred to as "wavies" or "beamers," these individuals are usually dismissed when asserting that they are the victims mind control weapons testing. In fact, at the "University of South Florida researchers have published a study showing that fears of the Internet are replacing the CIA and radio waves as a frequent delusion in psychiatric patients. In every case of Internet delusion documented by the researchers, the patient actually had little experience with computers."346 The problem is that it is difficult if not impossible to sort out which people might be victims and which are delusional. Attempts to determine the reality of the complaints are often the butt of jokes and fear. For example the "University of Albany has shut down the research of a psychology professor probing the 'X-Files' world of government surveillance and mind control. At conferences, in papers and research over two semesters, Professor Kathryn Kelley explored the claims of those who say they were surgically implanted with communications devices to read their thoughts."347 Since the release of our first book, *Angels Don't Play This HAARP: Advances in Tesla Technology,* we have heard from hundreds of people making such claims. We can not sort out what might be real experimentation from that which resides only in the minds of these people. We believe that the claims should be taken seriously and that people should have someplace to go in order to find the truth or gain the medical treatments they otherwise deserve. The history of the United States is littered with examples of people being exploited by scientists working under the cover of darkness provided by "black budgets" – could these reports have a factual basis? We believe that some do.

Government control of the mind in order to impose its will on people is best summarized on the wall of the Franklin D. Roosevelt Memorial where an inscription appears that reads, "They (who) seek to establish systems of government based on the regimentation of all human beings by a handful of individual rulers...call this a new order. It is not new and it is not order."348

345. Ibid.
346. AP. "Internet Feeds Delusions." July 5, 1999. EPI123
347. Brownstein, Andrew. "UAlbany Suspends implants research." *Times Union*, Aug. 25, 1999. EPI833
348. Lalli, Anthony N. "Human Research Subject Protection." http://www.pw2.netcom.com~allalli/BillSite_analysis/paper_web.html EPI619

Chapter 8

The End of Privacy

"Over the past 20 years Congress has encouraged the U.S. military to supply intelligence, equipment, and training to civilian police. That encouragement has spawned a culture of para-militarism in American law enforcement." [349]

Several factors are contributing to the restraint of freedom and the end of privacy. A great deal of emphasis has been placed on the need for higher levels of surveillance in order to combat crime, terrorism, welfare fraud, deadbeat dads, tax evaders and others. The rationale is centered on the need for safety and security at the expense of personal privacy. Over the years repeated attempts have been made by the government to gain access to personal information and, when the public is aware of these attempts, they have been largely thwarted.

The issue of terrorism is real and significant resources are being allocated at the federal level to fight the threat. A good part of these resources are for sophisticated surveillance systems. In a 1999 press report President Clinton said "that he has been persuaded by intelligence reports that a terrorist chemical or germ threat or attack in the United States is likely. Without being specific, Clinton made his comments as his administration disclosed plans to ask Congress to earmark $2.8 billion for programs to fight terrorists armed with such weapons. He is also considering a proposal from the Defense Department to establish a commander in chief for the defense of the continental U.S. According to Clinton, he has been spending a lot of time considering the security challenges facing the country and thinking of how to solve them."[350] This recurring theme has been fostered by the federal government, the military and others as justification for dissolving the otherwise well defined lines of privacy.

349. Weber, Diane Cecilia. "Warrior Cops: The Ominous Growth of Paramilitarism in American Police Departments." CATO Institute Breifing Paper No. 50, Aug. 26, 1999. EPI1121
350. Miller et Al. "Clinton Describes Terrorism Threat for 21st Century." *New York Times*, Jan. 12, 1999. Source: NLECTC *Law Enforcement & Technology News Summary*, Jan. 28, 1999. EPI1042

In 1999 ChoicePoint Inc. "won an exclusive contract with the Department of Justice's Telecommunications Services Staff (TSS) to help federal law enforcement agencies in their investigations by way of the company's CDB Infotek System. The technology serves as a repository of databases with more than 3.5 billion public records online. Government officials have been using the records to find hidden asset ownership, locate individuals, and provide leads for criminal and civil investigations."351 Later in the summer, the Department of Justice announced that "the National Crime Information Center 2000 (NCIC 2000) which helps police to find out in a matter of minutes after scanning a fingerprint if suspects are wanted, rather than waiting hours, days or longer for results. The system, which links the bureau to state and local police, can process fingerprints and mug shots, as well as matches names to a list of wanted people, criminal histories, missing and deported people, and stolen goods, including guns, vehicles, license plates, stocks, boats, and other articles."352 Not to be outdone, the legal research organization "LEXIS-NEXIS has created CertiFINDER, a tool that helps law enforcement officers find people that would rather remain hidden. LEXIS-NEXIS has compiled 2.3 billion documents online in almost 9,876 databases, and is expanding by 30 percent each year. CertiFINDER can search corporate and limited partnership registrations, doing-business-as and fictitious name registrations, professional licenses, aircraft and boat registrations, deed transfers, tax assessor records, bankruptcy filings, crime indices, and Social Security Administration death benefits limits."353 In other words, just about any public record, anywhere, can be accessed instantaneously without warrants, court orders or anything else but the interest of the person with access to the information. The idea that private companies have access to this information allows profiling and marketing manipulations to take place at unprecedented levels.

One of the recent doorways to these intrusions was adopted with the passage of the "Welfare Reform Act – formally titled the Personal Responsibility and Work Opportunity Reconciliation Act of 1996. It includes a provision that gives the government almost carte blanche authority to gather information on everybody, under the guise of tracking down those who owe child support payments."354 The access to information is not limited to "deadbeats," but allows access to everyone's personal records. "'We're heading toward the profiling of America,' says Robert Gellman, a privacy-rights consultant based in Washington, D.C. It's the creation of a consumption bureau to parallel the credit bureau – someone is going to keep track

351. *Business Wire*. "U.S. Dept. of Justice ChoicePoint Online Public Records Contract to Assist Federal Law Enforcement Investigations." April 6, 1999. Source: NLECTC *Law Enforcement & Technology News Summary*, April 8, 1999. EPI995
352. *Washington Times*. "FBI Able to Finger Suspects in a Hurry." July 16, 1999. Source: NLECTC *Law Enforcement & Technology News Summary*, July 22, 1999. EPI919
353. *Law Enforcement Technology*. "CertiFINDER Uses News, Public Records to Help Locate Suspects." Vol. 26, No. 7. Source: NLECTC *Law Enforcement & Corrections Technology News Summary*, Aug. 12, 1999. EPI907
354. *Anchorage Times*. "Snooping Sam." July 7, 1999. EPI169

of all the details of your life. Personal privacy is at risk as never before. However, this rush to electronic profiling is about more than just the digital invasion of privacy."355 The right to access this data has been extended by President Clinton to new financial conglomerates who have "the power to collectively exploit the information their varied affiliates have collected on their customers – health records, stock transactions and credit histories, for example – shredding the basic American right to privacy."356

These new technologies overlap into several areas. Doctors predicted in November 1999, "that brain scans would soon diagnose schizophrenia, toilets would analyze urine and sensors would smell out infection-causing bacteria...Implanted biochip photo sensors would restore sight to the elderly and wheelchairs would take the disabled up and down stairs."357 An interesting set of predictions. Consider the concept of using a public toilet while on medication. It would be possible to set off a silent alarm which analyzed a person's waste, identified the person, and determined the legality of the substances discharged from a person's body. The technology already exists and may soon be used to effect government and corporate policies. The confidentiality of medical histories and other matters will fall victim to these new invasions of personal privacy.

In the patent category an apparatus and method for remotely detecting super-low frequency (SLF) and extremely low-frequency (ELF) signals emanating from human subjects has been developed. "The SLF/ELF signals are composed of various wavelengths and amplitudes which correspond to the subjects internal physiological processes."358 This device will allow an observer access to your biological information which is unique to a person and private. Even a device for measuring blood flow and heart rates has entered the scene.359 "Other new devices will allow a number of other possibilities:

"Back-Scatter Imaging System for Concealed Weapons Detection. The low-energy x-rays delivered by this device, which are equivalent to about 5 minutes' exposure to sunlight at sea level, are reflected back, rather than penetrating the body...The major advantage this device has over conventional magnetometers is that it can detect nonmetallic as well as metallic weapons.

Ballistocardiogram Human Presence Detection Technology Demonstration. Ballistocardiogram technology detects the heartbeat of persons hidden in a large or small vehicle.

355. Mark off, John. "The Rise of Little Brother." *UpsideToday.com* News, Feb. 23, 1999. EPI838
356. Sheer, Robert. "Privacy Issue Bubbles Beneath the Photo Op." *Latimes.com,* Nov. 16, 1999. EPI1227
357. Ryan, Patricia. Technology Set to Transform Medicine. Reuters, Nov. 11, 1999. EPI1197
358. US Patent #4,940,058, July 10, 1990. Cryogenic Remote Sensing Physiograph. Inventors: Tiff et Al. EPI1125
359. US Patent #3,980,076, Sept. 14, 1976. Method For Measuring Externally Of The Human Body Magnetic Susceptibility Changes. Inventors: Woks, Jr. et Al. Assignee: The Board of Trustees of Eland Stanford Junior University. EPI1123

Electromagnetic Portal for Concealed Weapons Detection. The magnetometers detect anomalies in the earth's magnetic field caused by magnetic material in objects carried by individuals.

Handheld Acoustic System for Concealed weapons Detection. This device will be able to detect both metallic and non-metallic weapons concealed under an individual's clothing utilizing acoustic technology.

Thermal Imagers. Thermal imagers detect infrared radiation instead of visible light and allow officers to 'see' any heat-emitting object, even one hidden in total darkness.

Physical Gas Sampling. Remote monitoring of effluent by-products of illegal activities such as methamphetamine synthesis can be detected through the use of gas sampling devices.

Radar-Based Through-the-Wall Surveillance System. It employs a form of radar that can locate and track an individual through concrete or brick walls.

Speaker Identification. Automatic speaker identification technology determines the identity of a speaker solely from a speech sample, using as little as four seconds speech for system training, regardless of the speaker's language or choice of words."[360]

Each of these technologies deserves attention because they are all subject to abuse. Keep in mind that each of these signals is interpreted digitally and stored in this way, which could allow for computers to manipulate the data so that it would allow law enforcement to "prove" their cases against people. The courts are considering the ramifications of thermal imaging but still allowing evidence into court proceedings. As a result, "thermal imaging has become a hot issue in law enforcement and the judiciary system. Attacks on the admissibility of thermal imaging in court have been largely overturned. However, courts have ruled that thermal imaging cannot be used to broadly sweep an area in the absence of probable cause."[361]

The Militarization of the Police

The need for efficient and well armed police is a fact of life as criminals continue to gain access to ever increasing firepower and technology. Crime rates continue to rise and prison populations grow, largely because of the war on drugs. "According to a report released by the Justice Department's Bureau of Justice Statistics, the total prison population in the nation grew to 1.3 million inmates last year; however, the 4.8 percent rise is less than the average growth of 6.7 percent since 1990. There are an additional 592,462 inmates awaiting

360. National Institute of Justice. Technology Projects. http://www.nlectc.or/techproj/to_ltlw.html EPI767

361. Paynter, Ronnie L. "When the Gavel Goes Down..." *Law Enforcement Technology*, July 1999, Vol. 26, No. 7. Source: NLECTC *Law Enforcement & Technology News Summary*, Aug. 5, 1999. EPI914

trial and serving short sentences in local jails. In 1998, there were an estimated 461 inmates sentenced to at least a year in prison for every 100,000 U.S. residents, which is 169 more inmates per 100,000 residents than in 1990."[362] This trend needs to be evaluated as the United States, one of the "freest" countries in the world, jails an ever increase number of citizens at great public expense. Each new crime initiative costs taxpayers billions of dollars, has marginal impacts on crime, and great costs to personal liberty.

An increased federal presence in local law enforcement is also blurring the line between states' and federal rights. "The National Defense Authorization Act of 1994, section 1122, established a requirement that the U.S. Department of Defense (DoD) develop procedures to enable state and local governments to purchase law enforcement equipment suitable for counterdrug activities through federal procurement funds."[363] "Over the past 20 years Congress has encouraged the U.S. military to supply intelligence, equipment, and training to civilian police. That encouragement has spawned a culture of para-militarism in American law enforcement...nearly 90 percent of the police departments surveyed in cities with populations over 50,000 had paramilitary units..."[364] Closer to home, and with much less fanfare and public attention, the US Army has participated in an unprecedented number and type of domestic activities including: disaster relief operations, military support to law enforcement in the war against drugs, and discrete cases of military support to federal law enforcement agencies.

The next few pages draw on an article published in the fall of 1997 in the military quarterly, *Parameters*. The coming years will likely see an increase in the use of the military within the United States. Because of the potentially adverse effect of such uses, the relationship between the military and the public can only be maintained by strict conformance with the legal framework established by the constitution and federal law. Consequently, it is vital that national leaders and their staffs understand the legal aspects of these operations.

The Posse Comatitus Act prohibits the Army and Air Force from enforcing civil criminal law in the United States. This 1878 law stopped soldiers from interfering with voters during Reconstruction and is a foundational concept in democratic society.

The Posse Comatitus Act provides a broad proscription against soldiers enforcing the law. Later congressional acts liberalized the original restrictions of Posse Comatitus and have altered the way the armed forces can assist law enforcement. Congress began to liberalize the law in order to fight illegal drugs. The army can provide equipment, training, and expert military advice to law enforcement

362. Seper, Jerry. "U.S. Prison Inmate Population Swells to 1.3 Million." *Washington Times*, Aug. 16, 1999. Source: NLECTC *Law Enforcement & Corrections Technology News Summary*, Aug. 19, 1999. EPI896
363. Bureau of Justice Assistance. The State and Local Law Enforcement Equipment Procurement Program. BJA Fact Sheet. http://www.ojp.usdoj.gov/BJA/html/site1.htm EPI655
364. Weber, Diane Cecilia. "Warrior Cops: The Ominous Growth of Paramilitarism in American Police Departments." CATO Institute Breifing Paper No. 50, Aug. 26, 1999. EPI1121

agencies as part of this effort. In addition, troops are authorized to provide a wide range of support along the borders to stem the tide of illegal drugs entering the United States.

During the now famous siege at Waco a request was made to Operation Alliance for the military to support a BATF operation against a methamphetamine laboratory. The request asked for military training in a number of areas, including "room clearing discriminate fire operations," termed "close-quarter combat" by the military. The BATF requested more than just advice: they wanted army medics and communicators to accompany them on the actual mission.

The results of the BATF's botched attempt to serve a warrant at the Branch Davidian compound was disastrous. Initially, four BATF agents were killed and twenty were wounded, which was the greatest loss of life in the bureau's history. At that time six Branch Davidians were killed and four wounded. The resulting siege held the attention of the nation until its fiery conclusion two months later. Seventy-four Branch Davidians died in the final assault including twenty-one children under the age of fourteen.

The lessons for America's military are still being sorted out at the end of 1999. Major issues stand out. First, military decision makers cannot rely on the word of other federal agencies. Aggressive agencies will attempt to use the "war on drugs" mandate to drag in the military even when they know the circumstances don't warrant it. The drug issue was nonexistent, having been settled six years before, when David Koresh removed a resident involved in illegal drug activity and reported him to police.

The real issue for the BATF stemmed from their conclusion that the Branch Davidians were stockpiling weapons. That issue is what likely drove the BATF's request for Special Forces support. The lack of involvement of the federal drug laboratory reaction force and the extensive nature of the requested support should have sent up signal flags for the military. These should have provided strong indications that greater care should be given in evaluating the request.

The image of the Army's special operations forces joining BATF agents while storming a religious compound, however misguided its leader, could have seriously impacted public support of the military. The initial request of the BATF was approved but not visibly acted upon. If the request had been acted upon it would likely have resulted in greater losses of life in the initial attack and had a chilling effect on popular support of the military. This would have been a clear violation of the Posse Comatitus Act and would have been further complicated by the military's involvement in a case of religious freedom.

The next lesson for the military is to recognize that their enthusiasm to complete missions needs to be limited while supporting federal agencies in suppressing drugs. The military zeal that creates conformance and dedication to team efforts must be subordinated to healthy skepticism and critical review. Missions such as those described above are on the periphery of the role of the military. Any actual or perceived departure from applicable legal restrictions can lead to an unacceptable loss of confidence in the government.[365]

The military has been establishing regional centers for taking on the task of law enforcement normally reserved for other agencies. "The Department of Defense has emphasized, however, that its RAID (Rapid Assessment and Initial Detection) commanders would defer to the command of the local incident commanders, rendering assistance when needed. The 22 RAID teams are scattered strategically across the country, and will eventually grow in number, pending funding from Congress. The military is integral to law enforcement research because the Department of Defense spends much more money than the National Institute of Justice on related research and development."[366]

In other rehearsals for the future, the "largest operation of the controversial Urban Warrior maneuvers held to date – a week of joint Navy-Marine Corps exercises intended as a way to test new technologies, equipment and systems which the military says are necessary for warfare in the next decade,"[367] was initiated in 1999. "Criticism of the training event has centered around the use of civilian locations, exposing innocent bystanders to the use of live ammunition and explosives. The military has also been criticized for not providing people in the area of the exercises with advance warning."[368]

Many criminologists "fear that the trend toward militarization of policing will lead to the further hardening of police culture and practices. 'This fascination with the hardware of war flies right in the face of the traditional model of police as public servants,' says Tony Platt, co-author of the seventies radical classic on law enforcement *Iron Fist & The Velvet Glove*."[369] This attitude was exemplified in a recent press interview where "a Los Angeles-based federal tactical officer lamented the plight of American SWAT teams, which lack the development of a terrorist-fighting work ethic where 'due process of law meant that all the hostages were alive, and whoever stood in their way ended up with a 9mm round between the eyes.'"[370] While there is no doubt that the battle for the streets of America is a tough one and local law enforcement deserve our support however, we must remember what we are protecting. The rule of law must prevail and the Constitution which assures our continued liberty must be maintained.

365. Lujan, Thomas R. "Legal Aspects of Domestic Employment of the Army." *Parameters*, Autumn 1997. EPI826
366. *Law & Order Online*. "Technologies and Tools for Public Safety." July 29, 1999. Source: NLECTC *Law Enforcement & Corrections Technology News Summary*, Aug. 19, 1999. EPI900
367. Foster, Sarah. "War games come to Monterey." *WorldNetDaily*, March 15, 1999. http://www.worldnetdaily.com/bluesky_exnews/19990315_xex_war_games_co.shtml EPI350
368. Bresnahan, David M. "Press charges rough treatment." *WorldNetDaily*, Feb. 22, 1999. http://www.worldnetdaily.com/bluesky_exnews/19990222_xex_press_charge.shtml EPI651
369. Parenti, Christian. "The Robocop's Dream." *The Nation*, Feb. 3, 1997. pp. 22-24. EPI4
370. Katz, Samuel M. "SWAT Teams of Europe." *Law and Order Magazine Online*, Oct. 14, 1998. Source: NLECTC *Law Enforcement & Technology News Summary*, Nov. 5, 1998. EPI1085

In The Name of the Drug War

The programs for research and development of new technology for the "war on drugs" are huge. "The National Counterdrug Research and Development Program: The national program of enforcement-related counterdrug R&D consists of 82 projects being performed by six drug control agencies with counterdrug R&D programs. (Of the 21 Federal agencies with counterdrug missions, only six agencies have formal R&D programs).

Federal Drug Control Spending by R&D Function 1997-1999 ($M)

FY 97 actual	FY 98 Enacted	FY 99 Request
652.221	676.474	722.052 [371]

The Technology Transfer Pilot Program

The Department of Defense is increasing the rate of technologies transfers to civilian law enforcement agencies. The Technology Transfer Pilot Program matches existing technology systems with State or local law enforcement agencies in need of those technologies and funds the technology transfer.

Systems Available for Transfer

1. Miniaturized Covert Audio "Bugging" Device

2. Voice Intercept Monitoring and Recording System for Title III Investigations

3. Data Locator System

4. Surface Residue Drug Test Kit

5. Cellular Phone Billing Records Analysis

6. Portable Contraband Detection Kit

7. Money Laundering/Suspicious Financial Transaction Detection Software – This project focuses on providing a visualization tool used to help analysts understand complex data systems deployed in counterdrug interdiction efforts. This includes scenarios dealing with money laundering, telephone bills, trafficking events and drug networks. This CTAC sponsored effort is being conducted in conjunction with the State Attorney General's office in Arizona, Texas and Utah in addition to FinCEN, the Defense Information Systems Agency and others.

371. ONDCP, CTAC. Counterdrug R & D Blueprint Update: Introduction. http://www.whitehousedrugpolicy.gov/scimed/blueprint99/intro.htm#1_1 EPI306

8. Command and Control Vehicle Tracking – NAVTRACK Air-Ground Surveillance Management System

9. Secure Messaging and Investigative Image Information Transmission System

10. Command and Control Vehicle Tracking – Signcutter – CTAC is sponsoring Project Signcutter to provide law enforcement officers with an ability to track and locate both field units and suspects using a system based on Global Positioning System (GPS) technology and dedicated radio or cellular telephone infrastructures.

11. Suspect Pointer Index network

12. Tactical Video Communications System

13. Tactical Speech Collection and Analysis System – Voice sample sets are stored on the TASCAN's hard drive. The speaker source is collected from any audio input (radio, microphone, tape recorder, etc.) and fed directly into audio port on the TASCAN laptop computer. TASCAN's algorithms identify the speaker by performing comparisons between voice samples and speaker source.

14. Infrared (IR) imaging surveillance system -- Thermal Imagers

15. Handheld Narcotic Vapor Detection System

16. Surveillance Video Enhancement System

17. Computer-based Interagency Radio Communications Switching System.372

These are some of the technology transfers which are now taking place. These transfers began in the mid-1980s as law enforcement turned to the military for their expertise. At the same time, and shortly thereafter, the military and their research and development laboratories where seeking ways to increase the scope of their missions to justify continued funding. The war on drugs, terrorist threats and "subversive groups" were the targets of these new joint efforts.

372. ONDCP, CTAC. Appendix C - Current Technology Transfer Listing Needs. http://www.whitehousedrugpolicy.gov/scimed/blueprint99/app_c.htm EPI307

Communications

The use of computers is providing another avenue for emergency notification to civilian populations. Law enforcement agencies will increasingly rely on computers to disseminate important announcements as more Americans spurn television and radio in favor of the Internet. Early in July 1999, United Water asked 750,000 residents of New Jersey's Bergen and Hudson Counties to boil drinking water and refrain from outdoor use of water until a major leak at the Oradell Reservoir was repaired. According to the Internet research firm Jupiter Communications, thirty-seven percent of American households are now online.[373] Another variation on the same theme could also be very helpful in a number of emergency situations. "The Berkeley, Calif. police department is exploring the use of Reverse 911 in crime stopping. Reverse 911, produced by Sigma Micro, when combined with a GIS mapping program and local phone numbers, allows police to send generic alert messages to any chosen area. An officer records a public message, highlights an area in the GIS, and clicks the mouse. The message is sent out to all phone numbers in the highlighted area."[374]

Meanwhile, the Amherst Police Department in New Hampshire is now providing its field officers with laptops for instantaneous communication. "The new system, which has already resulted in arrests, but is used by fewer than 5 percent of American police departments, allows officers to tap into local, state, and federal law enforcement databases and automatically sends back-up to an officer who has stopped a wanted or dangerous criminal."[375] The use of these systems is increasing and in a few years will allow police access to everyone's personal information regardless of a person's criminal history. Now is the time to begin to raise the questions about access to information about people who do not have criminal histories or other legal reasons for government to investigate our private affairs.

Court proceedings are also beginning to take on the look of the information age. In May 1999, the Orlando, Florida, based Circuit Court "unveiled a courtroom with videoconferencing equipment and other technology to facilitate Internet broadcasts. For law enforcement officials, the technology means the long process of moving prisoners from jails to courts for arraignment or trial can be made more cost effective and efficient when court administrators present evidence through digital means. The system will have the capability to link to external systems and allow experts to testify from a videoconferencing locale anywhere in the world. In the future, this technology will enable the broadcast of live and archived trials via the court's Web site."[376]

373. *Bergen Record Online.* "E-Mail From the Police." July 27, 1999. Source: NLECTC *Law Enforcement & Technology News Summary*, Aug. 5, 1999. EPI912
374. Overton, Rick. "Taking a Byte Out of Crime." *Business 2.0*, Vol. 4, No. 7. Source: NLECTC *Law Enforcement & Technology News Summary*, Aug. 12, 1999. EPI908
375. Zimmerman, Emily. "Police Have a New Weapon." *Cabinet.Com*, July 28, 1999. Source: NLECTC *Law Enforcement & Technology News Summary*, Aug. 5, 1999. EPI913
376. Chen, Anne. "Court Summons IT." *PC Week*, June 14, 1999. Vol. 16, No. 24. Source: NLECTC *Law Enforcement & Technology News Summary*, July 1, 1999. EPI936

These innovations should be evaluated based on a number of factors including the value of personal appearances in front of judges and juries. Subtle body language, gestures and other actions communicating information to those sitting in judgement are lost in the video conference medium.

One of the innovations which most people can see the relevance of is the use of robots in situations where the risk of loss of life is high. A "robot can now enter a building with its camera, maneuverable arm, lights, and speaker system and 'negotiate' with a suspect...but in some instances, the robots have caused dangerous criminals to become angry or scared about coming into contact with the futuristic units."377 This use of technology provides safe visual access to otherwise unsafe situations and might even compel criminals to surrender – or shoot the robotic messenger rather than a living person.

Motion Detection

New technologies are also leading to better systems that will enhance public safety. For instance, a new type of radar has been patented which "could protect your house or apartment too. Built into a burglar alarm, it bounces radio waves off objects in a room to detect whether any are moving. Called micropower impulse radar, or MIR, the new technology is so versatile that it will soon be incorporated into dozens of products."378 This system could be used in developing senors for notifying drivers of objects outside of their field of vision and in other similar applications. "Proximity technology allows for the transmission of data from a card, keyfob, or other type of transmitter to a receiver for electronic access control...Among the applications for proximity technology are revenue collection and automatic billing; security authorization; private and commercial parking garages; building/area entry control; fire command applications; alarm system activation; guarded tour systems; ATMs; and automotive anti-theft."379

In another development, the use of specially designed tags and receivers allows the movements of individual people to be tracked. "The data in these tags can be picked up by radio waves from several feet away – without direct line of sight and without permission of the person carrying the cards or merchandise. Thus they could tell employers who is spending too much time by the water cooler, and let merchants peek inside purses and billfolds to see who is entering their stores."380 These technologies invade privacy in that merchants do not have a right to access the purses and billfolds of customers.

377. Moore, Russ. "Consider Robocops for the 21st Century." *Police*, Aug. 1999. Vol. 23, No. 8. Source: NLECTC *Law Enforcement & Corrections Technology News Summary*, Aug. 26, 1999. EPI894

378. Stover, Dawn. "Radar On a Chip, 101 Uses in Your Life." *Popular Science*, March 1995, pp. 107-110, 116-117. EPI5

379. O'Leary, Tim. "Proximity Technology: Today and Tomorrow." *Security Technology & Design*, Feb. 1999. Vol. 9, No. 2. Source: NLECTC *Law Enforcement & Technology News Summary*, March 11, 1999. EPI1012

380. Holcomb, Henry J. "Privacy may be cost of Technology," *Anchorage Daily News*, July 29, 1999. EPI162

Tracking employees, although it might be legal, certainly will cause employees to question their employer's level of trust in them.

"Burdened with overcrowded and expensive jails, state governments are increasingly relying on a new means of tracking prisoners: electronic monitoring...So far, more than 30 states use some form of electronic monitoring, and experts surmise that about 100,000 prisoners are involved in the program."381 This system for minor offenses will reduce costs, increase efficiency and limit criminal movement in ways that the public can appreciate. Yet, abuse is possible. "Pentagon satellites once reserved only for guiding nuclear missiles are now being used to monitor prison parolees and probationers as a way to reduce the soaring incarceration rates. The ConTrak monitoring system currently tracks 100 people in nine states using 24 Department of Defense satellites. Currently, 11,000 Americans are monitored under house arrest by a less sophisticated system that cannot track the movements once they leave the house. Critics worry about the implications of an Orwellian future and say the system is rife with possibility of misuse."382 Each new system must be evaluated in the light of civil liberties and basic privacy issues as they relate to the average person... or that *"Brave New World"* will arrive sooner than we think.

The Movement Police

Everything is going digital. Global Positioning Systems (GPS) for law enforcement, commercial purposes and personal use is the fastest area of technology transfer. The use of these systems in automobile designs, remote recreation and as standard safety equipment on boats and small airplanes is now common.

Government databases are being constructed which use this information not only for locating people, but also in conjunction with other information systems for mapping, for example, residence locations. "Geographic Information Systems (GIS) and the Global Positioning Systems (GPS) aid officials in addressing over 75 percent of Vermont residents who live in rural locations, or who have a box-type address. Satellite transmissions provide the longitude and latitude of a home, and then a GIS database stores the information. In the coming year, the system will also include a map for call-takers to help point emergency service providers toward their destination."383

One of the other interesting uses for this technology is in conjunction with other systems. In recent years significant attention has be focused on whether police should pursue or not pursue a fleeing criminal suspect. This issue is being confronted by police departments across the country as safety issues continue to be

381. Kahn, Michael. "Long Arm of the Law Wraps Around Offenders' Ankles." *Reuters Online*, Jan. 21, 1999. Source: NLECTC *Law Enforcement & Technology News Summary*, Jan. 28, 1999. EPI1041

382. Fields, Gary. "Satellite 'Big Brother' Tracks Parolees." *Gannett News Service Online*, April 9, 1999. Source: NLECTC *Law Enforcement & Technology News Summary*, April 15, 1999. EPI992

383. *Law and Order*. "9-1-1 System Most Advanced." Jan. 1999. Vol. 47, No. 1. Source: NLECTC *Law Enforcement & Technology News Summary*, Feb. 4, 1999. EPI1039

raised...More developments are on the way which include a GPS tracking system for stolen vehicles and a remote ignition shut down system. Both tools would require fleeing vehicles to be equipped with a sensor that could be remotely activated by the pursuing officer."384 This technology already exists, with a New York engineer and a sheriff's deputy having "developed a radio-controlled microchip, which they hope will be placed in every new car sold in the United States, allowing police to disable suspect cars' accelerators. Before the Frequency Activated Neutralization Generator System (FANGS) cuts off the gas pedal, the suspect's lights will flash and the horn will blow repeatedly to warn others in the area. Also, some fear that FANGS' use of radio waves opens it to manipulation by non-police individuals, although the technology incorporates an internal alarm to alert authorities to tampering."385 The idea that these systems could be activated by others is a significant issue. Almost any technology can be "one-upped" by the next generation of technology. A good example is what happened with radar guns used by police being superseded by improved radar detection devices by drivers. The need for these to be placed in "every new car" is also questionable, considering new cars are already vulnerable to electromagnetic pulse emitters, described in earlier chapters, which can totally shut down the electronic systems on new cars. This technology would then really be limited to the targets of police rather than imposing the system on everyone.

Another twist in the newtech investigation of crime was really quite innovative. A primary suspect in a murder case had his vehicles impounded for searches about a week after his child disappeared. Investigators returned the vehicles with some additional gear – monitoring devices linked to the high-tech Global Positioning System, a court document indicated. Sheriff's officials refused to discuss or describe the GPS equipment."386 Nonetheless, it led the police to the burial site of his victim. In another incident across the country, police in State College, Pa., "recently used Global Positioning System (GPS) technology to nab an arson suspect. After receiving tips from local residents who had seen David Beer, 39, near some dumpsters that later caught fire, officers obtained the necessary warrant and placed a small GPS receiver on Beer's vehicle. Officers were able to monitor Beer's vehicle from the station, and they obtained visual confirmation that Beer was indeed driving. Using laptops, officers tracked his movements, and eventually linked Beer to a fire."387 In these instances proper warrants were issued and the suspects caught.

---

384. Strandberg, Keith W. "Pursuit at High Speeds." *Law Enforcement Technology*, Sept. 1998. Vol. 25, No. 9. Source: NLECTC *Law Enforcement & Technology News Summary*, Oct. 16, 1998. EPI1094

385. Frio, Dan. "Radio Device Could Put FANGS Into Police Pursuit Abilities." *Police*, July, 1999. Vol. 23, No. 7. Source: NLECTC *Law Enforcement & Technology News Summary*, July 8, 1999. EPI930

386. Wiley, John K. "Girl's body found after satellites track father." *Anchorage Daily News*, Nov. 17, 1999. EPI1213

387. Kinney, David. "GPS Technology Used to Nab Suspect." AP, March 15, 1999. Source: NLECTC *Law Enforcement & Technology News Summary*, March 18, 1999. EPI1000

"The Rock Hill, S.C. police are using the Automated Vehicle Locator (AVL), a device which employs global-positioning technology so the speed and position of police cars can be tracked and displayed at police headquarters, says Lt. Glenn Robinson, commander of support services. This fall, the technology will also be used when assigning officer cars that are closest to the location of a 911 call and will list the nature of the problem, the quickest route for getting there, and the history of calls at that address."388 Commercial users are also accessing this technology as well. "Cost reductions in Global Positioning Satellite (GPS) has made automatic vehicle location technology (AVL) a cost-effective tool for all businesses who operate or manage more than one vehicle. Many companies have borrowed from the Car Trace Inc. patent that uses GPS, issued in 1996, to expand on the technology employed to automatically notify emergency response agencies and allow them to monitor vehicles that might be in danger from crime, accidents, or natural disasters."389

"The Maryland State Police are endeavoring to reduce aggressive driving on the Capitol Beltway, where traffic problems rival the turbid, smog-clogged throughways of Los Angeles. Similar to the movement to curtail drunk driving, Project Advance is part of a widespread effort to fight aggressive driving. Sgt. Janet Harrison supervises Project Advance, which employs a unique automated enforcement system that records computer images of aggressive drivers speeding while either tailgating or negligently changing lanes. From a mobile checkpoint parked on a Beltway shoulder, the system's laser fixates on vehicles that exceed a certain speed while cameras record front, side, and rear shots of the vehicle. Project officials mail warnings to the offenders after analyzing the data."390 Another innovation along this same line is a device which allows very quick license plate comparisons for on-the-road police work. "Armed with voice recognition software and laptop computers, law enforcement officers nationwide are finding less use for their guns...Software developers have solved the biggest obstacle to effective mobile computing – how to enter data. The software has also reduced the time spent checking license plates because officers no longer need the assistance of overburdened dispatchers. As a result, officers typically run 150 plates per night, compared with the 12 to 15 run through dispatch on a 12-hour shift."391

In another development, a patent has been issued which provides for an "improved method and apparatus for automatically monitoring objects, such as vehicles, for example, wherein signpost units are positioned at predetermined locations and each signpost unit

388. Collins, Jeffrey. "Technology to Help Police Serve, Protect." *Herald Online*, Aug. 2, 1999. Source NLECTC *Law Enforcement & Technology News Summary*, Aug. 5, 1999. EPI909
389. *Business Wire*. "Car Trace Inc. to Enforce Patent for System That Monitors Vehicles During Crisis Situations." July 19, 1999. Source: NLECTC *Law Enforcement & Technolofy News Summary*, July 22, 1999. EPI918
390. Battaile, Janet. "Mad as Hell, But Not Getting Away With It Any More." *New York Times*, May 19, 1999. Source: NLECTC *Law Enforcement & Technology News Summary*, May 20, 1999. EPI967
391. Paynter, Ronnie L. "Alpha, Bravo, Charlie, One, Two, Three..." *Law Enforcement Technology*, Feb. 1999. Vol. 26, No. 2. Source: NLECTC *Law Enforcement & Technology News Summary*, March 18, 1999. EPI1006

transmits a binary signpost code for reception via units installed in the objects being monitored and the objects being monitored receive the signpost codes and store object location information."392 This technology may soon be installed in some of the nations highways for tracking truck traffic initially, and later perhaps all traffic. "Since NAFTA took effect in 1994, leaders from all the states on the highway's path have been asked for an estimated $4.7 billion to add electronic tracking devices and other improvements so I-35 can handle more and heavier trucks."393 Think about this as your car zips by predetermined distance markers which record your vehicle's speed and then dispatch a traffic ticket via mail and perhaps electronically updates your driving record, or shuts your car engine down for nonpayment of a fine. Although the current technology would allow this to be the case, the reality has not yet materialized, but may in the near future.

The ultimate smart car and other "smart" technologies have arrived and are being tested now. A "'smart car' will memorize your route home and transport you there quickly based on real-time traffic data. Your telephone will automatically dial up your number when you tell it to 'phone home.' Tiny biochemical sensors implanted in your body will monitor medications and a computer will direct the release of the next dose."394 The idea of self dispensing drugs may seem like a nice convenience, however, consideration should be given to the possibility of forced implants of mental patients and others "for their own good." The violation on the human body without consent again pushes the envelope too far.

In another innovation, separating normal movements from "suspicious" activity is driving the search for new systems capable of sorting out the "good" from the "bad." "Differentiating between normal human activity and suspicious behavior is a difficult task, whether performed by a sensing device or a human observer. Such a human observer would find such a task tedious and costly to perform. Fortunately, a sensing system is not bothered by tedious tasks. Such a system could be implemented to prune out obviously normal behavior, and tag human activities which could be suspicious and would therefore need more attention by a human operator. However, such behavior recognition systems have not been developed due to the difficulty of identifying and classifying such motions. We propose the development of a system which not only identifies humans in the environment and their location, but can also classify and identify their activity, providing a threat assessment. The heuristics needed involve recognition of information bearing features in the environment and the determination of how those features relate to each other over time. This is gesture recognition. This proposal addresses the technology development necessary to create a gesture recognition sensor system,

392. US Patent #4,217,588, Aug. 12, 1980. Object Monitoring Method and Apparatus. Inventor: Charles C. Freeny, Jr. Assignee: Information Identification Company, Inc. EPI11
393. AP. "Congress Delays Funding For I-35 Improvements." *Tyler Telegraph*, July 7, 1997. EPI95
394. Lin-Eftekhar, Judy. *UCLA Today*. "Tiny Computers To Make Objects 'Smart'." http://www.today.ucla.edu/html/990727tiny.html EPI139

enhancing it to create a behavior recognition, which would perform the dual task of classifying objects in terms of threats as well as determining the behavior (state) of objects in motion."395 This area is also drawing on some of the best minds in the country: "researchers at the Massachusetts Institute of Technology's Media Lab are working on computers that can 'read" a person's mind by monitoring a person's body movements."396 This area is of great interest to the military as well. "Unattended surveillance of human combatants in urban environments is an important and challenging task in law enforcement and peace keeping missions. Traditional approaches, based either on temporal difference or pattern recognition using still frames, are inadequate. As a technology transfer effort based on prior research results of leading scientists, ImageCorp proposes in this Phase I effort to design a site model supported, PC-based surveillance system for automatic detection, tracking, and monitoring of human combatants in urban environments. Using a site model, information from multiple sensors and platforms, as well as prior knowledge such as locations of sensitive areas and typical combatant activities can be employed in image detection. Further, the history of moving targets can be effectively represented, enabling the tracking of targets even under concealment or cessation of motion."397 In combat situations there is no question that these systems will be of great use in protecting lives. The issue for civilian populations will again be focused on what is reasonable for balancing the peace and public order against the values of personal freedom.

Another interesting and very useful technology is a system which can pinpoint the location of gunfire and immediately dispatch officers to the scene even before the first telephone call comes into the police station. "A gunshot location system consists of a set of elevated acoustic sensors networked to a central computer in a dispatch center. When gunfire occurs the precise location is rapidly displayed on a computer map. Police response can be immediate or delayed."398 This is a great invention for stemming urban violence and locating the criminals who are responsible. Shooting guns in urban areas is not something that most law abiding citizens are involved in but when they are, the reinforcement of arriving police would surely be welcome.

The last and more insidious technology is also subject to abuse. The use of GPS systems for locating the locations of 911 calls by cellular phones was reported in September 1999. The article stated that "hundreds of lives are being lost because rescuers get to an accident too late, regulators moved forward Wednesday on plans to

395. Cybernet Systems Corp. Behavior Recognition System for Identifying and Monitoring Human Activities. Topic# AF 98-062. EPI1194
396. Noonan, Erica. "Computer experts take 'user-friendly' to whole new level." *Anchorage Daily News*, July 6, 1999. EPI569
397. Imagecorp, Inc. An Unattended System for Monitoring Human Combatant. Topic# Navy 97-113. EPI1193
398. Trilon Technology. Shot Spotter: The 9-1-1 Gunfire Alert System. http://www.shotspotter.com/g-index.html EPI493

make cellular 911 calls automatically provide a person's location."399 This technology, according to some reports, could even work when the phone is off. "Illegal phone use in correctional facilities can now be detected by Zetron's CellPhone Detector Plus, which senses the frequencies emitted by analog and digital cell phone, even when the phones are not in use."400 The abuse should be obvious – anyone with a unique phone number could be located within a few feet of their location, their movements tracked around the globe and their right to privacy stolen in the process.

Government Snoops

One of the objectives of military planners has been to create low-orbit permanent working platforms in space. The historic problems surrounding this technology have been in transferring enough fuel to operate these kinds of machines. During our initial research on HAARP, two patents were uncovered which facilitate this technological advance. In fact, the company owning the patents was able to prove the concept in tests conducted in Canada. "The goal is to provide a permanent unmanned platform for one or more of a variety of missions: telecommunications, including direct broadcast TV and mobile telephone service; military radar warning networks; surveillance of drug smugglers, illegal immigrants, and fishing marauders; and monitoring of atmospheric pollution and carbon dioxide levels – all at much less than the cost of a satellite."401 Certainly there is merit in the arguments for this new technology, in that it could further reduce communications costs, among other things. Surveillance systems should be subject to careful scrutiny and accountability standards in order to protect the rights of individuals.

Another promising project being developed at Sandia Laboratory "is the construction of an advanced explosives detection portal that is capable of identifying microscopic traces of chemicals and other materials used to construct bombs. In development since 1994, the portal shoots jets of air over the person inside and then collects the air sample through a vacuum mechanism so it can be analyzed for explosive materials...The device is able to find even the smallest trace of such materials by detecting the weak fluorescent emissions that are given off by all types of organic substances."402 This technology will prove very useful in airport security and other security applications where explosives are likely to be used. The technology would go far in protecting people without violating their rights.

399. Srinivasan, Kalpana. "FCC pursues plan to locate 911 calls from cellular phones." *Anchorage Daily News*, Sept. 16, 1999. EPI1205
400. *Law Enforcement Technology*. "Products & technolgy: Cell Phone Detector." Jan. 1999. Vol. 26, No. 1. Source: NLECTC *Law Enforceent & Technology News Summary*, Feb. 25, 1999. EPI1020
401. Fisher, Arthur. "Beam-Power Plane." *Popular Science*, Jan. 1988. EPI661
402. Partington, George. "Sandia Labs Developing Next Wave of Technology." *Access Control & Security Systems Integration*, Oct. 1998. Vol. 41, No. 12. Source: NLECTC *Law Enforcement & Technology News Summary*, Nov. 5, 1998. EPI1083

In a January 1999 report on new "see-through-your-clothes" technology officials found that people preferred to be stripped naked before they would allow customs officials to test out their new gadgetry. "Last October, the Customs Service started giving foreign visitors and returning Americans who were suspected of drug smuggling the option between a strip-search or an x-ray...Of the nine people at Kennedy and eight at Miami who chose between the search and x-ray within the first three months of the experiment, everyone opted for the strip search. Custom Service officials said the x-ray option could have been rejected because drug smugglers who swallow drugs would want to avoid the detection of an x-ray, innocent tourists could have had health concerns about an x-ray, and being driven to another building where the x-rays are located could have been seen as an inconvenience."403 It is interesting in that the technology was resisted, although one of the reasons for saying "no" to the new technology might have had more to do with privacy than inconvenience. Increasingly these technologies will be utilized and the option for a strip search discouraged or perhaps even disallowed.

"The next generation of weapons detectors is deadly accurate, able to look through clothes to find guns, explosives and even syringes and drug vials tucked into rolls of fat."404 These systems are being introduced in prison work release programs for returning inmates at the end of their work day.405

Throughout this text we have pointed out the risks and documented actual abuses. The Internal Revenue Service (IRS) is well known for abusing its power. This agency of the government was the most abusive until recent changes were forced on it by the U.S. Congress, which had grown tired of their overreaching approaches and disregard for due process. In July 1994, "Senator John Glenn, Ohio Democrat, released Internal Revenue Service papers showing that its employees were using IRS computers to prowl through the tax files of family, friends, neighbors, and celebrities. Since 1989, the IRS says, the agency has investigated more than 1,300 of its employees for unauthorized browsing; more than 400 employees have been disciplined."406 This is the typical approach to law breaking by government employees; they are "disciplined" while anyone else would be arrested. Invading the privacy of individuals in our cyber worlds should be considered no different than kicking in our front doors. The "front door" to your life is now extended beyond the place where you reside. We all exist, in a sense, in cyberspace, where the most personal and intimate details of our life are stored. Violating this space is more than a minor intrusion, it is a "break-in" and raw violation of each of us as people – not just detached datafiles.

---

403. Finder, Alan. "Customs: Strip-Search Is the Preferred Option." *Sacramento Bee*, Jan. 24, 1999. Source: NLECTC *Law Enforcement & Technology News Summary*, Feb. 11, 1999. EPI1030
404. Butterfield, Fox. "Technology Will Allow Camera To See Weapons At a Distance." *Anchorage Daily News*, April 8, 1997. EPI93
405. Bain, Jackson. "X-Ray Vision." *Corrections Technology & Management*, Jan/Feb 1999. Vol. 3, No. 1. Source: NLECTC *Law Enforcement & Technology News Summary*, Feb. 11, 1999. EPI1034
406. Ness, Eric. "Big Brother @ Cyberspace." *The Progressive*. December 1994, pp. 22-27. EPI18

A year after the IRS abuses were reported by the Senator, the ability to access even more personal information was made available to them. "In an effort to catch more tax cheats, the Internal Revenue Service plans to vastly expand the secret computer database of information it keeps on virtually all Americans. Likely to be included are credit reports, news stories, tips from informants, and real estate, motor vehicle and child support records as well as conventional government financial data."407 So here it is again, the passing of even more power to an agency that should be losing power rather than gaining it.

The IRS was not alone in the blatant disregard for the law or personal privacy. Over 900 FBI files on allies, political adversaries and others deemed by White House insiders as dangerous were turned over to the Clinton White House. This happened before the big scandals broke regarding the President and his administration. It is interesting to note that every time one of his opponents raised an issue in attempting to take the moral high ground some mysterious "leak" would occur and the adversary was brought down. This pattern was repeated several times, which leads one to wonder; would he, could he do such a thing or allow it to happen? "FBI Director Louis Freeh bluntly reported Friday that the Clinton White House had no justification for seeking sensitive reports on employees of previous administrations and the bureau had no excuse for providing them. 'The prior system of providing files to the White House relied on good faith and honor,' Freeh said. 'Unfortunately, the FBI and I were victimized. I promise the American people that it will not happen again on my watch.'"408 Good faith and honor? Are these attributes of the political leaders of today? We think that these attributes are the exception rather than the rule.

During the mid-1990s we were not the only government engaged in high level abuse of privacy rights. The Canadian government had their own scandal brewing. The top secret Canadian Security Establishment (CSE) "is accountable to Parliament through the minister of defense and uses high-tech eavesdropping equipment to intercept foreign communications."409 But it was doing a great deal more as far as surveillance was concerned. Organizations which are allowed to operate in secret will always be used by those in power to further their own political objectives. Systems which provide for whistle-blowing and accountability are needed in almost every country on the globe. "Canada's secret high-tech spy agency has thousands of files on ordinary Canadians and key separatist politicians including Premier Jacques Parizeau, Radio-Canada reported last night."410 "Liberal cabinet ministers did all they could to downplay allegations that Canada's most secret spy agency had targeted Quebec

407. Greve, Frank. Knight-Ridder Newspapers. "IRS Wants More Data About You." *Anchorage Daily News*, Jan. 20, 1995. EPI103
408. Lardner, George and Thomas, Pierre. "FBI Puts Brakes On Abuse Of Files." *Anchorage Daily News*, June 15, 1996. EPI30
409. Canadian Press. "High-tech agency spied on citizen, new book claims." Oct. 20, 1994. EPI1149
410. *Gazette/Southam News*. "Was Parizeau a spy victim?" Oct. 22, 1994. EPI1152

separatists. The first, in a book by Mike Frost, was that the top-secret Canadian Security Establishment eavesdropped on communications between the governments of France and Quebec while Rene Levesque was premier. The second, reported by Radio-Canada, was that the CSE has files on Levesque, current premier Jacques Parizeau and Quebec Intergovernmental Affairs Minister Louise Beaudoin."[411] This is another case where the separatists have, in Canada, the right to greater levels of self-determination. Yet, even though their political activities were probably legal, their rights were violated by their own government. The real question which should be asked is: did heads roll? What about the little guys in this super-secret fraternity? Didn't any of them have the courage, or foundational democratic beliefs, to "just say no" and challenge the system? Are these super-secret "guardians of democracy" so lost in their own world view that they have forgotten what our countries were founded on...or perhaps they never really knew, and, after all, it was just a job? But a Canadian said it best, "'If these allegations prove to be true it is something extremely serious,' said deputy premier Bernard Laundry at a news conference. 'It is a highly immoral act in a democracy for a central government to spy on one part of its constitutional federated regime, regardless of the circumstances.'"[412]

Company Snoops

The use of computer chips to allow law enforcement and commercial interests access to the information housed in personal computers was the subject of a great deal of debate during the 1990s and was thought to have been stopped. This proved not to be the case. In early 1999 it was reported that the Electronic Privacy Information Center (EPIC) was "planning to file a petition requesting that the FTC investigate Intel for consumer privacy violations regarding the ID numbers contained in the company's new Pentium III chips. The group is concerned that the chip's embedded serial numbers, which identify consumers to each web site they visit, will be used without the consumer's consent for marketing or law-enforcement purposes."[413] Apparently, industry thought better than the public and moved forward anyway without consideration for the privacy of its customers.

In another report, security expert Richard M. Smith of Brookline, Mass., found that one company's software secretly transmitted to the company's headquarters details about which music CDs each customer listens to and how many songs he copies, along with a serial number that could be used to identify him."[414] While this might be ok with customers, they should be permitted to choose if this data is to be collected or not. Grocery stores are even more well

411. Wells, Paul. "Those people are hiding something." *The Gazette*, Oct. 25, 1994. EPI1150
412. Authier, Philip. "PQ rakes Ottawa over spy-on-Levesque report." *The Gazette*, Oct. 21, 1994. EPI1148
413. Mosquera, Mary. "Privacy Group Wants FTC to Investigate Intel." *TechWeb*, Jan. 29, 1999. Source: NLECTC *Law Enforcement & Technology News Summary*, Feb. 11, 1999. EPI1031
414. AP. "Software secretly gathered user data." *Anchorage Daily News*, Nov. 2, 1999. EPI1163

known for tracking data and no matter what "promises retailers make, nobody can be sure who has access to their data warehouses, and that goes for all kinds of stores, not just supermarkets."415 It is now possible to define the shopping habits of individuals, keeping track of each detail of their consuming patterns.

In our home state, the poor are the targets of this new technology, with paper food stamps and welfare checks being virtually replaced in Alaska by electronic benefits through the Quest card, as well as automatic deposit."416 This kind of transaction tracking should not be made by government as it is not necessary for managing their programs.

Private companies use of technology in these ways is by no means limited to the United States. In a scandal which raised the hackles of the British, it was disclosed that British Petroleum "had set up 148 hidden microphones across the UK to record customers conversations, and without their knowledge, their workers. Despite the resignation of two members of staff who only found out about the hidden devices after an engineer refurbished a service station in Ayr, the company still plans to install them into all of its 1,600 petrol stations."417 Why would they do this? What values are traded to commercial interests that are worth such invasions of privacy?

The new technology is not limited to its obvious uses but can also create significant unintended problems. One such system was tagged as being hazardous to one's health and perhaps deadly if an individual was wearing a heart pacemaker. "A new study conducted by the North American Society of Pacing and Electrophysiology has found that anti-shoplifting systems using 'acoustomagnetic' technology pose a serious health risk to the approximately 1 million Americans that have surgically implanted pacemakers. Based on controlled testing with 75 volunteers, the study determined that acoustomagnetic systems affected 96 percent of the pacemakers tested. Researchers said that the high disruption rate can be attributed to the fact that acoustomagnetic security gates produce a more intense electromagnetic field than other types of anti-shoplifting systems. Acoustomagnetic gates are typically used to patrol retail establishments with wide entrances, such as mall stores, because they emit such a powerful pulsing capacity."418 Other systems include Radio Frequency Identification (RFID) which uses "radio or electromagnetic reproduction to penetrate materials and read an embedded tag. RFID systems utilize an identification tag, reader, and data-processing equipment. Electronic Article Surveillance (EAS) systems use RFID to detect goods that have not been authorized to leave a retailer. EAS systems are based on a tag which emits radio or

415. Deibel, Mary. "Grocery discount cards raise privacy issue." *Anchorage Daily News,* Oct. 31, 1999. EPI1164
416. Demer, Lisa. "Food stamps enter the electronic age." *Anchorage Daily News,* Oct. 31, 1999. EPI1165
417. BBC Online Network. "'Alarming' Threat to Workplace Privacy." Feb. 18, 1999. EPI101
418. Larson, Ruth. "Some Anti-Shoplifting Gates Disrupt 96 Percent of Pacemakers." *Washington Times,* Oct. 28, 1998. Source: NLECTC *Law Enforcement & Technology News Summary,* Nov. 5, 1998. EPI1082

magnetic frequencies, and a detector."[419,420]

The dehumanizing of populations, treating them as "criminals," "consumers" or "tax paying drones," is becoming even more prevalent as technology outstrips good sense and basic democratic ideas.

Who's Listening

"A number of key technological and political trends are converging to facilitate the creation of Total Surveillance Societies. Four factors facilitate that evolution; (1) the development of technical means, (2) the evolution of political will, (3) the existence of legal mandates, and (4) the formulation of common standards. Their subtle interplay slowly dissolves privacy."[421]

The desire and ability to manipulate the population of the United States through the media and total surveillance was on the political drawing boards as far back as 1971. When first put forward the idea was stopped, not because it was not possible, but because it had already been achieved. "The general efficiency of our surveillance system made it possible to shelve a program submitted to President Nixon in August 1971 to wire every house, car, and boat in America. The program included a blueprint for a government-operated propaganda system via a TV network that would have linked every state, city, and home. One channel would have been devoted entirely to children's programs."[422]

As technology advanced, so did the desire to revisit the issue, and new laws were passed to make it happen.[423] "In June 2000, the Communication Assistance for Law Enforcement Act (CALEA) will mandate telecommunication service providers to deliver data to law enforcement agencies for electronic surveillance. Pen-Link software allows users of LINCOLN to run a local area network, and automatically loads call contents into its databases, along with tags such as target, date, time, duration, and number dialed. Pen-Link databases may be searched by call event, part of the call, or keyword, and may be shared among any of the 4,000 federal, state, and local agencies that use them."[424] "Full implementation of the Communications for assistance for Law Enforcement Act (CALEA) has been postponed until the summer of 2000 due to disagreements between telecommunications companies and law enforcement agencies. Following implementation, CALEA will make available the technological methods to conduct surveillance, with previous court permission, on wireless communications and modem and pager

419. Harwood, Emily M. "RFID Systems As Security Solutions." *Security Technology & Design*, July, 1999. Vol. 9, No. 7. Source: NLECTC *Law Enforcement & Corrections Technology News Summary*, Aug. 26, 1999. EPI893
420. Trovan. Electronic Identification Systems. http://www.trovan.com/hometxt.html EPI648
421. Advanced Surveillance Technologies Conference. Conference Summary, Sept. 4, 1995. EPI134
422. Frazier, Howard. *Uncloaking The CIA*. Macmillan Publishing Co., 1978. EPI861
423. AP. "FCC moves to give FBI, police cell-phone wiretap capability." *Anchorage Daily News*, Aug. 28, 1999. EPI804
424. Kanable, Rebecca. "New Resources to Tap Into for CALEA Compliance." *Law Enforcement Technology*, June 1999. Vol. 26, No. 6. Source: NLECTC *Law Enforcement & Technology News Summary*, July 22, 1999. EPI926

transmissions. Although cellular and phones and pagers have enabled criminals on the go to partake in illegal activities, most criminals still count on the dependable hard-line phone. The 1997 Wiretap Report reveals that 69 percent of all wiretaps authorized by courts were made on low-tech hard-line phones."425 This sweeping change in the law will make electronic communications privacy a thing of the past. "Telecommunications and privacy groups object to the increased surveillance, arguing that the FBI is overstepping its authority...In its review of CALEA, the FCC has restricted its inquiry to 11 FBI proposed modifications. Already, the FCC has decided that at least 4 of the 11 modifications overstep the existing CALEA law."426 The law gives the government unprecedented access to information, yet there is still pressure from law enforcement authorities to go even further. Fortunately, there are still independent thinkers at the FCC who understand the law and are not afraid to evaluate requests in light of the law.

One arm of the surveillance matrix includes an organization which draws very little attention. "The secret Foreign Intelligence Surveillance Court approved 576 government applications for domestic electronic surveillance of suspected foreign agents last year under the Foreign Intelligence Surveillance Act (FISA) of 1978, up from 509 approvals in 1993, according to annual reports obtained by S&GB from the Justice Department under the Freedom of Information Act. The integrity of the little-known FIS Court and its utility as an oversight body have been called into question because of the Court's failure to ever deny a single application for electronic surveillance."427

Several systems have been designed to improve the flow of information to police. One of these systems was brought on line in 1995. "The long-awaited upgrade of the National Crime Information Center, a 21-year-old online law-enforcement database maintained by the FBI, was completed Monday at a cost of $106.2 million. The upgraded system, NCIC 2000, can handle mugshots and fingerprints, in addition to the textual information previously available for the system's nearly 39 million records. Sun Prairie, Wisconsin, police chief Frank Sleeter estimates a $2,000 retrofitting cost for each patrol car to include NCIC 2000 equipment, including digital cameras and fingerprint scanners. The upgrade was originally slated to debut in 1995 at a cost of $77 million."428

Each of these issues is significant when considered alone. However, when considered together, the magnitude of the surveillance ability of the United States goes far beyond any normal person's wildest dreams – and the revelations continue.

425. Muscoplat, Richard. "Biting a Smaller Bullet on Wiretap Costs." *Sheriff*, Nov/Dec 1998. Vol. 50, No. 6. Source: NLECTC *Law Enforcement & Technology News Summary*, Jan. 21, 1999. EPI1044
426. Radcliff, Deborah. "Crime in the 21st Century". *InfoWorld*, Dec. 14, 1998. Vol. 20, No. 50. Source: NLECTC *Law Enforcement & Technology News Summary*, Dec. 31, 1998. EPI1049
427. *Secrecy & Government Bulletin*. "Secret Court Increased Surveillance in 1994." Issue No. 52, Sept. 1995. EPI869
428. Fields, Gary. "Upgraded database to Aid Patrol Officers." *USA Today*, July 12, 1999. Source: NLECTC *Law Enforcement & Technology News Summary*, July 15, 1999. EPI927

You're on Camera

"With today's cheap, tiny video cameras and the connectivity of the Internet, it is possible that someone – or everyone – may be watching at moments you consider private. Live video feeds are streaming across the World Wide Web from atop highway light poles and outside restaurants, from the bedrooms of voyeurs and the cribs of babies, from uninhabited corners of the earth and the most crowded urban subways."429 The use of cameras is growing across the country, "but privacy advocates worry about the introduction of yet another piece of technology to monitor the everyday activities of Americans. The cameras, long used in Europe and other countries, were slow to catch on in the United States." Now, "police surveillance cameras are used in at least 12 cities across the nation and over 200,000 private surveillance systems are set up in and around homes, as well as in banks, convenience stores, and school buses."430 "...Using built in telephone links, digital cameras allow for easy remote monitoring when contacted by modem or ISDN. A permanent link can be established over the Internet where users can access the camera with a click of the mouse."431 "A video surveillance system installed by a Neighborhood Watch group in a quiet cul-de-sac in Sacramento has recorded the blurry image police believe is the man suspected of murdering an elderly neighborhood woman and raping her granddaughter. The case highlights the escalation of video surveillance in today's society. Neighbors believe the cameras have helped keep crime down, and say the feel safer because of the system. Some crime experts worry about the potential for abuse, with cameras being surreptitiously used to look in a neighbor's bedroom. However, residents of the cul-de-sac in Sacramento say the benefits outweigh the potential for abuse."432

"A new study by the American Civil Liberties Union (ACLU) found that there are 2,380 surveillance cameras on public streets, sidewalks, and parks on Manhattan island. Volunteers for the group conducted a block-by-block survey of the cameras. Most of the cameras were found on Wall Street and in midtown, with the largest concentration in the three blocks around Madison Square Garden...Mayor Rudolph W. Guiliani endorses the use of public surveillance as a measure to increase public safety, while Police Commissioner Howard Safir says people 'have no right to privacy in a public place' and that 'no court order is required' for camera use."433 This first quote really does frame the issue – Privacy -vs- Security. We believe that it does not have to be an either/or question

429. *Anchorage Daily News.* "Someone May be Watching." Sept. 5, 1999. EPI1097
430. *CNN Interactive Online.* "Neighborhood Spycam Helps Catch Murder Suspect." Feb. 9, 1999. Source: NLECTC *Law Enforcement & Technology News Summary,* Feb. 18, 1999. EPI1026
431. *Intersec.* "Video: Towards a Digital Future." Dec. 1998. Vol. 8, No. 12. Source: NLECTC *Law Enforcement & Technology News Summary,* Dec. 31, 1998. EPI1047
432. *Los Angeles Times.* "Residents Video Could Solve Murder." March 8, 1999. Source: NLECTC *Law Enforcement & Technology News Summary,* March 11, 1999. EPI1007
433. Lambert, Bruce. "Rise of Secret Surveillance Cameras Criticized." *New York Times,* Dec, 13, 1998. Source: NLECTC *Law Enforcement & Technology News Summary,* Dec. 17, 1998. EPI1057

but suggest that there is a balance which can be found. The problem is that many of the discussions are driven by limited law enforcement budgets rather than fundamental human and civil rights. As the United States continues to escalate the war on drugs and others crimes, questions of balance must be raised. At what point do we evaluate our attempts to regulate the "abuse of self" from drug use, prostitution and other crimes where the only victims are the persons engaged in the choices they make as adults? Do we prosecute these crimes and spend billions of dollars trying to win a war that man has fought with himself for thousands of years without success? Is abuse of oneself an issue between men and men or between men and God? The real cost of trying to prosecute everything is the loss of freedom and privacy for the rest of the population. We are not suggesting that we look the other way, but that we more carefully evaluate the real costs associated with a continuing increase in control of people through invasions of privacy and other means.

"Closed Circuit TV (CCTV) is proving an effective tool for police in Britain. Except for high-profile areas, such CCTV centers are monitored by municipal authorities rather than police. CCTV is effective at catching criminals on tape as they are leaving the scene of a theft. Security employees who monitor the action screen may zoom in and out, as well as pan around a scene."434 "We are leading the world in closed-circuit television technology and its use," said John Stevens, police chief of the Northumbria region here in northern England. Other European countries are studying the British successes, but what alarms some is that while the technology is cutting-edge, rules for its use are scant."435 "All vehicles entering or leaving the City of London or British seaports are being watched by robot automatic number plate scanners (ANPS), which feed the data to the Police National Computer (PNC) in Hendon. The PNC replies within five seconds if the vehicles are 'of interest' to police."436 "Police at the Newham precinct in East London are testing a system of closed circuit television cameras that are positioned around the borough and designed to pinpoint known muggers and shoplifters. "Mandrake" uses facial recognition software to compare live images with file photos that have been scanned into the computer. In the event of a match, the system automatically highlights the suspect's face with a reddish halo and activates a hotline to police in the immediate vicinity."437 "Using neural network technology to collect, store, and process a person's physical characteristics, the Mandrake camera can pick out a suspect even when the person is disguised...The system ignores easily changeable features like hairstyles, eyeglasses and facial

434. *Toronto Globe & Mail.* "Patrolling With TV Like Flying a Helicopter." May 25, 1999. Source: NLECTC *Law Enforcement & Technology News Summary,* June 3, 1999. EPI958
435. Montalbano, William. "Big Bobbie Keeps Unblinking Eye On British City." *Anchorage Daily News,* June 15, 1996, p.1. EPI29
436. Campbell, Duncan. "Cops call the shots." *The Guardian Online.* http://www.gn.apc.org/duncan/875112275-phone.html EPI477
437. Quinn, Paul. "Big Brother's Got His Eye on You." *ID Systems,* Feb. 1999. Vol. 19, No. 2. Source: NLECTC *Law Enforcement & Technology News Summary,* March 18, 1999. EPI1005

hair."438 The advance of camera use in Britain was so rapid that there was little or no debate until it was already widespread. Britain may become the first fully camera-monitored country in the world and should be viewed as a bad example of public policy being shaped by technology rather than technology being used to enforce democratic values. The United States, Canada, Germany, France and every country should begin to examine the questions of surveillance now, before the technology overruns us. Whether, when and where these technologies should be used must be debated in the face of the array of technology now available.

"A compact (less than one-tenth cubic inch) covert camera with digital data capture is proposed to satisfy military, DEA, and FBI surveillance requirements...Anticipated Military Benefits/Potential Commercial applications of the Research or Development: The camera can be carried covertly by law enforcement officers or emplaced in rooms, parcels, luggage or vehicles. The FBI has an airport surveillance requirement. Military special operating forces have many relevant applications and will be the primary performance drivers."439 This camera will allow for the monitoring of huge areas because their cost will be cheap when mass produced. This type of camera was predicted years ago by military think tanks and would allow the policing of countries from border-to-border. Such systems can be monitored from anywhere in the world at very little cost. "According to C-Phone, the Sentinel provides highly reliable, worldwide, real-time remote video and audio within a user-friendly configuration. The Sentinel is a stand-alone device that operates over analog or ISDN phone lines and connects with either a PC or TV for remote viewing. An infrared remote control and on-screen menus facilitate the simple operation of the Sentinel, while a serial port enables both local and remote control through the use of other hardware or a variety of software applications. The recent addition of Internet/IP connectivity removes the assessment of long distance phone charges on communication between surveillance headquarters and sites being monitored."440 This kind of technology combined with others could be used in creating the ultimate "Big Brother" and will likely evolve to this point through the zeal of fear-based security planning, moving from war zones and high crime areas to everyone's neighborhoods.

Miniaturization, coupled with thermal imaging, offers even greater possibilities. "Airborne cameras and thermal imaging company FLIR Systems plans to unveil a state-of-the-art new airborne camera system – called the UltraMedia LE – designed for law enforcement. Some of the features of the new camera system include super low light requirements that allow video color clearly to be captured even during nighttime, long-range optics that permit

438. *Futurist.* "Cameras See Through Digital Disguises." Jan. 1999. Vol. 33, No. 1. Source: NLECTC *Law Enforcement & Technology News Summary*, Feb. 11, 1999. EPI1032
439. Irvine Sensors Corp. "Covert Compact Camera." Topic# DARPA 95-005. EPI1192
440. *Business Wire.* "C-Phone Introduces Sentinel Video Surveillance System." March 12, 1999. Source: NLECTC *Law Enforcement & Technology News Summary*, March 18, 1999. EPI1001

helicopter pilots to keep a far distance between their vehicle and the subject, proprietary gimbal technology that offers picture balance stability good enough to enable chopper pilots to maintain faster speeds while filming, and a versatile digital control system."441 Another version of this camera genre includes cameras that "work much like night-vision goggles, detecting differences in temperature and translating them into a picture that appears on a small screen. Hot objects show up as white, while cold objects are dark."442 In yet one more version of specialized cameras one can "see through walls." "The RADAR flashlight was designed to detect the respiration of a human subject behind a wall, door or an enclosed space with non-conductive walls. The use of the system as a foliage penetration radar has also been explored."443

Enhanced imaging for law enforcement surveillance cameras is a powerful addition to their capabilities. Through technology transfers incredible possibilities now exist in this area. "Using satellite surveillance and industrial imaging technology, scientists at Oak Ridge National Laboratory have developed a tool that enhances images captured by a security camera. By employing algorithms, the new device – Video Imaging Tool for Aiding Law Enforcement (Vitale) – increases pixels by combining elements from two closely related frames, taking certain features from one to enhance the other, thereby increasing overall clarity. A computer using technology designed for automated inspection of semiconductors compares multiple images for minute changes. Vitale-enhanced camera images enables law enforcement officials and security personnel to obtain more evidence from surveillance cameras."444

Cameras in jails are standard in many places. They can protect both security personnel and prisoners. The call for cameras in such places are in fact a very good use of the tools and can create safer environments for those working and those incarcerated in these places. As an example, "The Nassau County District Attorney urged the County legislature to install video cameras in the jail on Thursday, after an inmate serving a 90-day term for a traffic violation was beaten to death on Jan. 7. Thomas Pizuto, 38, a recovering heroin addict, was beaten to death in his cell by guards when he began calling out for Methadone the day after his sentence began. Law enforcement officials confirmed that guards forced Pizuto to sign a statement saying his injuries were accidental. Federal Agents and prosecutors are investigating the death."445 Uses could include daycare facilities and

441. *Business Wire*. "FLIR Systems Launches Breakthrough Airborne Camera for Law Enforcement Market." Feb. 16, 1999. Source: NLECTC *Law Enforcement & Technology News Summary*, Feb. 18, 1999. EPI1024
442. Komarnitsky, S.J. "Thermal camera aids firefighters." *Anchorage Daily News*, Aug. 28, 1999. EPI805
443. Greneker, Eugene F. Radar Flashlight For Through-The-Wall Detection of Humans (abstract). Radar Systems Division, Surface Systems Branch, Sensors and Electromagnetic Applications Laboratory. Georgia Tech Research Institute, Georgia Institute of Technology. EPI125
444. Schlesinger, Hank. "Souped-Up Security Cam." *Popular Science*, May 1999. Vol. 254, No. 5. Source: NLECTC *Law Enforcement & Technology News Summary*, May 20, 1999. EPI969
445. *New York Times*. "Video Cameras Are Sought for Jail Where Inmate Died." Jan. 29, 1999. Source: NLECTC *Law Enforcement & Technology News Summary*, Feb. 4, 1999. EPI1036

school rooms with entry codes held by supervisors, parents and others with access needs. Perhaps the public needs to install surveillance systems to keep track of government operations and keep them accountable to the public they serve?

Another helpful use of this technology involves the use of cameras when apprehending criminals. "Beginning as a foolproof method to establish probable cause in traffic stops, video in patrol cars has become a standard feature, perhaps to the point of creating liability in its absence. Also, a portable system available through Video Systems Plus allows the law enforcement officer to detach the camera to film away from the patrol car."446 "The Cook County, Ill., sheriff's department has announced it will expand its pilot video program to include 12 squad cars and will later apply for federal grants to equip 50 to 75 more cars. Harrison said the cameras, capable of recording audio from 1,000 feet with a personal microphone, are particularly helpful with drunk driving prosecutions."447

Speeders and red light runners are the subject of totally remote photo-ticket issuing systems. In many communities these are accepted with minimal debate while in others they are hotly contested. An attempt to institute electronic speed traps around school zones was made in Anchorage, Alaska, and was perceived as "smacking of big-brother." It was fought in the courts and in several public forums where it was defeated and removed. Politicians involved in the issue learned by the public outcry that "big-brother" had its limits. In other areas these technologies are readily integrated without significant public attention. "A 40-camera system, introduced by the Beverly, Mass. manufacturer EDS, will be installed in Baltimore County, MD., as a way to deter drivers who run red lights. The county, under a one-year contract, is trying to make highway intersections safe, while not charging taxpayers for the technology."448 "A growing number of communities already have implemented red light photo enforcement programs or are planning to install them to prevent red light running...The agencies that will manage these programs must choose a publicity campaign determined by the program's motivation: does the municipality want to deter red light running or catch and penalize those who commit the transgression? Also, while the IIHS reports that 66 percent of Americans polled favor photo enforcement, the other third of the population could very vocally complain, and a strategy must be developed to handle that situation."449 Think about that ...for a moment.

446. Paynter, Ronnie L. "Patrol Car Video." *Law Enforcement Technology*, June, 1999. Vol. 26, No. 6. Source: NLECTC *Law Enforcement & Technology News Summary*, July 1, 1999. EPI937

447. O'Brien. "Cook Sheriff's Police to Expand Videotaping." *Chicago Tribune Online*, Aug. 9, 1999. Source: NLECTC *Law Enforcement & Corrections Technology News Summary*, Aug. 12, 1999. EPI904

448. *PR Newswire*. "Baltimore County Selects EDS for Red-Light Camera Program; Six Maryland Jurisdictions Will Use EDS System to Make Streets Safer." July 20, 1999. Source: NLECTC *Law Enforcement & Technology News Summary*, July 22, 1999. EPI924

449. *Sheriff*. "You Oughta Be in Pictures: Red Light Photo Enforcement." Jan/Feb 1999. Vol. 51, No. 1. Source: NLECTC *Law Enforcement & Technology News Summary*, Feb. 25, 1999. EPI1022

Eyes in the Sky

Eyes in the sky are being introduced in a number of areas with very different requirements for surveillance assistance. "...The grueling geographic challenges of Pima County have motivated Sheriff Dupnik to begin testing yet another CTAC-sponsored technology: a remotely controlled, unpiloted aircraft not much bigger than a model airplane...equipped with a live video camera and transmitter. It is designed for undetected surveillance of drug dealers who often take to the wide open spaces to defend against routine surveillance by cops in cars."[450] This use gives a limited number of officers a greatly enhanced ability to monitor rural areas and boarder regions. "NASA and the U.S. Army have developed a remote-controlled helicopter that could be used for a wide range of tasks, including precision crop spraying, border patrol, hazardous spill inspection, fire surveillance, crowd security and emergency medical delivery."[451] This technology represents another significant military technology transfer area.

On another scale, the technology for local policing is moving into outer space. "Astronauts pointing a HERCULES unit out of a space shuttle window can take a high-resolution picture of a house at a specific street address. The HERCULES unit is capable of taking video and still pictures at night with a resolution of less than 10 meters."[452] The idea that the space program would get involved in these areas was well beyond anyone's expectations of what going to the moon would result in, but here it is. "NASA IG Roberta Gross said in a November 3, 1998 interview that lawyers and special investigators have been considering the use of satellite imagery to identify and prosecute crimes, as well as enforce contractor compliance. Investigations would compare historical data with the satellite imagery, enabling forensic scientists to detect a change in soil composition or vegetation in a suspect area. Gross wants NASA to lead an inter-agency effort to use space-based remote sensing platforms to fight crime; the agency already uses remote sensing devices on airplanes, as well as hand-held devices."[453] Who ever thought that NASA would be involved in "crime fighting?" "And yet, there are even more sophisticated and powerful eyes in the sky – pointed not at the outer cosmos, but toward Earth. Their pictures are rarely if ever seen by the public, and very little is ever revealed about the billion dollar orbiting spy gear that regularly looks in on the secret affairs of various nations."[454] In September 1999, "the first commercial satellite capable of taking pictures approaching the detail

450. ONDCP, CTAC. Advanced Surveillance & Wiretap Technologies. http://www.whitehousedrugpolicy.gov/ctac/page13.htm EPI304
451. NASA Press Release. Robotic Helicopter May Offer New Option For Public Safety. http://www.robotbooks.com/robot-helicopter.htm EPI1103
452. Privor, Cheri. Army Imager Flashes In Test from Space. *Defense News*, July 24-30, 1995. EPI777
453. Berger, Brian. "NASA Explores Using Crime-Fighting Satellites." *Federal Times*, Dec. 7, 1998. Vol. 33, No. 44. Source: NLECTC *Law Enforcement & Technology News Summary*, Dec. 28, 1998. EPI1055
454. Abatemarco, Fred. The Secret World of Spy Satellites. *Popular Science*, April 1997. EPI821

of spy satellites roared into orbit, five months after an identical spacecraft was destroyed in an unsuccessful launch."[455] Space technology companies now have a new set of customers – law enforcement agencies.

BIOMETRICS

Biometrics is one of the fastest growing areas of security for both government and private organizations. Biometrics involves scanning varies parts of the human body with machines and computers which can make minute measurements in order to determine a person's identity. "Biometric technology captures individual characteristics that can be converted into a mathematical code and stored as a template on a database, smart card, or bar-code. Depending on budget and security needs, users can select from an array of individual's unique characteristics, including fingerprints, hand geometry, facial maps, iris and retina patterns, vein structure, and voice recognition."[456,457] This type of scanning system provides a way to catalog each person's unique characteristics like fingerprints, facial image, eyes (iris), voice and even the energy patterns which radiate from the body.

"Biometrics can solve the problem of securing an individual's identity in a world increasingly dependent on computer networks, especially in the U.S., where more and more government services, such as taxation and social services are using networks, writes Saul Schiffman, Senior Consultant of IBM's Network Software Computing Division. Schiffman says that biometrics technology, which includes measures such as fingerprints, face geometry, voiceprints, and retinal scans, has the potential to provide a simple, inexpensive, and reliable way to confirm identities and protect against fraud and abuse of benefits."

The use of digital information storage is increasing as well. The storage of information on computers in this manner increases the amount of information which can be stored in less space. This technology also produces better images and is cheaper and much faster to produce than other systems. "Digital imaging is being seriously considered by many law enforcement agencies for its ability to produce convincing evidence, its efficiency, and its ability to save money. Instead of waiting to develop film, an investigator can download images from camera to computer; in addition, these images can be shared rapidly, and provide access for those with mobile computers. However, many are concerned about the ease with which one may tamper with digital images. One way to ease these concerns is to prioritize standardization in procedures involving digital imagery. The initial 'point of reference' for standardization lies in storing the

455. AP. Satellite launch succeeds. *Anchorage Daily News*, Sept. 26, 1999. EPI829
456. Siuru, Bill. "Who Are You." *Corrections Technology & Management*. Nov/Dec 1998. Vol. 2, No. 10. Source: NLECTC *Law Enforcement & Technology News Summary*, Dec. 28, 1998. EPI1053
457. Schiffman, Saul. "On Using Biometrics as a Basis for an Individual Identifier." *E-Gov Journal*, Jan. 1999. Vol. 2, No. 1. Source: NLECTC *Law Enforcement & Technology News Summary*, Feb. 25, 1999. EPI1019

image immediately, before the case begins to unfold. In addition, digital 'fingerprinting' of images is becoming available through technologies like Epson's Image Authentication System (IAS). Creating a forgery which evades detection from such software would take 300 years using a 1000 MHz CPU."458 Another important detail reveled in the above quote really explains why digital and biometric information is so desirable – it is presently claimed to be nearly impossible to forge.

In the future, computers will become faster and criminals even more sophisticated, making this technology obsolete. By the time this occurs there will likely be such widespread use of these technologies and success in controlling populations that people will accept the next level of "fraud and identity management" – the biological-mechanical implant.

666 – The Mark?

The suggestion that implants would be used to track and identify people has been the nightmare of many. Government control of all transactions, work opportunities and virtually all access to commercial or governmental goods and services is not far off. Consumer fraud, personal safety and identity theft will be the excuses for this final step in a totally controlled society. The miniaturization of technology, nanotechnology, will soon make it possible to install in humans a very sophisticated class of implants. An object the size of a grain of rice will store all of your vital personal information, a GPS locator, a transmitter/bug (only to be activated by court order) and perhaps even something which can create pain if you cross unseen security lines.

The first indication that we are moving in the direction of totally monitored citizens appeared in 1989 in a publication by the US Army War College. "In the near future every American at risk could be equipped with an electronic individual position locator device (IPLD). The device, derived from the electronic bracelet used to control some criminal offenders or parolees, would continuously inform a central data bank of the individuals' locations. Eventually such a device could be permanently implanted under the skin, with automatic remote activation either upon departure from the U.S. territory (while passing through the security screening system at the airport, for example) or by transmission of a NEO alert code to areas of conflict. Implantation would help preclude removal of the device (although, of course, some terrorists might be willing to remove a portion of the hostage's body if they knew where the device was implanted). The IPLD could also act as a form of IFFN (identification friend, foe or neutral) if U.S. military personnel were equipped with appropriate challenge/response devices."459 These devices are already

458. Biehl, Craig. "Ensuring the Integrity of Digital Images." *Police Chief*, June, 1999. Vol. 66, No. 6. Source: NLECTC *Law Enforcement & Technology News Summary*, June 17, 1999. EPI945
459. Metz, Steven and Kievit, James. "The Revolution in Military Affairs and Conflict Short of War." Strategic Studies Institute, U.S. Army War College, July 25, 1994. EPI516

available and growing in complexity with each passing year of technological advance.

The most likely people to initially receive implants are military personnel, who will be told that this will help rescue them if they are captured or they might just be injected along with the normal injections of various vaccines without anyone's knowledge. The military may be the first, setting the stage for the rest of the country. Will our military personnel, business travelers and others object, seeing this as an invasion of their private lives? Will more objections matter?

A crude version of this device, used for pet identification, was available as far back as 1990. "For nearly a year, the Marin Humane Society has been implanting every dog and cat adopted from the shelter with an Infopet microchip I.D...a high-tech answer to the age-old problem of permanently identifying your beloved pet?"460

In 1996 the idea hit the mainstream with this report: "A tiny chip implanted inside the human body to send and receive radio messages, long a popular delusion among paranoids, is likely to be marketed as a consumer item early in the next century. Several technologies already available or under development will enable electronics firms to make implantable ID locators, say futurists, and our yearning for convenience and security makes them almost irresistible to marketers."461

Personal locator devices are being considered for any number of things. Some of the "other applications are being contemplated and some are in hand. The systems could check up on medical patients with, say, heart trouble or Alzheimer's disease, and find where wounded soldiers lie on a battlefield. Keeping track of personnel, vehicles, and inventory, as well as providing security on college campuses, are being pursued."462

A larger version of this technology is being used for Pittsburg area children in troubled homes. They are being given an "electronic emergency device, a plastic pendant the size of a yo-yo, that they'll be able to use to alert authorities when they're being mistreated."463 This is a great deal of power for a small child who with such a device can literally dominate an entire household. Is this a good thing? If the child is at such great risk to need such a device should he even be in the home of the abuser? Who will sort out the real calls from the false alarms? More questions and less answers seem to be available. Is it time for a break, a "time-out" from the technology race before the race becomes a curse?

Fingerprints

The ability to sort through fingerprints in order to get a match was impossible until the last few decades. Today, "live-scan stations

460. Wright, Mary. "An ID Tag That Won't Get Lost." *Pacific Sun*, week of May 4, 1990. EPI28
461. Van, Jan. "In Future, Tiny Chip May Get Under Skin." *Chicago Tribune*, May 7, 1996. p.1. EPI21
462. Hoshen et al. Keeping Tabs on Criminals. *IEEE Spectrum*, Feb. 1995. EPI113
463. Stanley, Bruce. "Experimental Device Offers Children Protection from Domestic Abuse." *Anchorage Daily News*, June 14, 1996. EPI99

are being used to process bookings and applications for background checks – including tenprints – as the time required for the task has shrunk from weeks to days. Both small and large police departments nationwide are using scan technology to process tenprints for gun regulation, criminal bookings, as well as job-seekers in both public and private industries, including those seeking work as day-care providers, bankers and stockbrokers. Scan systems should be used with the Automated Fingerprint Identifications Systems (AFISs), according to vice president of worldwide marketing for Printrack International Steve Yeich. San Bernardino County, California's 32 live-scan stations and 10 card scans will be used to process the 180,000 prints of suspects as well as job applicants; once San Bernardino is connected to AFIS, it will save 50 percent to 60 percent of the time spent processing prints, according to Stg. Gary Eisenbeisz."[464] This news was released in May 1999.

Between May and August 1999, the speed of processing went from days to hours, again illustrating the speed of technological advances. "With the new $680 million Integrated Automated Fingerprint Identification System (IAFIS), which is currently available to several federal agencies and 15 states, law enforcement officers will cut down the amount of time it takes to compare suspects' fingerprints to those in the FBI database. The old system, which required mailing the fingerprints to the FBI, took up to two weeks for processing; the new system uses technology that can get results within two hours."[465,466] While the ink had not even dried on the above quoted article the technology had advanced again, this time reducing the search to a few minutes. "Advanced Precision Technology has received notice of allowance on patent application No. 08/853,850 for Real Time Fingerprint Sensor and Verification System. The package offers APT's patented holographic optical fingerprint sensor, integrated with a real-time optical fingerprint correlation system. The system uses holographic optics to capture a perfect fingerprint image, which can then be compared to ink-based prints or filters stored as holograms."[467]

In the 1980s, requirements for fingerprinting school personnel, in all school districts receiving federal money, became the law. I was working in education at the time and remember many of the concerns which were raised in terms of what would happen with the information. The fingerprints were collected and sent to law enforcement personnel for comparison with FBI records. The person

464. Kanable, Rebecca. "Live-Scan Is Making its Print." *Law Enforcement Technology,* April 1999. Vol. 26, No. 4. Source: NLECTC *Law Enforcement & Technology News Summary,* May 13, 1999. EPI974
465. Lipton, Beth. "FBI Debuts Enhanced Fingerprint Database." *CNET,* Aug. 10, 1999. Source: NLECTC *Law Enforcement & Corrections Technology News Summary,* Aug. 12, 1999. EPI903
466. *Business Wire.* "Litton PRC Goes Live With Identification Tasking and Networking for FBI Fingerprinting System." Aug. 11, 1999. Source: NLECTC *Law Enforcement & Corrections Technology News Summary,* Aug. 19, 1999. EPI898
467. *Business Wire.* "Advanced Precision Technology Inc., Announces Another Innovative Fingerprint Patent." July 28, 1999. Source: NLECTC *Law Enforcement & Technology News Summary,* Aug. 5, 1999. EPI911

would then start work and six months later the report of the fingerprint checks would arrive. People were assured that the information was not being retained by the government, but, given the abuses of the past, who really knows?

These invasions of our privacy, for our "own safety," do not stop with school employees. New gun laws are forcing the collection of information on every person who purchases a gun. This information will likely be included in government computers as part of their cataloging of the American people.

These databases are being integrated with other records in order to compile increasingly more complete information on people around the world. Even in third world countries like El Salvador, these systems are being used. "Identix Incorporated, a manufacturer of live-scan and biometric verification systems, announced earlier this month that its IT Security Division, Indenticator Technology, has been contracted by the government of El Salvador to implement a system for driver's license and vehicle registration. The system captures prints and creates fingerprint templates that are stored on the smart card license registration, enabling the system to identify and authenticate new and renewing applicants. Two thumbprints, along with any additional data like demographics, physical characteristics, and a photograph, are stored on an associated smart card. The system includes fingerprint scanning hardware, fingerprint quality control software, and fingerprint matching algorithms."[468] El Salvador, a country with a repressive history, where hundreds of people were murdered by death squads, now will have the ability to search out future victims using American technology.

Fingerprint security is not limited to law enforcement. These technologies will soon be introduced in consumer products. "Researcher's at Siemens' Perlach R&D facility in Germany recently unveiled Biometric Sensor Fingertip for distribution in cell phones and computer boards this year. The company is also studying other applications for use in handguns, starting cars, and conducting electronic commerce."[469] The use of this technology to protect computers is also about to be introduced and it is raising significant privacy concerns. "The use of fingerprint scanners to protect computer files and data is considered by some experts to be a great advancement in high-tech security, but privacy advocates say the technology is susceptible to abuse. 'The fingerprint identification raises the difficult tension between security and privacy,' says Marc Rotenburg, director of the Electronic Privacy Information Center. 'Security gives more authentication, and that's good...but on the privacy side we create the risk of a new global identifier.' Bank customers, for example, have different passwords for different accounts, but using a fingerprint gives a person a single, universal ID

468. *Business Wire.* "Identix Expands International Use of Biometrics." June 15, 1999. Source: NLECTC *Law Enforcement & Technology News Summary*, June 24, 1999. EPI941

469. Buderi, Robert. "Fingerprints Don't Lie." *Upside.Com*, March 1, 1999. NLECTC *Law Enforcement & Technology News Summary*, March 11, 1999. EPI1008

for all accounts. That ability means all of a person's data is related to a single ID and thus puts more of that data at risk."470

Faceprints

Technology now exists which can pick a person out of a crowd by the unique characteristics of a person's face. "In attempting to teach computers not only to ask a question but to answer it, one researcher has captured thousands of faces in a photographic computer database – he calls it his facebase – and has developed software for searching through this collection and picking one face out of the multitude."471 This technology is of great interest to intelligence organizations, which could use this information in a number of applications. The U.S. Secret Service was caught gathering their own "facebase." A small New Hampshire company that was working to "build a national database of driver's license photographs received nearly $1.5 million in federal funds and technical assistance from the U.S. Secret Service last year, according to documents and interviews with officials involved in the project."472

"ImageWare Software has released Face ID version 2.0, a program which will allow law enforcement officers to quickly identify suspects from their image. The program will compare an image, be it a mugshot or a frame capture from a surveillance camera, to a vast database of photographic criminal composites. ImageWare plans, in future releases, to enable the program to check through databases at the time of booking, allowing an officer to identify those suspects with aliases or outstanding warrants. ImageWare developed Face ID using technology from Visionics Corporation, the leading developer of facial recognition technology worldwide."473

"The premier development of face recognition technology, Visionics Corporation, announced on Aug. 4 (1999) that GTE Corp. has incorporated the FaceIt engine into its Internet-based search engine, the Bastille. The first such Internet-based system designed for investigators and intelligence analysts, the Bastille will now allow users to enter new images and match them against existing records."474 In other developments, "Viisage technology has announced that its facial recognition system used by the Illinois Driver Services Department and the Illinois State Police exceeds 2 million images. The database, which is expected to grow beyond 25 million images, is thought to be the largest in the world. Viisage's Data Mining product

470. Wayner, Peter. "Signing On With Your Fingerprints." *New York Times*, March 18, 1999. Source: NLECTC *Law Enforcement & Technology News Summary*, March 25, 1999. EPI997
471. Schwartz, Evan. "A Face One's Own." *Discover*, December 1995. EPI19
472. O'Harrow Jr., Robert and Leyden, Liz. "U.S. Helped Fund License Photo Database." *Washington Post*, Feb. 18, 1999, pp. A1. EPI88
473. *Business Wire Online*. "Image Ware Announces Several New Orders for Revolutionary Facial Recognition Product." May 13, 1999. Source: NLECTC *Law Enforcement & Technology News Summary*, May 20, 1999. EPI966
474. *Business Wire*. "GTE Adds Visionics Face Recognition Technology Into Its On-Line Law Enforcement Database Search Engine." Aug. 9, 1999. Source: NLECTC *Law Enforcement & Corrections Technology News Summary*, Aug. 12, 1999. EPI906

searches large databases for duplicate images, and identifies individuals."475

"InSpeck Inc. has developed a 3D digitizer that may revolutionize recognition and identification technology. InSpeck 3D is a computer-assisted non-contact 3D digitizer that projects white light from a halogen bulb on a target, and uses video to record the 3D coordinates from a person or object. The Royal Canadian Mounted Police hopes to use InSpeck 3D for skull reconstruction, and is considering using it to create 3D images of Canadian school children. In addition, InSpeck could be used for age progression purposes, and for generating 3D line-ups. Viewers of a 3D line-up could see images at any angle, and under lighting conditions that occur in a specific incident."476 Why Canadian school children? Is it so that they can be identified in case they are missing or is it to begin to build their future database files beginning with their images?

Facial imaging does not end with sorting out mugshots and 3D imaginings. It is also being engineered to read the emotions and the truthfulness of people by interpreting the images. "A Salk-led team has developed a computer program to discern certain facial expressions and determine which are genuine or false. The researchers say the program is as accurate as a psychologist trained to read faces and performs much better that human non-experts. Developers believe the program may be useful to law enforcement officials, as well as mental health professionals. Paul Eckman, professor of psychology at the University of California, San Francisco and author of the study, notes that when an individual is lying, the real expressions move across the person's face in a micro-expression; the person, however, masks true feelings with a posed expression. While the movements can be discerned on a video tape, such analysis is time consuming. The program developers say their program can analyze a minute of tape in roughly five minutes."477

Eye Scans

The romantic notion of "let me look into your eyes" or the idea that the "eyes are the window of the soul" gets lost in the new technologies. The eyes are considered a source of information for keeping track of people and confirming who they are. Commercially, "Texas' largest financial institution, Bank United, uses biometrics technology in the form of iris scanning at its ATMs – the first company to apply such technology to a consumer tool. Customers need only look into a camera mounted on top of the machine for a

475. *PR Newswire.* "Viisage Accelerates Deployment of Facial Recognition Technology." Aug. 13, 1999. Source: NLECTC *Law Enforcement & CorrectionsTechnology News Summary*, Aug. 19, 1999. EPI899
476. Paynter, Ronnie L. "Imaging Technology: 3D I.D.s." *Law Enforcement Technology*, June, 1999. Vol. 26, No. 6. Source: NLECTC *Law Enforcement & Technology News Summary*, July 8, 1999. EPI935
477. *Science Daily Online.* "Computer Program Trained to Read Faces Developed by Salk Team." March 22, 1999. Source: NLECTC *Law Enforcement & Technology News Summary*, March 25, 1999. EPI998

few seconds, and their scanned iris is checked against an earlier scanned image within the bank's database of previously scanned irises. Iris scanning, fingerprints, voice prints, hand and face geometry, and signature authentication are all forms of biometrics technology that have become cheaper, easily deployed, and secure, for a variety of applications."478 However, "One of the most promising developments in biometric identification is iris recognition. Everybody has his or her own unique iris pattern that can function like a human barcode...To access a machine such as an ATM, an individual peers into an IriScan reader, and if his or her iris matches the valid code, access is given. The IriScan resembles a camcorder and is capable of reading the iris from a distance."479 "...Officials at N.J.-based IrisScan say their eyeball-identifying technology is more reliable than hand and fingerprint identification technology. While the fingerprint-sensing community struggles to get past the police and high security markets, other technologies, like iris scanning, are likely to remain stagnant until prices subside."480

These systems are becoming more widespread and will continue to grow in importance as technology improves and prices decline. In prisons these new systems are being developed for use on both prisoners and visitors. "The new system includes two databases, one each for inmates and visitors. When an inmate is admitted to the prison, he is enrolled in the system. Once the best of three iris scans are chosen, the operator inputs a name and PIN number for the inmate, which is stored with the code. Visitors, such as lawyers, doctors, ministers, and family, once enrolled in the system, will be scanned on subsequent visits and will not have to present any other ID."481

Thermal Imaging

New systems also include those which can read the heat of the body and from that information computers are able to create thermal images. "While thermal imagers – equipment that reads thermal radiation between 7.5 and 13.5 microns – have a wider range of sanctioned uses, image intensifiers help law enforcement officers perform search and rescues, look for fugitives, and conduct surveillance. Sometimes officials use the two technologies together; for example, officers in Texas and Indiana note that thermal technology helps detect suspects, while image intensifiers help identify them. The Law Enforcement Thermographers Association has indicated 11 sanctioned uses of thermal images, including perimeter

478. Cugliotta, Guy. "Bar Codes for the Body Make It to the Market." *Washington Post,* June 21, 1999. Source: NLECTC *Law Enforcement & Technology News Summary,* June 24, 1999. EPI938
479. *Law and Order Magazine Online.* "Iris Recognition: The Ultimate in Biometric Identification." Dec. 7, 1998. Source: NLECTC *Law Enforcement & Technology News Summary,* Dec. 28, 1998. EPI1054
480. Costlow, Jerry. "Fingerprint Security lacks Punch." *TechWeb Online,* April 22, 1999. Source: NLECTC *Law Enforcement & Technology News Summary,* April 29, 1999. EPI983
481. Carey, Carol. "Iris Recognition Gives Positive IDs of prison Inmates." *Access Control & Security Systems Integration,* Jan. 1999. Vol. 42, No. 1. Source: NLECTC *Law Enforcement & Technology News Summary,* Feb. 4, 1999. EPI1040

surveillance, vehicle pursuits, air and marine safety, structure profiles, disturbed surfaces, hidden compartments, and environmental and officer safety."482 "Indeed: A security system consisting of an infrared camera and a computer is now using the heat patterns emitted by facial blood vessels to identify people quickly and precisely. Developed by Technology Recognition Systems of Alexandria, Virginia, the system's camera takes a picture of the heat radiating from a person's face. The computer then compares this picture, called a thermogram, with an earlier image stored in its memory."483 This is another technology which will likely be integrated with several of those described in this book. Thermal imaging is already in field use in a number of areas across the country.

Voice Recognition

A new device was patented which will allow for the manipulation and storage of voice patterns. This system will provide a way to create a synthetic voice which sounds natural but is altered in order to hide the person's real voice and take on a new identity. The device is described as, "An apparatus for transforming a voice signal having characteristics of a different person provides apparatus for separating the talker's voice into a plurality of voice parameters including frequency components, a neural network for transforming at least some of the separated frequency components into those characteristic of the different person, and apparatus for combining the voice parameters for reconstituting the talker's voice signal having characteristics of the different person."484

Voice recognition is also being used in law enforcement. "The Dane County, Wisconsin., Sheriff's Office says its program to monitor non-violent offenders using SpeakerID voice verification from SecurityLink has saved $2.8 million since 1994. SpeakerID tracks the activity of offenders via random calls to their home or work. Offenders must repeat a series of number pairs during the call, while the computer compares their speech pattern to a digitized voice print. Experts say the voice verification program accurately differentiates between twins and will even identify a voice that is affected by a cold."485

Regulation and Control

"The Defense Department's Smart Card Office is expected to recommend later this summer that all military identification cards be replaced with smart cards. Smart cards have the potential to replace not only military ID cards, but also government driver's licenses, weapons cards, library cards, meal cards and others, Neal said. What

482. Paynter, Ronnie. "Images in the Night." *Law Enforcement Technology*, May 1999. Vol. LXVI, No. 5. Source: NLECTC *Law Enforcement & Technology News*, June 3, 1999. EPI962
483. Stover, Dawn. "In Your Face." *Popular Science*, Sept. 1995. EPI1101
484. US Patent #5,425,130, June 13, 1995. Apparatus For Transforming Voice Using Neural Networks. Inventor: Morgan, David P. Assignee: Lockheed Sanders Inc. EPI795
485. Maier, Ed. "Phoning It In." *Corrections Technology & Management*, Jan/Feb 1999. Vol. 3, No. 1. Source: NLECTC *Law Enforcement & Technology News Summary*, Feb. 18, 1999. EPI1027

makes smart cards attractive is that 'the technology for most DOD functions is commercially available."486 This is the first step in most electronic monitoring systems. It is expected that the loss of these cards will lead to breaches of security which will create the arguments the military will eventually present in favor of increased uses of biometric devices including the possibility of body implanted devices.

"Correctional facilities around the country are bolstering security with biometrics. Such places are using the high-tech systems to scan body parts or behaviors that are unique to an individual. David Banisar, legal counsel for the Electronic Privacy Information Center, warns of the day when employers will scan the retinas of job candidates to ascertain criminal history."487 The use of these systems for employees is not in the distant future and again demonstrates how systems designed for criminals will eventually be used on the entire population. "Barry Babler, a special agent in the FBI's Milwaukee office, says that in the near future officers will be able to check a criminal's background and outstanding arrest warrants in as little as 10 minutes. Chris Ahmuty, the American Civil Liberties Union executive director for Wisconsin, believes the technology can be used to exonerate innocent suspects more rapidly... but should not be abused by officers, which means strong regulations need to be set."488 Where are the regulations that experts in privacy and civil rights are demanding? The only changes in the law we have seen require increasing segments of the population to the invaded by the new technology rather than limiting its use. First these technologies are used to contain the villains, then the poor and, after that, the rest of us. Such is the creeping pattern of tyranny in the surveillance society.

The poor in one state are now being expected to give up aspects of their personal privacy in order to obtain welfare funds so they can buy food and the basics of life. Again, the motivation is for security in separating criminals from ordinary citizens. "The Connecticut Department of Social Services (DSS) is among several state welfare agencies, departments of motor vehicles, and military groups using a $5.2 million biometrics-base fraud prevention system. It uses a Identix Inc. fingerprint capture device – which goes for $2,2000 – that recalls a person's face on-screen from a database...Researchers at DSS say 11 states have adopted biometrics in their social service programs."489

Internet Control

The control and regulation of the Internet is the result of the government's fear of identity theft, terrorists, drug dealers, hackers

486. O'Hara, Colleen. "DoD may use smart cards as ID." *Federal Computer Week*, June 11, 1999. EPI1139

487. Beiser, Vince. "Biometrics Break Into Prisons." *Wired Online*, Aug. 21, 1999. NLECTC *Law Enforcement & Corrections Technology News Summary*, Aug. 26, 1999. EPI889

488. Burnett, James H. III. "System Lets Police Check Prints Quickly with FBI." Aug. 10, 1999. Source: NLECTC *Law Enforcement & Corrections Technology News Summary*, Aug. 12, 1999. EPI902

489. Radcliff, Deborah. "Busted!" *Computerworld*, Dec. 14, 1998. Vol. 32, No. 50. Source: NLECTC *Law Enforcement & Technology News Summary*, Dec. 31, 1998. EPI1050

and computer coding (encryption), which keeps the government out of a persons communications. "The increasing use of electronic records from credit card transactions over the Internet to electronic medical records gives identity thieves more and more opportunities to claim victims."490 Another development is the issuing of a specialized "chip" which allows easy access to information which communicators wish to keep private. "The chip is being launched as privacy advocates are organizing a boycott of it. They object to a feature that allows the chip to send an identification code over the Internet, arguing that it could allow operators of web sites to easily track and trace consumers."491 One of the things about using the Internet is that many people do not want to give information on who they are until they are ready for two-way communication or a purchase. The tracking of on-line shoppers is analogous to providing a name and mailing address in order to get into a store and "window shop" or into a library to read a book. How long would stores stay in business if they were intruding in this way? How would a person feel if the information was used to send piles of junk mail, or sold to telemarketers and door-to-door salespeople?

Pornography on the Internet is a problem, particularly child pornographers and the assortment of perverts preying on children. "Sgt. Joe Duke, and Oakland County, Michigan, sheriff's detective, is part of one of the ever-growing departments fighting child pornography on the Internet. These detectives often spend hours online portraying young children to catch pedophiles."492 The use of Internet police will increase as human activity increases on the Internet. However, it should be recognized that people are not electronic, they are real people who are all entitled to both protection from electronic stalking or random government intrusions without cause. In protecting children we must be aggressive while at the same time making sure we are respecting our Constitutional safeguards. These are not easy things to accomplish – the balance needs to be found with civil libertarians, politicians, the public and police seeking a real solution together.

"Cookies" are a coding system built into computers to identify who a user is. What happens is that when visiting locations on the Internet, an electronic message is left which can be followed back to your home location. After your address is located, the electronic junk mail begins to flow your way. "People feel that on the Internet they're fairly anonymous, but cookies are one of the ways that are whittling down that anonymity," says Ethan Preston, a legal researcher for the Electronic Privacy Information Center in Washington, D.C.493

490. Yip, Pamela. "Identity theft one of country's fastest-growing crimes." *Anchorage Daily News*, Oct. 25, 1999. EPI375
491. Segal, David. "Intel Unwraps New Chip." *Anchorage Daily News*, Feb. 2, 1999. EPI26
492. Zemen, David. "More Cops are Policing Cyberspace in Hunt for Pedophiles." *Detroit Free Press Online*, July 20, 1999. Source: NLECTC *Law Enforcement & Technology News Summary*, July 22, 1999. EPI920
493. Sutel, Seth. "Cookies don't have to crumble Internet privacy walls." *Anchorage Daily News*, Oct. 31, 1999. EPI1169

Encryption & Security

Encryption, or the coding of information, is as old as human communication. Hiding information by coding is a person's right and the way business is done on the Internet. Love letters, medical records, financial information and other communications on the Internet are susceptible to being intercepted by many different kinds of people with unknown motives. Coding information is like the sealing of an envelope before mailing, but now that envelope is electronic.

"Illegal activity on the Internet increases as the number of people using the Web increases, concludes a three-year study from the UK National Criminal Intelligence Services (NCIS). NCIS says crimes being committed on the Internet include pedophilia, pornography, hacking, fraud, and software piracy, as well as use by criminals for secure communications. Britain's trade and Industry Select Committee says a national computer crime organization, similar to London Metropolitan Police computer crime unit, is necessary to combat criminals who might use encryption to commit crimes."494 What else could a person expect? As cities grow, crime increases, so why would it be any different for electronic communities? The motivation behind such studies is to create larger budgets for policing organizations and as a way to justify the violation of civil rights.

"Rep. Bob Goodlatte (R-VA), chief sponsor of the Security and Freedom Through Encryption Act (SAFE), says the legislation, which would lax U.S. encryption export rules, has a good chance of passing this year. Law enforcement groups and the FBI have expressed concern over the export of strong encryption, fearing it would facilitate terrorists and criminals. But SAFE's main goal is to prevent electronic crime, Goodlatte said, by providing a level of confidentiality."495 "One Capitol Hill staffer had some concerns. 'I think they are really trying to hobble how people use encryption,' said Ellen Stroud, spokeswoman for Rep. Bob Goodlatte (R-VA), sponsor of the Security and Freedom Through Encryption Act, which would relax controls on the export of encryption and prohibit the government from requiring a back door into people's e-mail and computer files."496 This law is an attempt to find a balance between protection and privacy. In most of these cases, law enforcement attempts to restrict privacy, while others fight to secure it. These issues also extend beyond the United States, as the Internet is the most "globalized" technology on the planet. By 1999, the necessity of privacy for electronic, and, what people now call "snail mail" (regular postal mail), is finally being recognized and the laws changed. "Although international encryption policies are restrictive

494. Nuttall, Chris. "Net Crime prompts Cyber Squad Call." *BBC News Online*, June 26, 1999. Source: NLECTC *Law Enforcement & Technology News Summary*, June 24, 1999. EPI940
495. Mosquera, Mary. "SAFE Encryption Bill Re-Introduced." *TechWeb*, Feb. 25, 1999. Source: NLECTC *Law Enforcement & Technology News Summary*, March 4, 1999. EPI1014
496. Brown, Doug and Tillet, L. Scott. "Bill reopens encryption access debate." *Federal Computer Week*, Aug. 16, 1999. EPI656

and outdated, these policies may be relaxed or made obsolete in the near future, writes Jim Reavis. Several countries have loosened their encryption laws this year. France, for example, has significantly relaxed its encryption policies, now requiring simple declaration rather than prior authorization to supply and use encryption systems. Germany also has changed its stance on encryption, encouraging the technology as a way of protecting personal liberties."497 The transition into the next age will not be without the tension of debates, with the same energy as those waged over two hundred years ago during the forming of our constitutional democratic republic. "While public encryption is limited, the U.S. military faces no such restrictions, which created a degree of skepticism when military-grade encryption was used to keep President Clinton's Aug. 17 televised grand jury testimony secure. The U.S. policy seems to be that insecure encryption is OK for the public but not for politicians involved in scandals."498 Is information security only to be allowed for criminals and politicians who need to hide the truth?

The sad story in the debate is that the United States speaks with a voice which moves in opposite directions, or as American Indians once said, they speak with a "forked tongue," which moves in two different directions. "The most contentious source of trench-coat contretemps among trans-Atlantic allies: Internet encryption. The United States is trying to persuade the European Union to allow only Internet codes for which law enforcement and national security agencies would have a 'key.' That would help to combat terrorists and drug smugglers. But it would also give U.S. officials potential access to the commercial secrets of foreign companies."499

"The Department of Justice (DOJ) is seeking to add 55 assistant U.S. attorneys under the proposed budget for fiscal 2000. The attorneys would be involved in developing a global response to computer attacks and in prosecuting computer and high-tech crimes."500 "Internet service providers and message boards are attracting lots of attention from the legal community, which has increased its number of information requests drastically since the shooting rampage at Colombine High School. Those who think they are hiding behind a curtain of electronic anonymity should be warned: law enforcement can learn the user's true identity, according to Ron Horac of the Loudon County Sheriff's office in Leesburg, VA. The use of electronic evidence is becoming more and more popular, and accounts for 30 percent of evidence in legal cases, according to Ontrack lawyer Deborah Schepers in Minnesota."501

497. Reavis, Jim. "Trends in Government Encryption Policies." *Network World*, Aug. 20, 1999. Source: NLECTC *Law Enforcement & Corrections Technology News Summary*, Aug. 26, 1999. EPI892

498. Joseph, Regina. "The Encryption Imperative." *Editor & Publisher*, Jan. 1999. http://www.icij.org/about/joseph.html EPI466

499. Ford, Peter. "What's a little spying between friends?" Nando Media, Sept. 6, 1999. EPI1140

500. *National Law Journal*. "DOJ Requests Funding For More AUSAs." Feb. 15, 1999. Vol. 21, No. 25. Source: NLECTC *Law Enforcement & Technology News Summary*, Feb. 25, 1999. EPI1018

501. Weise, Elizabeth. "Electronic Evidence Hot New Tactic." *USA Today*, May 10, 1999. Source: NLECTC *Law Enforcement & Technology News Summary*, May 13, 1999. EPI970

"The FBI has organized two separate task forces, one that specializes in computer crimes and another, the Computer Analysis Response Team (CART), that assists FBI agents in handling computer evidence. Looking to the future, while it is clear that computer and other types of digital evidence are becoming increasingly important to criminal investigators, what remains cloudy is the definition of what constitutes 'digital evidence,' and how it can be used."[502]

The irony is that at the same time that authorities are expressing concern over Internet crime we are training inmates in the skills needed for using computers. "An increasing number of prison inmates are being used for computer data entry jobs, as prison industries go high tech. One prison industry, UNIGROUP, is using inmates to provide critical data conversion services. Meanwhile, UNICOR trains the inmate population of the Federal Bureau of Prisons for computer-oriented services ranging from word processing to converting images, maps, charts, and drawings into digital formats. UNICOR then markets the prisoners' skills to other Federal government agencies."[503] An attempt to use the services of inmates to catalog public information was stopped in Alaska as it was quickly realized that they were gaining access to the personal information of private citizens which they might use later in criminal activities.

Various security measures continue to grow as the technology expands. "A recent technological 'breakthrough' may be the union of smart card technology and public key cryptography. The combination of these technologies should allow for persons and businesses to conduct electronic commerce and electronic business securely. Smart cards can provide secure storage of a user's private key that will allow for a signature verification process to occur without exposing the key to the PC's hacker-prone network. When the user wishes to sign an electronic message, he inserts his smart card into a reader installed on the PC, and uses a PIN to access his private key."[504]

The proposed Federal Intrusion Detection Network (FIDNET) is a "netted" intrusion detection monitoring system for non-Department of Defense government computers. Intrusion detection monitors installed on individual systems or networks will be "netted" so that an intruder or intrusion technique used at one site will be automatically known at all sites. The FBI will be at the center of the system: "raw/filtered" data from the network of sensors will be provided to the National Infrastructure Protection Center (NPIC) that has been created at the FBI. Ultimately, the plan states, it is the goal to have similar monitoring sensors installed on private sector information

502. Pilant, Lois. "Electronic Evidence Recovery." *Police Chief*, Feb. 1999. Vol. 66, No. 2. Source: NLECTC *Law Enforcement & Technology News Summary*, March 11, 1999. EPI1011
503. Siuru, Bill. "Prison Industries Go High Tech With Low Cost Labor." *Corrections Technology & Management Magazine Online*, Oct. 27, 1998. Source: NLECTC *Law Enforcement & Technology News Summary*, Nov. 11, 1998. EPI1076
504. Williams, John. "Strengthening Smart Card Security with the Public key." *Card Technology*, May, 1999. Source: NLECTC *Law Enforcement & Technology News Summary*, June 17, 1999. EPI946

systems."505 This would allow for identifying those who are trying to illegally enter private data bases.

Executive Intrusion

"The White House has created a working group to review illegal activities on the Internet, including the sale of guns, explosives, controlled substances, and prescription drugs; fraud; and child pornography. The plan appears to favor the use of key recovery technology, a tool that the FBI and law enforcement agencies have been clamoring for. The working group also would explore the possibility of using other technology tools to help protect children on the Internet. Attorney General Janet Reno will head the group."506 This working group was created based on the following order:

"EXECUTIVE ORDER 13133
Working Group On Unlawful Conduct On The Internet

> By the authority vested in me as President by the Constitution and the laws of the United States of America, and in order to address unlawful conduct that involves the use of the Internet, it is hereby ordered as follows:
>
> Section 1. Establishment and purpose. (a) There is hereby established a working group to address unlawful conduct that involves the use of the Internet ("Working Group"). The purpose of the working group shall be to prepare a report and recommendation concerning:
>
> 1) The extent to which existing Federal laws provide a sufficient basis for effective investigation and prosecution of unlawful conduct that involves use of the Internet, such as the illegal sale of guns, explosives, controlled substances, and prescription drugs, as well as fraud and child pornography.
>
> (2) The extent to which new technology tools, capabilities, or legal authorities may be required for effective investigation and prosecution of unlawful conduct that involves the use of the Internet; and
>
> (3) The potential for new or existing tools and capabilities to educate and empower parents, teachers, and others to prevent or to minimize the risks from unlawful conduct that involves the use of the Internet."507

505. Center for Democracy & Technology. "Initial Comments on Draft National Plan for Information Systems Protection," July 27, 1999. http://www.cdt.org/policy/terrorism/fidnet/ EPI481
506. *Newsbytes*. "White House Working Against Internet Crime." Aug. 10, 1999. Source: NLECTC *Law Enforcement & Corrections Technology News Summary*, Aug. 12, 1999. EPI905
507. Executive Order, Working Group On Unlawful Conduct On The Internet. The White House, Office of the Press Secretary, Aug. 6, 1999. EPI275

The balance between privacy and safety is the central issue. Their conclusions support that balance, but achieving that balance requires an open public debate, not decisions made entirely by the executive branch of government.

> "Other proposals, however, have been offered that are not voluntary in nature and that raise privacy concerns. The Presidential commission recommended the establishment of an 'early warning and response capability' to protect telecommunications networks against cyber-attack. The commission said that such a capability should include a means for near real-time monitoring of the telecommunication infrastructure, the ability to recognize and profile system anomalies associated with attacks, and the capability to trace, reroute, and isolate electronic signals that are determined to be associated with an attack."508 "For Americans to trust this new electronic environment, and for the promise of electronic commerce and the global information infrastructure to be fully realized, information systems must provide methods to protect the data and communications of legitimate users. Encryption can address this need, because encryption can be used to protect the confidentiality of both stored data and communications. Therefore, the Administration continues to support the development, adoption, and use of robust encryption by legitimate users.
>
> At the same time, however, the same encryption products that help facilitate confidential communications between law-abiding citizens also pose a significant and undeniable public safety risk when used to facilitate and mask illegal and criminal activity. While cryptography has many legitimate and important uses, it is also increasingly used as a means to promote criminal activity, such as drug trafficking, terrorism, white collar crime, and the distribution of child pornography.
>
> A sound and effective public policy must support the development and use of encryption for legitimate purposes but allow access to plaintext by law enforcement when encryption is utilized by criminals. This requires an approach which properly balances critical privacy interests with the need to preserve public safety. As is explained more fully below, CESA provides such a balance by simultaneously creating significant new privacy protections for lawful users of encryption, while allowing law enforcement to preserve existing and constitutionally supported means of responding to criminal activity."509

---

508. O'Neil, Michael J., and Dempsey, James X. "Critical Infrastructure Protection: Threats to Privacy and Other Civil Liberties and Concerns with Government Mandates on Industry." Center for Democracy & Technology," Feb. 18, 1999. EPI482
http://www.cdt.org/policy/terrorism/oneildempseymemo.html EPI482
509. Executive office of the President, Office of Management and Budget. JUSTICE REVISED Draft Bill on Cyberspace Electronic Security Act, Aug. 5, 1999. EPI480

"Litton Industries' TASC subsidiary has initiated a joint effort with the U.S. Air Force to develop sophisticated computer forensics to root out computer and Internet criminals. Hackers, terrorists, and pornographers are among those that TASC hopes to help investigate by using an automated Integrated Media Analysis Tools kit and by establishing a computer forensics lab."510 The tools are here and being used.

Government Break-ins?

The worst hackers are the national governments, which assure privacy when collecting information, but then use that private knowledge to violate people. "In particular, there was 'consternation, horror and outrage' earlier this year when charities providing support services to people with HIV and Aids learned that information they had supplied to health authorities was being used to identify their clients. Department of Health staff had lied, said Anderson, when they reassured clinicians that personal health data being fed into the NHS network was not even linkable, let alone identifiable."511 It doesn't stop with electronic hardware and traffic through the ether of the Internet. "The Clinton administration reportedly plans to ask Congress to give police authority to secretly go into people's homes and crack their security codes. Legislation drafted by the Justice Department would let investigators get a sealed warrant from a judge to enter private property, search through computers for passwords and override encryption programs."512 Draw your own conclusions – privacy or hypocrisy?

Wiretap Dancing

Wiretaps have been around since the telephone came into general use. Wiretaps have been used just as long by politicians in order to further their own political ends. In the 1960s the issue was the center of much debate when it was revealed that, "From May, 1966 to January, 1969, the FBI provided President Lyndon B. Johnson with biweekly reports on conversations by or about anti-Vietnam-war senators or congressmen overheard by the bureau agents wiretapping foreign embassies, according to information obtained by the Senate intelligence committee."513 Even the highest level of elected representatives were subject to the whims of powerful people who, holding the public trust, chose to violate it rather than honor privacy and honest government.

---

510. *Law Enforcement Technology*. "Examining Evidence of Illegal Computer Use." April 1999. Vol. 26, No. 4. Source: NLECTC *Law Enforcement & Technology News Summary*, April 29, 1999. EPI988
511. Campbell, Duncan. "Hypocritic Oaths: Medical records have lost their way, says Duncan Campbell." *The Guardian Online*, Nov. 5, 1997. http://www.gn.apc.org/duncan/878749662-priv.html EPI478
512. AP. "Feds want authority to secretly crack personal computer codes." Aug. 20, 1999. http://www.cnn.com/TECH/computing/9908/20/computer.codes.ap/ EPI385
513. Pincus, Walter. "LBJ Got FBI Wiretap Data." *Washington Post*, Dec. 10, 1975, pp. 1-2. EPI54

President Johnson was not alone. President Reagan in 1991 "eased many of the restrictions imposed on U.S. intelligence agencies since the mid-1970s when he signed an executive order that he said is designed 'to remove the aura of suspicion and mistrust that can hobble' their work. The 17-page order, which supplants one signed by president Carter three years ago, authorizes the Central Intelligence Agency to use secret means to collect 'significant foreign intelligence' from unsuspecting Americans here and abroad."514 These restrictions were imposed when the Congress discovered that the CIA had opened mail, tapped phones, infiltrated domestic political groups and were experimenting on Americans in violation of the law. These restrictions were eased, essentially giving them the license to begin their abuse of power again.

Today, "Motivated by the public's increasing reliance on wireless communication, the Federal Communications Commission (FCC) passed a rule in 1996 requiring that Public Safety Answering Points (PSAP) be equipped to identify the location of cellular phones used to make 911 calls by October 1, 2001."515 These locator questions do not end with 911 emergency calls because these technologies will allow tracking cell phones users not only by their calls but also by their locations. This was confirmed when news reports began to link the systems in this way. "Space-based telephone systems, which can be used to operate a wireless phone anywhere on earth, has FBI and other U.S. law enforcement agencies concerned that they will not be able to trace criminal activities and locations through cell phones. The new technology, being produced by Globalstar, Iridium, and other companies, is being held up by the 1994 U.S. Communications Assistance for Law Enforcement Act, which gives law enforcement agencies access to digital call information when tapping and determining locations of users."516 The intrusions continue even today, with periodic examples of abuse of the law.

The Los Angeles Police Department seems to be one of those places which continues to be a source of examples of abuse. They have been nailed to the wall in the past for violating the rights of suspects and beating people when they were apprehended, as was the case with Rodney King. Electronic surveillance, although it does not inflict the same pain as a beating, does deliver an even stronger blow against democratic principles. In a news story released in 1999, "The Los Angeles Police Department is facing charges that it has been improperly concealing a nine-month wiretapping operation and the information derived from the operation."517

514. Lardner, Jr.,George. "President Eases Restrictions On Intelligence Gathering." *Washington Post*, Dec. 5, 1981. EPI85
515. Dees, Tim. "Locating Cellular Users: Mandates and Solutions." *Law and Order*, Jan. 1999. Vol. 47, No. 1. Source: NLECTC *Law Enforcement & Technology News Summary*, Feb. 25, 1999. EPI1023
516. Borland, John. "FBI Wiretap Worries Slow Satellite Phones." *CNET*, Aug. 3, 1999. Source: *Law Enforcement & Technology News Summary*, Aug. 5, 1999. EPI910
517. Berry, Steve. "Wiretap Inquiry Shines Light on Police Tactics." *Los Angeles Times--National Edition*, Dec. 3, 1998. Source: NLECTC *Law Enforcement & Technology News Summary*, Dec. 10, 1998. EPI1061

Roving Wiretaps

"In a closed-door maneuver, controversial 'roving wiretap' provisions were added to the Intelligence authorization bill and passed by Congress. Previously, the federal wiretapping law allowed roving taps only under very strict limits. The new provisions dramatically expanded that authority by allowing taps on any phone used by, or 'proximate' to, the person being tapped, no matter whose phone it is."518 What this means is that entire areas could be secured and tapped. Using voice recognition software, sophisticated computers could immediately record conversations over the entire area. As this combination of technologies advances, it will be used globally. The idea of "proximate" is changing as the technology advances and its real meaning may soon extend to all global communications. In fact, the evidence thus far presented indicates that the technology, computing power and political will already exists for a global communications monitoring system which will violate the rights of all people equally – a much different view of what the phrase "equal protection under the law" really means.

"In Pima County, Arizona, the Sheriff's Department, in partnership with CTAC, installed and adapted Borderline, a system that makes court ordered wiretap much more effective. In a state-of-the-art facility funded by HIDTA, the High Intensity Drug Trafficking Area program, wiretaps from all over the huge county are digitally recorded and transcribed, making them digitally accessible for rapid search and retrieval."519 The area networks are already in place and the combination already being used in various regions of the world. "So-called roving wiretaps work through a digital switching system linked to voice analyzers that can record and identify a conversation instantly. The analyzer produces a numerical pattern of a suspect's voice. Then the switching system constantly scans outgoing phone lines for that pattern."520

"In July 1992, we reported to Congress that law enforcement's ability to execute court-approved wiretaps was challenged by advanced telecommunications technologies. These discussions revealed that although some technological solutions have been developed to facilitate law enforcement agencies' wiretap efforts, other technology changes have made it more difficult for them to use traditional wiretap methods. Because the details of law enforcement agencies' problems and the specific technological challenges are classified, I cannot elaborate on them at the moment. S. 2375 specifies that the Attorney general, subject to the availability of appropriations, shall reimburse telecommunications carriers for reasonable costs

518. Center For Democracy & Technology. CDT's Counter-Terrorism issues page. http://www.cdt.org/policy/terrorism/ EPI555
519. ONDCP, CTAC. Advanced Surveillance & Wiretap Technologies. http://www.whitehousedrugpolicy.gov/ctac/page13.htm EPI304
520. Johnson, Chip. "Techno-cops: Police Tools of the 90's Are Highly Advanced, But Privacy Laws Lag." *Wall Street Journal*, Nov. 12, 1990. EPI118

incurred in carrying out the provisions of the bill, and details the nature of the costs that can be reimbursed...However, it is virtually impossible to precisely estimate the reimbursement costs discussed in this bill because costs will depend on evolving law enforcement requirements."521 What? This means that all of the nation's telecommunication systems must provide a way for the government to intrude on a massive scale. These laws will allow many government agencies the ability to easily hack through the safeguards and illegally tap the phones of millions of people with or without court orders. Current events in the media certainly confirm that when the opportunity for government abuse is there it will be taken by those who forget the underlying mission of all government activity – to represent and protect the rights of citizens and uphold the law.

In other countries which remember the losses of freedom in recent history this issue has become very important. "The wiretapping law is similar to those in other countries. But many Japanese, remembering secret police brutality during World War II and crackdowns on radical students and labor unions in the 1950s and 1960s, have long been reluctant to hand police greater powers."522 Recent memory of wholesale abuses by government should be the reminder we all need to insist on the security of the sacred trust between the government and those who are governed. Privacy rights are the foundation of that trust.

Echelon – Big Brother Goes Global

"Members of Congress, the European Parliament and civil liberties groups have begun to ask tough questions about the National Security Agency's interception of foreign telephone calls, faxes and electronic mail, the most intense scrutiny of NSA operations since the so-called Church committee probed the spy agency 24 years ago."523 Echelon has become one of the biggest exposés of this century in terms of the most significant example of abuse of control ever instituted. This level of surveillance surpasses all of the combined efforts of our adversaries throughout history. "Moreover, the level of intrusion has reached new heights in recent years with the development by America's NSA of a secretive system called ECHELON, which has the capability of intercepting all the world's communications at any given time."524 "Virtually invisible to the American Public, NSA runs the nation's most ambitious spying operation, eclipsing the Central Intelligence Agency in budget and personnel. Its operations cost nearly $1 million an hour, $8 billion a year. It oversees tens of thousands of eavesdroppers in listening posts from Alaska to Thailand, and with a Maryland work force of 20,000 is

521. GAO. "Electronic Surveillance: Technologies Continue To Pose Challenges." GAO/T-AIMD-94-173, Aug. 11, 1994. EPI119
522. Coleman, Joseph. "Japan Arms Law Enforcement With Wiretapping Powers." *Anchorage Daily News*, Aug. 13, 1999, p. A-13. EPI100.
523. Loeb, Vernon. "Critics Questioning NSA Reading Habits." *Washington Post*, Nov. 13, 1999. EPI1210
524. Frost, Mike. "Big Brother Really Is Listening." *Anchorage Daily News*, Nov. 22, 1998. EPI102

that state's largest employer."525 The European Parliament had the following to report on this system:

"PART 1: INTERIM STUDY
Part B: Arguments and Evidence

General: *(.......The NSA is one of the shadowiest of the US intelligence agencies. Until a few years ago, it's existence was a secret and its charter and any mention of its duties are still classified. However, it does have a web site (www,nsa.gov.8080) in which it describes itself as being responsible for the signals intelligence and communications security activities of the US government. One of its bases, Menwith Hill, was to become the biggest spy station in the world. Its ears – known as radomes – are capable of listening in to vast chunks of the communications spectrum throughout Europe and the old Soviet Union.*

In its first decade the base sucked data from cables and microwave links running through a nearby Post Office tower, but the communications revolutions of the seventies and eighties gave the base a capability that even its architects could scarcely begin to imagine. With the creation of Intelsat and digital telecommunications, Menwith and other stations developed the capability to eavesdrop on an extensive scale on fax, telex and voice messages. Then, with the development of the Internet, electronic mail and electronic commerce, the listening posts were able to increase their monitoring capability to eavesdrop on an unprecedented spectrum of personal and business communications.
(Simon Davis report: http://www.telegraph.co.uk)

Led by Germany and the Scandinavians, the EU has been generally distrustful of key escrow technology. In October 1997, the European Commission released a report which advised: 'Restricting the use of encryption could well prevent law-abiding companies and citizens from protecting themselves against criminal attacks. It would not, however, totally prevent criminals from using these technologies.' The report noted that privacy considerations suggest limiting the use of cryptography as a means to ensure data security and confidentiality.

Technical File
1. Introduction
Surveillance and Privacy

The recent years have seen the emergence and refinement of a new form of surveillance no longer of the real person, but of the person's data shadow or digital persona. Data

525. Shane, Scott and Bowman, Tom. "SHHHHH! They're Listening." *Anchorage Daily News*, Jan. 28, 1996. EPI96

surveillance or dataveillance is the systematic use of personal data systems in the investigation or monitoring of the actions or communications of one or more persons. Dataveillance is significantly less expensive than physical and electronic surveillance, because it can be automated. As a result, the economic constraints on surveillance are diminished and more individuals and larger populations are capable of being monitored. Like surveillance, more generally, Dataveillance is of two kinds: 'personal Dataveillance', where a particular person has been previously identified as being of interest, 'mass Dataveillance', where a group or large population is monitored, in order to detect individuals of interest, and/or to deter people from stepping out of line.

Surveillance technology systems are mechanisms, which can identify, monitor and track movements and data.

During the last few decades since information technology has become immensely sophisticated real benefits have been achieved in the development of surveillance technology systems...

People often think of privacy as some kind of right. Unfortunately, the concept of a 'right' is a problematic way to start, because a right seems to be some kind of absolute standard. What's worse, [it] is very easy to get confused between legal rights on one hand and natural or moral rights on the other. It turns out to be much more useful to think about privacy as one kind of thing (among many kinds of things) that people like to have lots of.

Privacy [is] the interest that individuals have in sustaining a 'personal space' free from interference by other people and organizations.

To a deeper level privacy turns out not to be a single interest but rather has several dimensions:

- privacy of the person
- privacy of personal behavior
- privacy of personal communications
- privacy of personal data.

Risks inherent in Data Surveillance

In mass surveillance

a. Risks to the individuals:

- arbitrariness
- a contextual data merger
- complexity and incomprehensibility of data
- witch hunts

– ex - ante discrimination and guilt prediction
– selective advertising
– inversion of the onus of proof
– covert operations
– unknown accusations and accusers
– denial of due process

b. Risks to society

– adversarial relationships
– focus of law enforcement on easily detectable and provable offenses
– inequitable application of the law
– decreased respect for the law and law enforcers
– reduction in meaningfulness of individual actions
– reduction in self-reliance and self-determination
– stultification of originality
– increased tendency to opt out of the official level of society
– weakening of society's moral fiber and cohesion
– destabilization of the strategic balance of power
– repressive potential for the totalitarian government"526

"While EU is aware that Echelon may be a useful tool for tracking down global terrorists, drug barons, and international criminals, Ford said the parliament is concerned that the system may also be used for espionage, spying on peaceful nations, or gaining unfair economic advantage over non-member nations."527 Despite the European concerns about Echelon, they are creating their own tapping system in cooperation with the very nations that are violating their citizens communications. Is this the trend? To yell and scream and give the appearance of outrage and then make other agreements giving the same countries another ticket to the game? "Under the plan, Enfopol 98, European telecommunication companies will be required to build tapping connection into their systems. Each EU country's 'interception Interface' must be capable of allowing member states to tap communication throughout the EU. The US, Canada and Australia are likely to participate in the network, giving the FBI and other non-European security agencies access to communications in Europe."528

"Echelon is NSA's cold war vintage global spying system, which consists of a worldwide network of clandestine listening posts capable of intercepting electronic communications such as e-mail,

526. European Parliament. *Development Of Surveillance Technology And Risk Of Abuse Of Economic Information.* Parts 1-4. April, May 1999. STOA. EPI1158
527. McKay, Niall. "Eavesdropping on Europe." *Wired News*, Sept. 30, 1998. EPI105
528. Campbell, Duncan. Revealed: secret plan to tap all mobile phones. *Observer*, Dec. 6, 1998. http://www.gn.apc.org/duncan/Enfopol_98_obs.htm EPI464

telephone conversations, faxes, satellite transmissions, microwave links and fiber-optic communications traffic. However, the European Union last year raised concerns that the system may be regularly violating the privacy of law-abiding citizens."529 "Over the past 10 years, the Echelon system has been automated and many of the hundreds of thousands of listening operators formerly working in communication intelligence have been replaced. Interception and analysis is now done automatically, through global networks of computers that can sift unattended."530

The system was developed to share data between the United States, Great Britain, Canada, Australia and New Zealand after WWII. "Australia, one of five countries running the controversial Echelon global surveillance network, has become the first to admit it. The Australian government has confirmed that the system spies on the international communications of it's own and other countries' citizens."531 The concept was that since it was illegal to spy on one's own citizens participating countries would allow other countries to do it for them and then pass the information on to the country in control of the citizen that was spied on. This provided a way around the privacy laws which each nation had in place for its citizens. The net result was establishment of the largest spy network in the world, which now has the ability to intercept every form of electronic and telephonic communication, and sort the data by voice recognition, key words used or any number of other criteria. "The NSA patent, granted on 10 August, is for a system of automatic topic spotting and labeling of data. The patent officially confirms for the first time that the NSA has been working on ways of automatically analyzing human speech."532 Once selected material is captured, the largest computers in the world sort the data and forward it to tens of thousands of intelligence organizational workers in the five countries for further analysis. The number of people required to do the work significantly drops as the technology advances. It is clear that when the full scope of advancing technologies is available, there will be nothing left of privacy, unless these trends are reversed and properly contained. The power of government has gone beyond the humanity of the planet and its people.

Unfortunately, the government which wants to keep track of all of our communications can not keep track of our money, or is it that they are so good at hiding the money that no one can find it? "The National Reconnaissance Office, the secret agency that builds spy satellites, lost track of more than $2 billion in classified money last year, largely because of its own internal secrecy, intelligence officials say. Critics of the reconnaissance office said Monday that the money

529. Verton, Daniel. "Congress, NSA butt heads over Echelon." *Federal Computer Week,* June 3, 1999. EPI1138
530. Campbell, Duncan. "EC concern about US phone spies." *The Age*, Sept. 12, 1998. http://theage.com.au/daily/980912/news/news29.html EPI465
531. Campbell, Duncan. Australia first to admit "we're part of global surveillance system." http://www.heise.de/tp/english/inhalt/te/2889/1.html EPI794
532. Dreyfus, Suelette. "This is just between us (and the spies)." *Independent News,* Nov. 15, 1999. EPI1215

was hidden in several rainy-day accounts that secretly solidified into a 'slush fund.'[533] That's a lot of slush in that fund...What they are hiding is the truth, which is that these innovations are a greater threat to our democracy and freedom than any rogue nation with a "nuke" or any terrorist with a threat. These "intelligence" organizations supposedly protecting the free world have created systems to enslave us to their own warped version of freedom.

Fortunately, there are a few champions of freedom left in the world like European Parliament Member Glenn Ford of Great Britain and United States Representative Bob Barr. "That the National Security Agency intercepts American's missives is clear. 'I have a problem with what the program appears to be doing, and that is, invading the privacy of American citizens without any reason, any court order, without any reasonable cause, without any probable cause, almost a dragnet invasion of privacy,' says Rep. Bob Barr, R-Georgia., one of the NSA's most outspoken critics."[534]

Many over the years have found fault with the American Civil Liberties Union. However, it is important to remember that their only charge is to defend the Constitution and the Bill of Rights. We personally object to some of the cases they bring forward, but, in the end, judges decide, law is clarified and our governing documents are upheld. It is not always perfect, but it is a system which works. On this issue they have not been silent either, "The American Civil Liberties Union has focused its eye on an international electronic surveillance system that allegedly eyeballs regular citizens. The civil liberties watchdog launched Echelon Watch, a site designed to prompt governmental investigation...of a global electronic surveillance system said to be code-named 'Echelon.'"[535]

The technology is still expanding and all the agencies are looking for skilled people to assist them in both their legitimate and illegitimate programs. These agencies, from the CIA to the FBI, and "Other members of America's covert intelligence community, including the National Security Agency, which conducts electronic eavesdropping around the world, and the Defense Intelligence Agency, which studies foreign military forces, also have gone public in the hunt for fresh faces."[536] They are recruiting from our colleges and universities the best minds they can get for the defense of freedom and, unfortunately, their other programs. We, the writers of this book, hope that people connected to this work will consider these words. We hope that during their careers, as they think about their mission, they will remember the mission of our forefathers when they laid their lives down for freedom. We can have security – but there is no security unless we have freedom, liberty and honor.

533. Weiner, Tim. "Spy agency finds big bucks in own coffers." *Anchorage Daily News*, Jan. 31, 1996, pp. A3. EPI326
534. Ruppe, David. "Big Brother is Listening." *ABCNEWS.com*, July 27, 1999. EPI272
535. Oakes, Chris. "ACLU to Spy on Echelon." *Wired News*, Nov. 17, 1999. EPI1224
536. Drogin, Bob. "CIA stakes out colleges for new recruits." *Anchorage Daily News*, Nov. 17, 1999. EPI1214

The problem of privacy extends beyond the government as well. We have allowed corporate cultures to create intrusive, inhumane environments for both customers and employees, treating both as common criminals rather that honest people. "Two-thirds of U.S. businesses eavesdrop on their employees in some fashion – on the phone, via videotape and through e-mail and Internet files – according to a 1999 survey by the American Management Association International."537 So, a great deal must be done to reclaim a balance between those who would control and those who seek a small measure of privacy in an increasingly complex world.

European Perspectives on Surveillance

The news in the United States is often quite limited. O.J. Simpson, Bill Clinton and other scandalous figures dominate the news, to the exclusion genuine information. The European perspective is an important one because their view is one from the vantage point of people who have been subjected to huge restrictions on freedom in recent decades. People who know repression understand what Americans have learned to take for granted. The time for all of us to pay attention is now. The European Parliament has the following to say in one of their reports:

"2. Developments in Surveillance Technology

> Surveillance technology can be defined as devices or systems which can monitor, track and assess the movements of individuals, their property and other assets. Much of this technology is used to track the activities of dissidents, human rights activists, journalists, student leaders, minorities, trade union leaders and political opponents. A huge range of surveillance technologies has evolved, including night vision goggles; parabolic microphones to detect conversations over a kilometer away; laser versions, can pick up any conversation from a closed window in line of sight; the Danish Jai stroboscope camera can take hundreds of pictures in a matter of seconds and individually photograph all the participants in a demonstration or march; and the automatic vehicle recognition systems can track cars around a city via a Geographic Information System of maps.
>
> New technologies which were originally conceived for the Defense and Intelligence sectors have after the cold war rapidly spread into the law enforcement and private sectors. It is one of the areas of technological advance, where outdated regulations have not kept pace with an accelerating pattern of abuses. Up until the 1960s, most surveillance was low-tech and expensive since it involved following suspects around from

537. Stevens, Liz. "Employer snooping's legal in most states." *Anchorage Daily News*, Nov. 15, 1999. EPI1211

place to place, using up to 6 people in teams of two working 3 eight hour shifts. All of the material and contacts gleaned had to be typed up and filed away with little prospect of rapidly cross checking. Even electronic surveillance was highly labor intensive. The East German police for example employed 500,000 secret informers, 10,000 of which were needed just to listen and transcribe citizens' phone calls.

By the 1980s, new forms of electronic surveillance were emerging and many of these were directed towards automation of communications interception. This trend was fueled in the US in the 1990s by accelerated government funding at the end of the cold war, with defense and intelligence agencies being refocused with new missions to justify their budgets, transferring their technologies to certain law enforcement applications such as anti-drug and anti-terror operations. In 1993, the US department of defense and the Justice department signed memoranda of understanding for 'Operations Other Than War and Law Enforcement' to facilitate joint development and sharing of technology. According to David Banisar of Privacy International, 'To counteract reductions in military contracts which began in the 1980's, computer and electronics companies are expanding into new markets – at home and abroad – with equipment originally developed for the military. Companies such as E Systems, Electronic Data Systems and Texas Instruments are selling advanced computer systems and surveillance equipment to state and local governments that use them for law enforcement, border control and Welfare administration. 'What the East German secret police could only dream of is rapidly becoming a reality in the free world.'

2.1 Closed Circuit Television (CCTV) Surveillance Networks: In fact the art of visual surveillance has dramatically changed over recent years. Of course police and intelligence officers still photograph demonstrations and individuals of interest but increasingly such images can be stored and searched. Ongoing processes of ultra-miniaturization mean that such devices can be made to be virtually undetectable and are open to abuse by both individuals, companies and official agencies.

2.2 Algorithmic Surveillance Systems: The revolution in urban surveillance will reach the next generation of control once reliable face recognition comes in. It will initially be introduced at stationary locations, like turnstiles, customs points, security gateways etc. to enable a standard full face recognition to take place.

Such surveillance systems raise significant issues of accountability, particularly when transferred to authoritarian regimes. The cameras used in Tiananmen Square were sold as advanced traffic control systems by Siemens Plessey. Yet after the 1989 massacre of students, there followed a witch hunt when the authorities tortured and interrogated thousands in an effort to ferret out the subversives. The Scoot surveillance system with USA made Pelco cameras were used to faithfully record the protests. The images were repeatedly broadcast over Chinese television offering a reward for information, with the result that nearly all the transgressors were identified. Again democratic accountability is only the criterion which distinguishes a modern traffic control system from an advanced dissident capture system. Foreign companies are exporting traffic control systems to Lhasa in Tibet, yet Lhasa does not as yet have any traffic control problems. The problem here may be a culpable lack of imagination.

2.3 Bugging and Tapping Devices: A wide range of bugging and tapping devices have been evolved to record conversations and to intercept telecommunications traffic. In recent years the widespread practice of illegal and legal interception of communications and the planting of 'bugs' has been an issue in many European states. However, planting illegal bugs is yesterday's technology. Modern snoopers can buy specially adapted lap top computers, and simply tune in to all the mobile phones active in the area by cursoring down to their number. The machine will even search for numbers 'of interest' to see if they are active. However, these bugs and taps pale into insignificance next to the national and international state run interceptions networks.

2.4 National & International Communications Interceptions Networks: The Interim Study set out in detail, the global surveillance systems which facilitate the mass supervision of all telecommunications including telephone, email and fax transmissions of private citizens, politicians, trade unionists and companies alike. There has been a shift in targeting in recent years. Instead of investigating crime (which is reactive) law enforcement agencies are increasingly tracking certain social classes and races of people living in red-lined areas before crime is committed - a form of preemptive policing deemed dataveillance which is based on military models of gathering huge quantities of low grade intelligence.

Without encryption, modern communications systems are virtually transparent to the advanced interceptions equipment which can be used to listen in. The Interim Study also explained how mobile phones have inbuilt monitoring and tagging dimensions which can be accessed by police and intelligence agencies. For example the digital technology required to pinpoint mobile phone users from incoming calls, means that all mobile phone users in a country when activated, are mini-tracking devices, giving their owners whereabouts at any time and stored in the company's computer. For example Swiss police have secretly tracked the whereabouts of mobile phone users from the computer of the service provider Swisscom, which according [to] SonntagsZeitung had stored movements of more than a million subscribers down to a few hundred meters, and going back at least half a year.

However, of all the developments covered in the Interim Study, the section covering some of the constitutional and legal issues raised by the USA's National Security Agency's access and facility to intercept all European telecommunications caused the most concern. Whilst no one denied the role of such networks in anti-terrorist operations and countering illegal drug, money laundering and illicit arms deals, alarm was expressed about the scale of the foreign interceptions network identified in the Study and whether existing legislation, data protection and privacy safeguards in the Member States were sufficient to protect the confidentiality between EU citizens, corporations and those with third countries.

Since there has been a certain degree of confusion in subsequent press reports, it is worth clarifying some of the issues surrounding transatlantic electronic surveillance and providing a short history & update on developments since the Interim Study was published in January 1998. There are essentially two separate systems, namely:

(i) The UK/USA system comprising the activities of military intelligence agencies such as NSA-CIA in the USA subsuming GCHQ & M16 in the UK operating a system known as Echelon;

(ii) The EU-FBI system which is linked up to various law enforcement agencies such as the FBI, police, customs, immigration and internal security;

2.4.1 NSA Interception Of All EU Telecommunications: The Interim Study said that within Europe, all email, telephone and fax communications are routinely intercepted by the United States National Security Agency, transferring all target information from the European Mainland via the strategic hub of London then by satellite to Fort Meade in Maryland via the crucial hub at Menwith Hill in the North York Moors of the UK.

The Echelon system forms part of the UKUSA system but unlike many of the electronic spy systems developed during the cold war, Echelon is designed for primarily non-military targets: governments, organizations and businesses in virtually every country. The Echelon system works by indiscriminately intercepting very large quantities of communications and then siphoning out what is valuable using artificial intelligence aids like Memex to find key words. Five nations share the results with the US as the senior partner under the UKUSA agreement of 1948, Britain, Canada, New Zealand and Australia are very much acting as subordinate information servicers.

Each of the five centers supply 'dictionaries' to the other four of keywords, phrases, people and places to 'tag' and the tagged intercept is forwarded straight to the requesting country. Whilst there is much information gathered about potential terrorists, there is a lot of economic intelligence, notably intensive monitoring of all the countries participating in the GATT negotiations. But Hager found that by far the main priorities of this system continued to be military and political intelligence applicable to their wider interests.

Indeed since the Interim Study was published, journalists have alleged that Echelon has benefited US companies involved in arms deals, strengthened Washington's position in crucial World trade organization talks with Europe during a 1995 dispute with Japan over car part exports. According to the *Financial Mail* on Sunday, 'key words identified by US experts include the names of inter-governmental trade organizations and business consortia bidding against US companies. The word 'block' is on the list to identify communications about offshore oil in area where the seabed has yet to be divided up into exploration blocks'...It has also been suggested that in 1990 the US broke into secret negotiations and persuaded Indonesia that US giant AT&T be included in a multi-billion dollar telecoms deal that at one point was going entirely to Japan's NEC.

2.4.4 EU-FBI Global Telecommunications Surveillance System: In February 1997, *Statewatch* reported that the EU had secretly agreed to set up an international telephone tapping network via a secret network of committees established under the 'third pillar' of the Mastricht Treaty covering co-operation on law and order. Key points of the plan are outlined in a memorandum of understanding, signed by EU states in 1995. (ENFOPOL 112 10037/95 25.10.95) which remains classified. According to a *Guardian* report (25.2.97) it reflects concern among European Intelligence agencies that modern technology will prevent them from tapping private communications. 'EU countries it says, should agree on international interception standards set at a level that would ensure encoding or scrambled words can be broken down by government agencies.' Official reports say that EU governments agreed to co-operate closely with the FBI in Washington. Yet earlier minutes of these meetings suggest that the original initiative came from Washington. According to *Statewatch*, network and service providers in the EU will be obliged to install 'tappable' systems and to place under surveillance any person or group when served with an interception order.

These plans have never been referred to any European government for scrutiny, nor to the Civil Liberties Committee of the European Parliament, despite the clear civil liberties issues raised by such an unaccountable system. The decision to go ahead was simply agreed in secret by 'written procedure' through an exchange of telexes between the 15 EU governments. We are told by *Statewatch* that the EU-FBI Global surveillance plan was now being developed 'outside the third pillar.' In practical terms this means that the plan is being developed by a group of twenty countries – the then 15 EU member countries plus the USA, Canada, Norway and New Zealand. This group of 20 is not accountable through the Council of Justice and Home Affairs Ministers or to the European Parliament or national parliaments. Nothing is said about finance of this system but a report produced by the German government estimates that the mobile phone part of the package alone will cost 4 billion D-marks.

Statewatch concludes that 'It is the interface of the Echelon system and its potential development on phone calls combined with the standardization of tappable communications centers and equipment being sponsored by the EU and the USA which presents a truly global threat over which there are no legal or democratic controls.' (Press release 25.2.97) In many respects what we are witnessing here are meetings of operatives on a new global-intelligence state. It is very difficult for anyone to get a full picture of what is being decided at the executive meetings setting this 'Transatlantic agenda.' Whilst *Statewatch* won a ruling from the Ombudsman for access on the grounds that the Council of Ministers 'misapplied the code of access', for the time being such access to the agendas have been denied. Without such access, we are left with 'black box decision making'. The eloquence of the unprecedented Commission statement on Echelon and Transatlantic relations scheduled for the 16th of September, is likely to be as much about what is left out as it is about what is said for public consumption. Members of the European Parliament may wish to consider the following policy options:

2.5 Policy Options:

(i) That a more detailed series of studies should be commissioned on the social, political, commercial and constitutional implications of the global electronic surveillance networks outlined in this Study, with a view to holding a series of expert hearings to inform future EU civil liberties policy. These studies might cover:

a. The constitutional issues raised by the facility of the US National Security Agency (NSA) to intercept all European telecommunications, particularly those legal commitments made by member states in regard to the Maastricht Treaty and the whole question of the use of this network for automated political and commercial espionage.

b. The social and political implications of the FBI-EU global surveillance system, its growing access to new telecommunications mediums including e-mail and its ongoing expansion into new countries together with any related financial and constitutional issues;

c. The structure, role and remit of an EU wide oversight body, independent from the European parliament, which might be set up to oversee and audit the activities of all bodies engaged in intercepting telecommunications made within Europe;

(ii) The European Parliament has the option of urging rejection of proposals from the United States for making private messages via the global communications network (Internet) accessible to US intelligence agencies. Nor should the Parliament agree to new expensive encryption controls without a wide ranging debate within the EU on the implications of such measures. These encompass the civil and human rights of European citizens and the commercial rights of companies to operate within the law, without unwarranted surveillance by intelligence agencies operating in conjunction with multinational competitors."[538]

You Might Do It Again – Jail Forever

What is happening here? I truly understand the feelings of people when it comes to victimization of children, having five of my own and coming from a large family. Even in prisons, these predators of the innocent are beaten and isolated for their crimes by their fellow prisoners who have the self-respect to know – we must not violate children. However, "In a 5-4 decision, the justices said that communities could go much further than they do in Arizona, keeping

538. The European Parliament. *An Appraisal of The Technologies of Political Control*, STOA Interim Study, Updated Executive Summary, Sept. 1998. EPI250
http://www.europarl.eu.int/dg4/stoa/en/publi/166499/execsum.htm#text1 EPI250

sexual predators locked up even after they have served their sentences if professionals predict that they may be dangerous in the future."[539] I feel very, very strongly about these perverted and depraved people and fear what I would do in the instant of discovery if I found that one of my children had been victimized in this way. However, the solution in Arizona is a wrong one because it will not end with these kinds of crimes, but continue to be expanded to other crimes until nobody once put in jail would have a reasonable hope for freedom. Who will test the prisoners, who will make the judgments, who will decide when they can be let free? If the sentences are too short, then change the laws to allow for these types of situations, but do not cage a man based on professional opinion – jail him on the merits of the law and severity of the crime. We cannot allow fear to dominate our futures. We must find the punishment that fits the crime and deliver it fairly, swiftly and justly.

What do you have Under Your Skin?

"Proposed devices that would be worn on the skin would detect such substances as methamphetamine, morphine, tetrahydrocannabinol (THC), and transmit radio signals in response to computer queries. Such a device, called a Remote Biochemical Assay Telemetering System (R-BATS), commonly referred to as a 'drug badge,' could be attached to the wearer by use of an adhesive wristband, for example."[540] This device makes possible significant abuse of medical privacy as its use will eventually extend beyond drug abusers. When people who are not criminals are forced to wear such technological controls, safeguards for privacy and individual choice must be secured. This technology could be helpful for the elderly and those unable to care for themselves but, again, we must respect the whole person and not impose our choices on them without careful safeguards, accountability for authorities and fair evaluation by advocates for those in need.

Government "funded medical scientists at the National Institute on Drug Abuse are using state-of-the-art Position Emission Tomography scanner technology to learn more about the mechanisms of action of drugs on the human brain, safely exploring the brains of living human subjects, including drug abusers. With this new knowledge we hope to make rapid advances in our quest for counterdrug medications.

At Columbia University's College of Physicians and Surgeons in New York City, an international research team led by Don Laundry, M.D., has created artificial antibodies that attach to cocaine molecules as they enter the bloodstream, rendering the cocaine inert in laboratory rats. While human subject trials are not expected for several years, Dr. Laundry believes that if he can make additional progress it

539. Tharp, Mike. "Tracking Sexual Impulses: An Intrusive Program To Stop Offenders From Striking Again." U.S. News & World Report. July 7, 1997, pp. 34-35. EPI97
540. *NASA Tech Briefs*. "Devices Would Detect Drugs in Sweat." May 1996. pp. 87-88. EPI98

will lead to an historic breakthrough: the development of a medication that would be an effective treatment for cocaine overdose, serve as a vaccine against cocaine addiction and provide physicians with a drug that would give cocaine addicts powerful support if they should choose to quit."

Anti-drug drugs may soon be a reality for substance abusers. "Designer antibodies may someday be used to immunize people against cocaine and other drugs to block the rush that users crave. 'Our goal would be to protect against the sudden unexpected urge to use, so that if the patient used it, he wouldn't get the effects.'"541 This is interesting technology but raises other questions regarding medical privacy. This is the same thinking that if applied to other types of medical care could be tragic. Drugs or procedures which deal with the health of an individual, and not the life of another, are not decisions others should make for us regardless of our status in life. These decision are ours alone unless we are incapacitated to the point where these decisions can not be made. Care must be taken to assure freedom of choice in what happens to our bodies.

"With the knowledge researchers are acquiring, we're going to end up talking about compulsive drug taking behavior as a function of neuro-chemically rewiring a person. It's an altered mind state; don't talk about spiritual or moral aspects, or will, or discipline, or family background. You should talk about a person who has a different neurochemical function.' ONDCP Director McCaffrey."542 What it normal? Was Einstein, Edison, Washington, Churchill, Kennedy, Noah, Mohammed or Gandhi "normal?"...We think not. And we hope the next great humanitarians on the planet don't run into social engineers when they are attempting to change the world for us and for our great grandchildren.

Electronic Control of Your Assets

"The federal government has announced a new regulation requiring some 5,000 to 8,000 companies to register with the Treasury Department by the end of 2001. The government is creating a national database of companies that exchange currency, cash checks, and issue money orders, trying to hinder drug traffickers who launder money through these entities. The companies will have to keep internal lists of their outlets and agents, and monitor which locations handle over $100,000 in transactions per month."543 The monitoring of all transactions is already here. The last few pieces to assure tracking and to penalize those who do not spy for the government are almost all in place.

541. McConnaughey, Janet. "Vaccine against cocaine, other drugs in works." *Anchorage Daily News,* Aug. 24, 1999. EPI562
542. ONDCP, CTAC. CTAC Research & Development: Behind the Headlines, Ahead of the News. http://www.whitehousedrugpolicy.gov/ctac/page03.htm EPI302
543. Fields, Julie. "U.S. Will Monitor Check Cashing Firms." *Bergen Record Online,* Aug. 19, 1999. Source: NLECTC *Law Enforcement & Corrections Technology News Summary,* Aug. 26, 1999. EPI891

Another somewhat chilling approach to the bill paying world is automatic payments. "Customers who come to a payment center once a week get a six-digit code when they pay their bills. They punch in the code on the keypad and if it matches the proper code, the light turns green and the car can be started. If more than a week passes without a new code, the light stays red, a buzzer sounds and the car won't start."544 What if you live in Alaska, where we do, and your car stalled at 40 below zero? We need to think ahead of our technology. Because technology gives us new things does not make them either good, right or truly useful when everyone's interest is considered.

Remote Body Surveillance

Now even the body can by monitored at a distance. A new technology was recently patented which is described as follows:

> "An apparatus for measuring simultaneous physiological parameters such as heart rate and respiration without physically connecting electrodes or other sensors to the body. A beam of frequency modulated continuous wave radio frequency energy is directed towards the body of a subject. The reflected signal contains phase information representing the movement of the surface of the body, from which respiration and heartbeat information can be obtained. The reflected phase modulated energy is received and demodulated by the apparatus using quadrature detection."545

Looking into your toilet and house are one in the same for the new technology mix. It is funny that toilets are one of the top targets of law enforcement when raiding a house. "Today, police here are able to use the CTAC-developed Data Locator System to check connected government agencies' public documents for information on suspect locations and persons. Burlington police lieutenant Darren Grimshaw directed the planning of the drug house raid pictured here and made routine use of the Data Locator. 'As a result we knew the number of rooms and the number of toilets in the place and were able to make sure we had officers assigned to hit those locations so fast that it was impossible for the suspects to flush the evidence.'"546

DNA Dog Tag

"Computers and information technology may be bringing average people increased comfort and convenience, but they are also quickly making the concept of privacy extinct. Although many believe that the greatest threat to privacy is the media, the real culprit is the long line of recorded transactions that occur every time a purchase, phone call, or email is made or written. Not only is there a

544. Hyde, Justin. "No pay, no start." *Anchorage Daily News*, Aug 14, 1999. EPI143
545. US Patent #4,958,638, Sept. 25, 1990. Non-Contact Vital Signs Monitor. Inventors: Sharpe et al. Assignee: Georgia Tech Research Corporation. EPI1126
546. ONDCP, CTAC. Brownsville, Texas and Burlington, Iowa: "Tech Transfer Stories from the Mean Streets." http://www.whitehousedrugpolicy.gov/ctac/page14.htm EPI301

record of most everything we currently do available to employers, law enforcement, and government, but there will soon be a record of who we are. America, Britain, Canada, and Australia are all in the process of starting national DNA databases of convicted criminals; however, most analysts believe this will slip over into the general population as the desire to make crime virtually obsolete is too strong for any government to ignore. On that note, the article says that tiny, powerful microphones and video cameras already have the capability to conduct surveillance at any location at any time. Video cameras are already filming public streets in most large cities, and the field of biometrics allows for an inexpensive and fool-proof method of personal identification from virtually any measurable physical characteristic, such as a person's eyeballs. Market solutions such as infomediaries (companies that protect consumer privacy by brokering information between a person and various companies) and encryption will not be enough to stop widespread surveillance and government databases. There will be no one solution to reverse the slow death of privacy, as technology is most likely too powerful and widespread to save the already decaying concept of being left alone."547

Sworn to uphold the law, "The leadership of a worldwide association of police agencies has called for collecting DNA samples from all crime suspects when they are arrested. The International Association of Chiefs of Police unanimously endorsed a resolution calling on the federal and state governments to pass broad DNA collection plans."548 Suspects? What will happen when mass arrests of government protesters occurs or when suspects are found to be innocent? What happened to the presumption of innocence? Should law enforcement be permitted this intrusion into a person's body without consent? "The International Association of Chiefs of Police's (IACP) executive committee has voted unanimously to approve New York City police commissioner Howard Safir's resolution, which calls for states and the federal government to fund DNA sampling of criminal suspects upon first arrest, much in the same vein that arrestees are fingerprinted. At the moment, the FBI coordinates a DNA sample database contributed to by 14 states, but sampling is done on a state-by-state basis, and generally only from those individuals convicted of crimes."549,550 It seems that government never stops attempting to establish a state control of everyone through invasive means or by any means which are possible for technology. In 1999 a new bill was introduced in Congress to provide "over US$45 million in federal funds to spur police into taking more DNA samples and using them in criminal investigations. The FBI in late 1998 finished constructing a

547. *Economist.* "The Surveillance Society." May 1, 1999. Vol. 351, No. 8117. Source: NLECTC *Law Enforcement & Technology News Summary,* May 13, 1999. EPI973
548. *New York Times.* "Police association joins in call for earlier DNA sampling." *Anchorage Daily News,* Aug. 16, 1999. EPI201
549. Chen, Hans H. "Police Chiefs Call for DNA Testing." *MSNBC Online,* Aug. 18, 1999. Source: NLECTC *Law Enforcement & Corrections Technology News Summary,* Aug. 26, 1999. EPI890
550. Commonwealth Biotechnologies, Inc. "Genetic and CODIS Identity Analyses." http://www.cbi-biotech.com/genetic.htm EPI1221

massive database with about a million entries..."551

In 1999, "As a number of states begin to broaden the use of DNA technology in law enforcement, New York Governor George Pataki proposed legislation Wednesday to increase the state's ability to take DNA samples from individuals convicted of a felony or attempted felony...Forty-three states already take some DNA samples, the majority from sex offenders, but more and more states have been building DNA databases. Last October, the Federal Bureau of Investigation launched its DNA Index System that allows nine states to share information, with 21 preparing to participate. Some oppose the breadth of potential information available in a DNA sample, including New York Civil Liberties Union executive director Norman Siegel, who questions whether the practice conflicts with the Fourth Amendment's constitutional protection from unreasonable search and seizure."552 The issues are many, but the important one is: should these samples be taken before or after conviction and if a person is later exonerated what should happen? "Using FBI-created software, states have matched DNA left at crime scenes to 425 other crimes in the last six years. State crime labs have collected 609,000 DNA samples, but fewer than half are ready to be used, according to the FBI."553

The pressure for expanded DNA databases comes from law enforcement agencies around the country and the United States Department of Justice has yielded to that pressure. "Establishing DNA sampling as a standard arrest procedure could protect the innocent as well as convict the guilty, members of a Federal commission to study DNA databases said Tuesday. Attorney General Janet Reno requested the 22-member Justice Department commission after repeated nudging from law enforcement officials across the country. Indeed, Louisiana recently adopted a law authorizing DNA testing of anyone arrested, and the New York City Police Commissioner has expressed support for the idea."554

"Unlike 'mug shot' photos and ink-and-paper fingerprinting, DNA identification involves taking human tissue. But few safeguards for it exist: of the 50 US states that demand DNA from criminals, only three prohibit the unauthorized release of that genetic material. The privacy implications are staggering."555 In 1999, The FBI's DNA advisory panel during a meeting "questioned whether sufficient safeguards exist to prevent the unauthorized access and release of DNA samples all 50 states collect from crime scenes and some criminals."556

551. McCullagh, Declan. "Expanding the DNA Database." *Wired News*, Nov. 17, 1999. EPI1222
552. Anderson, Lisa. "Pataki Seeks DNA Collection." *Chicago Tribune Online*, Jan. 7, 1999. Source: NLECTC *Law Enforcement & Technology News Summary*, Jan. 21, 1999. EPI1045
553. Willing, Richard. "Nationline: FBI and DNA." *USA Today*, Nov. 20, 1998. Source: NLECTC *Law Enforcement & Technology News Summary*, Nov. 30, 1998. EPI1064
554. *New York Times*. "Panel Offers Case For More DNA Tests." March 2, 1999. Source: NLECTC *Law Enforcement & Technology News Summary*, March 4, 1999. EPI1013
555. McCullagh, Declan. "The Marker of a Criminal." *Wired News*, Nov. 19, 1999. EPI1226
556. McCullagh, Declan. "What to Do with DNA Data." *Wired News*, Nov. 18, 1999. EPI1225

The use of the technology for instant information at the scene of a crime is also now possible. "Lawmakers (sic) may no longer have to wait weeks for labs to analyze DNA information. The National Institute of Justice has unveiled new technology, the Forensic DNA Chip, that will allow law enforcement officials to use DNA information at crime scenes. Law enforcement officials will be able to determine who was at a crime scene by inserting on-scene evidence into one end of the chip, which would be placed in a small computer mounted within a police vehicle. The computer will have access to a national criminal database, and will be able to search the DNA profile identified on the chip."557 The technology will be introduced into the field within a few years. "A computer-linked microchip device smaller than a credit card may soon allow DNA analysis to be conducted at crime scenes within seconds, the National Institute of Justice announced at the Conference of the Future of DNA on Monday. "This vision is very close to reality."558 "A team of engineering and genetic specialists at the University of Michigan announced their invention of a chip that automatically analyzes DNA samples and reports the results electronically...Analysis using the chip would require no lab work."559

It is the reach which is of great concern as well. The direction of the technology is such that it may not be far into the future when a DNA sample is required at birth. "Although the technology presents enormous possibilities in law enforcement, civil libertarians warn against practices such as mass DNA testing in crime scene areas and the maintenance of a DNA database on all citizens regardless of their criminal record. The American Civil Liberties Union recently offered to support the employees of a Massachusetts nursing home where a near-comatose woman was raped and impregnated earlier this year. The district attorney's office began collecting voluntary samples from men on staff, and indicated that non-cooperative employees might be subject to a warrant."560

In other developments, a tool for self protection has also been designed to make use of this technology in conjunction with pepper spray. "Pepper spray has its advantages, but imagine a self-defense tool that enables its user to both ward off an assailant and collect a DNA sample from the thug for law enforcers. Defender DNA, manufactured by Apik Inc. in Melrose, Florida., is a pocket-sized personal protection device that appears much like a small flashlight and works using a combination of electronics and biotechnology. When users strike an assailant with the device, it emits a loud alarm and extends a small needle-like probe to extract a small amount of

557. *Law Enforcement Technology*. "New DNA Technology." July 1999. Vol. 26, No. 6. Source: NLECTC *Law Enforcement & Technology News Summary*, July 15, 1999. EPI929
558. Willing, Richard. "Science at the Crime Scene: Device Will Test DNA in Seconds." *USA Today*, April 4, 1999. Source: NLECTC *Law Enforcement & Technology News Summary*, May 6, 1999. EPI977
559. *United Press International*. "Breakthrough in DNA Testing Announced." Oct. 22, 1998. Source: NLECTC *Law Enforcement & Technology News Summary*, Nov. 5, 1998. EPI1086
560. *Business Wire*. "Identicator, Key Tronic Sign Deal to Market Peripherals With Fingerprint ID Technology as a Network Security Device." Nov. 10. 1998. Source: NLECTC *Law Enforcement & Technology News Summary*, Nov. 19, 1998. EPI1067

blood or tissue from the attacker. Apik officials say that any purchasers who actually use their Defender DNA in a criminal case can have it replaced free of charge to avoid messy DNA mix-ups."561

At the end of 1999, "The International Biometric Industry Association (IBIA) recommended regulations for biometric data including retina scans, voice recognition files, fingerprint files, and DNA information Friday. IBIA spokesman Fred Norton said the group welcomes government regulation, and has not come up with specific suggestions because of 'too many different scenarios.' Simon Davies, a privacy and data protection instructor at the London School of Economics, says that a variety of scenarios will make regulation ineffective. In California, however, there are at least 12 bills pending that concern biometric privacy, including one by State Assembly-woman Liz Figueroa (D-Fremont) that would fine insurance companies and others $1000 for using, selling, or sharing medical information, including biometric data, without consent. Many worry about the potential misuse of biometric data, but Norton believes the problem can be solved."562 $1000 dollars for breaking into a person's body seems like a bargain for the industry in the few instances they are likely to get caught abusing people's trust. A person's DNA signature should have the same legal protections as his private home. What could be more private than the DNA which forms the chemical essence of who we are as individuals? "California governor Pete Wilson recently expanded the list of crimes that warrant a blood draw for the state's DNA database. Official estimates report about 60,000 more felons, including those convicted of manslaughter, kidnapping, mayhem, torture, and felony spousal abuse, will be required to provide blood samples."563

At the same time there have been instances where old cases should be reexamined to determine if the right decision was made when people were originally convicted. At the same time the use of this kind of evidence does have value in getting the "bad guys." "DNA analysis has taken center stage in the robbery, false imprisonment, and rape charges against 39-year-old Charles Peterson. Sgt. Mike Puetz was working undercover when he spotted Peterson expectorating from his motorcycle, and leapt to wipe the shady character's spittle from the St. Petersburg, Florida, street. Puetz turned the evidence over to the state police for DNA testing, where the samples matched those taken from victims in a brutal sexual battery case."564

DNA is our own personal signature and should remain personal. Occasions for DNA extraction should be strictly limited by law to suspects of serious crimes. Justice and security can have balance

561. *CIO*. "When Pepper Spray Isn't Enough." Sept. 15, 1998. Vol. 11, No. 23. Source: NLECTC *Law Enforcement & Technology News Summary*, Oct. 16, 1998. EPI1095
562. Kirtz, Heidi. "Boosting Biometric Privacy." *Wired News*, March 30, 1999. Source: NLECTC *Law Enforcement & Technology News Summary*, April 8, 1999. EPI996
563. *Government Technology*. "California's Criminal DNA Database Expanded." Dec. 1998. Vol. 11, No. 16. Source: NLECTC *Law Enforcement & Technology News Summary*, Dec. 31, 1998. EPI1051
564. *Houston Chronicle*. "Spittle on Street Leads to Slaying Suspect's Arrest." Oct. 24, 1998. Source: NLECTC *Law Enforcement & Technology News Summary*, Nov. 5, 1998. EPI1081

in a technological world while assuring a very basic element of democracy – privacy.

Drug Wars

"U.S. Customs Service agents are outsmarting cross-border drug smugglers by using the latest technology. Recently, by employing an X-ray matching the size of a car wash, agents near the border town of Laredo, Texas, noticed 5,600 pounds of marijuana worth $7.8 million stashed discreetly in a tractor trailer. Another tool law enforcement agents find useful is a thermal imaging camera – the size of an average camcorder – that can detect temperature changes as slight as one-quarter of a degree and creates an image of light and dark contrasts. This device can be used to detect marijuana growing – given away by grow light heat – in a person's home, or locate potential snipers during nighttime border patrols."565 "Police often use thermal imaging devices to detect grow lights used in indoor marijuana-growing operations to establish probable cause for search warrants, but use of the devices has implications regarding Fourth Amendment limitations on police activities, writes attorney Douglas A. Kash. Some thermal imagers are as small as 35-millimeter cameras and can detect slight temperature differences up to a half mile away."566 These technologies again must be treated with care in their use so that the innocent are not held hostage by the guilty. The increased use of technology can cause the innocent to be subjected to "looks into their homes" which can lead to serious mistakes. Aggressive systems designed for use against criminals should not be targeted against ordinary citizens without clear causes of legal action.

"Prison officials are buying six drug-detection machines, called Itemisers, that will allow guards to wipe an inmate or visitor with paper, place the sample into the machine, and find out whether the person tests positive for drug residue. According to officials, around 6 percent of the state's 16,500 inmates tested positive for drugs in 1999. The machines will cut back on the flow of drugs into prisons."567 There is an old saying when someone is bothering a person, "get out of my hair." It never made sense until the government decided to get into it. "However, hair can be used in law enforcement investigations; the National Institute of Justice and the Pennsylvania Department of Corrections (PADOC) have teamed up to include hair testing of inmates, as well as use the test to evaluate other ways of keeping prisons drug free. Ion-scan testing of visitors and employees, use of drug dogs, and random searches seem to be working, according to PADOC director of research Andt Keyser. However, hair tests might be better for probation tests, which are easier to falsify, Keyser explained. Hair is also good for pre-employment and random drug

565. Sprinivasan, Kalpana. "Technology Deployed in Drug War." AP, June 2, 1999. Source: NLECTC *Law Enforcement &Technology News Summary,* June 3, 1999. EPI955
566. Kash, Douglas A. "Thermal Imaging Devices Heat Up 4th Amendment Issues." *Police,* July 1999. Vol. 23, No. 7. Source: NLECTC *Law Enforcement & Technology News Summary,* July 8, 1999. EPI934
567. *USA Today.* "Tennessee: Nashville." July 19, 1999. Source: NLECTC *Law Enforcement & Technology News Summary,* July 19, 1999. EPI922

checks because it will show a pattern of use, according to Psychemedics' Mike Lamb, rather than miss the candidate who abstained specifically for the employment screening. Retests usually have the same result because they test a long period of time."568

"The assassination of major international drug traffickers is under consideration by senior U.S. officials devising aggressive new measures against them. Encouraged by the President, Congress and a new drug czar demanding stepped-up efforts to deal with narcotics control, a National Security Council-convened team is seriously considering 'military operations' including such actions against 'a number of high-value targets,' according to a senior administration official."569 This violates every value of our nation. We can not act as international vigilantes, invading the sovereign territories of countries to kill their citizens, without desecrating our own heritage. All men and women are innocent until proven guilty in a court of law by a jury of their peers. This is our system and it is not perfect, but we must not ignore our foundational laws. Everyone, whether Jew, Moslem, Christian, Democrat, Republican, poor, rich or, yes, even a suspected drug dealer, is entitled to due process under the law. If we lose sight of the law and our foundational beliefs, then democracy is indeed lost. What we will put in its place is a new kind of slavery. A slavery which removes every grain of our privacy and personal freedom from us and makes us instead conforming drones of a world order which will define "normal" in some dark closet and impose that "normal" on us all. Is this freedom? Is this justice? Is this what we have become – so filled with fear that we will give up the essence of our souls in exchange for a false security? All new technologies will be overtaken by better ones. All safeguards will be worked around by those with the will to do so. In the process of realizing this, are we prepared to give up our privacy? We hope your answer is the same as ours – a resounding No!

Lock Down

"'Domestic Terrorism: Is Your Corrections Facility Ready?' High-risk inmates pose special problems for correctional facilities, and steps must be taken to ensure the safety of staff, facility, and the inmate. High risk individuals should be housed in a high security area that is closely monitored, and attempts to keep the inmate isolated should be a priority."570 There is an increasing problem of violence and horror within our prisons. "Three Corcoran State Prison inmates infiltrated on Saturday a recreation yard for the facility's most famous and protected prisoners, beating up mass murderer Juan Corona, destroying Charles Manson's guitar and causing the staff to

568. Hoffmann, John. "Beware Hair." *Corrections Technology & Management*, March/April 1999. Vol. 3, No. 2. Source: NLECTC *Law Enforcement & Security Technology News Summary*, May 6, 1999. EPI982

569. Greve, Frank. U.S. weighs assassination of foreign drug traffickers. *Philadelphia Enquirer*, June 10, 1989. EPI572

570. Webb et al. "Domestic Terrorism: Is your Corrections Facility Ready?" *Sheriff*, March/April 1999. Vol. 51, No. 2. Source: NLECTC *Law Enforcement & Security News Summary*, April 29, 1999. EPI986

deal with the latest in a number of security violations. Shortly after noon on Saturday, the three inmates scurried through an unlocked emergency door that separates two recreation areas...Corrections Department spokesman Tip Kindel said the emergency exit sensor malfunctioned."571

Rehabilitation is a sad joke, with up to 85% of serious offenders returning for a second sentence. Crime requires justice: punishment, and if it is possible, a change in behavior, not "rehabilitation." Change can be imposed by drugs, mind control techniques or by a person's sincere desire to change. We can not impose change no matter how bad the situation is, because real change is, at its root, a spiritual matter that can only last if it occurs from within a person's very soul. We can mask the soul in a cloud of "rehabilitation" drugs or other invasions but it will not last. The prison of the soul is the dark heart of crime. The inner thoughts which are acted out in the world is the point where change can occur. In each personal disaster there is great opportunity to refocus energy. When people are at their worst they can move on toward their best. They have to want to change for it to happen.

Computers Cry for Help

We really like the following story but, once again, what can be used to catch a thief can be used to keep track of an innocent person using their portable computer as they travel around the world. "Law enforcement officials are having success recovering stolen computers, thanks in large part to new technology from Absolute Software Corporation. The company has introduced CompuTrace, a computer tracking system that allows a PC to call a monitoring center and leave its address. The Know County, Tennessee, Sheriff's Department was involved in a case in which a PC owned by First Tennessee Bank in Memphis was stolen out of the car of an employee on Dec. 3, 1997. Five days later, the monitoring center received a silent call from the computer, and tracked the location of the PC."572

Other Strange Engagements

"The CIA and U.S. intelligence community will start monitoring the environment more closely as a predictive measure for global conflict. 'Environmental trends, both natural and man-made, are among the underlying forces that affect a nation's economy, its social stability, its behavior in world markets and attitudes toward neighbors,' John Deutch, director of central intelligence, said July 25," (1996).573 This will become an increasingly important issue in conjunction with the destabilization of economies because of climate

571. Arax et al. "Prisoners Attack Safe Unit's Inmates." *Los Angeles Times,* March 16, 1999. Source NLECTC *Law Enforcement & Technology News Summary,* March 18, 1999. EPI1004
572. Verdeyen, Vicki. "Changing the Criminal Mind." *Corrections Today Online,* March 9, 1999. Source: NLECTC *Law Enforcement & Technology News Summary,* March 11, 1999. EPI1009
573. *Defense News.* "CIA Views Environment As Conflict Barometer." July 29-Aug. 4, 1996. EPI605

and weather changes which are accelerating globally. These factors will become significantly pronounced within the next two decades. These forces will create new instabilities as well as drive the search for technological and scientific solutions. Controlling the weather will become a factor as the science advances. As the next chapter will show – that science is advancing.

"Frustrated by restrictions on using military force against terrorists, the United States is turning to a lower-profile tactic. The CIA calls it 'disruption" – working with foreign law enforcement services to harass and hamper terrorists around the world before they can pull off major attacks."574 These are old tactics which are being brought into the drug wars and the war on terrorism but they have been around since wars began.

"France freely admits that it conducts intelligence operations against its allies and its enemies on the 'economic front.' As we move into the information age, distinction between enemies and friends begins to blur. The Japanese place a high priority on intelligence gathering through both 'official' and corporate intelligence networks. Japan's international telecom carrier, NTT, routinely cooperates with Japanese intelligence to tap the phone lines of competitors."575 The United States does the same. Allies and adversaries become confused when a new kind of "state" emerges – the multinational corporations which use governments as tools for their self interests at the expense of real public policy. Governments are elected by the money of special interests around the world. These interests are confusing what is right and what is wrong and substituting greed as the controlling factor of international politics. It is time for the revolution of the heart to replace the decay of greed.

The CIA used to hide behind others in setting up their business enterprises to support both their legal and illegal activities. Today they go to Silicon Valley and create companies in the open light of day and no one blinks an eye. "Temporarily forgoing its clandestine ways, the CIA is setting up a venture-capital firm in California's Silicon Valley to invest in young companies that are developing technologies with potential uses for spying."576

Smart Guns?

"Philadelphia City Councilman Angel Oritz and Mayor Edward Rendell have proposed a bill which would prosecute all owners of guns not equipped with 'smart-gun technology.' Oritz and Rendell propose that the bill go into effect in 2002, when they believe that gun manufacturers will have, and be able to use, smart-gun technology. As of yet, however, manufacturers say that smart-gun

574. Diamond, John. "CIA wields new secret anti-terrorist weapon: Disruption." *Anchorage Daily News*, March 1, 1999. EPI629
575. Littleton, Matthew J. *Information Age Terrorism: Toward Cyberterror.* Dec. 1995. http://ndcweb.navy.mil/concepts/terror/terror.htm#Chap1 EPI785
576. Raum, Tom. "CIA starts venture-capital firm to invest in new technologies." *Nandotimes.com*, Sept. 29, 1999. EPI828

technology is unreliable due to erratic recognition devices."[577] "Gun control activists in New Jersey are promoting legislation requiring that all handguns sold in the state be loaded with owner-controlled technology. The Smart Gun, developed by Colt Manufacturing Co., is inoperable without a coded transmission from a transponder in a wristband worn by the registered owner to a receiver in the grip of the gun."[578] "New Jersey Senators. Richard Codey (D-West Orange) and Joseph A. Palaia (R-Ocean Township) are pushing a bill requiring all guns in the state to use 'smart technology.' So far, opponents have had a field day with the issue, pointing out that no such technology exists. However, Colt is producing such a gun – which may be available in a matter of days – that utilizes a radio transmitter worn on the wrist or jacket."[579] 1999 has been an threshold year for new technology. When the villain breaks into your home we hope everyone remembers their code, has their radio transmitter close by and that the technology doesn't "jam."

New World Order

"The FBI report analyzes 'the potential for extremist criminal activity in the United States by individuals or domestic groups who attach special significance to the year 2000,' the bureau said in a written statement. 'The significance is based primarily upon apocalyptic religious beliefs or political beliefs concerning the New World Order conspiracy theory.'"[580] There will probably be extremists among such people, just as there are in all aspects of human interaction. The true paranoids are not the individuals in these groups that do not trust the government. The true paranoids are found in the governments which hatch oppressive, inhuman schemes. Governments have lost sight of their true missions in their rush to serve their real modern masters – the economic interests which harvest the wealth of modern nations at the expense of the people.

577. Sprenger, Polly. "Philly Shoots for Smart Guns." *Wired Online*, May 1, 1999. Source: NLECTC *Law Enforcement & Technology News Summary*, May 6, 1999. EPI978
578. Greider, William. "Will The Smart Gun Save Lives?" *Rolling Stone*, Aug. 6, 1998. No. 792. Source: NLECTC *Law Enforcement & Technology News Summary*, Oct. 16, 1998. EPI1096
579. Aseltine, Peter. "'Smart Gun' Could Blast Hole in Critics Argument." *New Jersey Times Online*, May 31, 1999. Source: NLECTC *Law Enforcement & Technology News Summary*, June 3, 1999. EPI960
580. Sniffen, Michael J. "FBI to dish out advice against attacks." *Anchorage Daily News*, Oct. 21, 1999. EPI1172

Chapter 9

Strange Technology

"Others are engaging even in an eco-type terrorism whereby they can alter the climate, set off earthquakes, volcanoes remotely through the use of electromagnetic waves."[581]

Technology revolutions have impacted this generation unlike any other in tremendous ways. The Revolution in Human affairs is the bigger revolution. As we consider the future, we can look forward to changes in the direction of human energy and creativity. Each new innovation has incredible possibilities in terms of its impacts on people and future events. We have the potential to create change as we each gain an understanding of the possibilities along with the wisdom to act on the knowledge in responsible ways.

This chapter pulls together a number of technologies and military approaches to these innovations. While previous chapters bring forward the detail of many of the technologies which deserve public discussions, this chapter presents a number of other areas of importance which are in need of consideration.

The framework within which some of these technologies will be used is embedded in the concept of Special Operations and other limited military operations. These operations present new, and different, challenges to overcoming the limits of the conflict while shifting from war fighting to policing. Are these the appropriate uses of the technology? Are these the directions we want our militaries to go? What technologies should be used? Who will decide the nature of weapons deployments into the battlespace?

"Special Operations and Military Operations Other Than War – *SO/MOOTW*

4.7 Special Operations and Military Operations Other Than War; *Operational Task: Acquire, maintain, and disseminate information to plan and execute SO/MOOTW.*

<u>4.7.1 Description of</u> Operational Task. The special operations

581. DoD News Briefing, Secretary of Defense William S. Cohen. Cohen's keystone address at the Conference on Terrorism, Weapons of mass Destruction, and U.S. Strategy. April 28, 1997. EPI317

mission, more than any other military mission, must be carried out successfully the first time. As history has shown, the failure of such a mission, usually a one-of-a-kind mission carried out by small elements, becomes a national embarrassment and international news at a minimum. Thus the forces must be fully equipped, very well informed, completely trained, and thoroughly rehearsed. In an age of many rapidly evolving technologies on both sides, this is a significant challenge. As an example, the Special Operations forces typically had the night as their friend. That is, they could operate with relative impunity at night. Advances in night vision equipment and in radar have given our adversaries the capability to detect special operations at night and in poor weather, threatening the success of those operations and the lives of the special operations forces (SOF).

The SO/MOOTW mission area includes an ever-increasing series of roles in the modern world. The Air Force missions include:

• successfully, survivably, and perhaps covertly infiltrate and exfiltrate personnel, equipment, and supplies in denied territory.

• Apply precise firepower in small areas to support military entry/exit, and combat non-combat rescue.

• Support counter-terrorism operations.

• Provide counter-drug surveillance and intercept.

• Assist in humanitarian operations.

In fact, SO/MOOTW is the 'now' war, 'fought' every day in many locations around the world. In the future, SO/MOOTW will have an increasing role in the world. with particular emphasis on counter-terrorism and counter-proliferation. One-of-a-kind missions will be the rule, rather than the exception.

4.7.2 Sensor Challenges. The sensors envisioned for the SO/MOOTW missions vary from small, covert unattended sensors up to airborne (on-board and off-board) sensors and even space-borne sensors with a wide range of capabilities. Concentrating on the SO/MOOTW unique sensors, the needs tend to be peculiar to a particular mission, rather than the broad applications associated with other Air Force elements.

4.7.4 Enabling Technologies. The enabling technologies supportive of the Special Operations and Military Operations Other Than War operational tasks include miniaturized versions of many of the conventional technologies.

• Miniature focal plane arrays combined with acoustic/seismic/tilt transducers, and GPS in very small packages.

• New technologies for UGS, including new transducer fabrication and integration methods such as MEMS technology.

• Chemistry lab on a chip technology to produce a micro-miniature effluent and agent detector-reporter.

• Tags made even smaller than current designs (micro-tags) could map water, sewer, and air flows which, in turn, could help map areas and buildings, and even help locate humans and equipment.

• Multispectral apertures (e.g., infrared (IR) with millimeter wave (MMW) and multicolor IR) are required, as are low cost, moderate power, tunable, coherent lasers, and analog-to-digital converters with both higher sample rates and higher linear dynamic ranges.

5.2.3.2 Sensor Concept. This concept brings a variety of sensor and platform technologies together to deal with the difficult and growing problem of buried and hardened facilities. The surveillance of underground facilities is key to monitoring and targeting the activities of enemy leadership, detecting and classifying weapons manufacture and storage (especially NBC materials and weapons), and locating and classifying command and control sites. It is also essential to tracking transfers of personnel (including captives), weapons, and other equipment and to dealing with many counter-terrorism, drug interdiction, and similar tasks.

RF/Seismic Tomography System. The radio frequency/seismic tomography system uses a bistatic approach in which either or both electromagnetic and seismic waves are transmitted from the UAV into the ground to be received by implanted UGS in a tomographic scheme to construct a three-dimensional map of the electromagnetic and seismic properties of the volume observed."582

The above capabilities are immense and will change the shape of conflicts into a future that brings new and an increasingly widely dimensionalized vision of the meaning of Reagan's "Star Wars." The above description establishes the mission scope and then, in very technical generalities describes the many technologies we have discussed in the preceeding chapters. These developments will allow these devices to become smaller and smaller until you could place the power of the computer writing these pages into a device small enough to inject under a person's skin...These possibilities are on the horizon and in the near term will arrive. It is time for the serious discussion to begin.

Micro Machines

"Imagine a machine so small that it is imperceptible to the human eye. Imagine working machines with gears no bigger than a grain of pollen. Imagine a realm where the world of design is turned upside down, and the seemingly impossible suddenly becomes easy – a place where gravity and inertia are no longer important, but the effects of atomic forces and surface science dominate. Welcome to the

582. USAF Scientific Advisory Board. *New World Vistas: Air And Space Power For The 21st Century - Sensors Volume.* 1996. EPI401

microdomain, a world now occupied by an explosive new technology known as MEMS (MicroElectroMechanical Systems) or, more simply, micromachines. Sandia National Laboratories, motivated by a guiding vision for MEMS, has become a recognized leader in this emerging field."583

"Tiny machines no bigger than a fingernail, a grain or rice or a red blood cell have been twirling, buzzing and slithering across the pages of science fiction and research laboratory benches for years. Enthusiasts say they are the advance wave of a technological revolution comparable to the introduction of computer microchips. 'Imagine a machine so small that it is imperceptible to the human eye,' said Al Romig, director of the Microsystems Science, Technology and Components Center at Sandia National Laboratory in Albuquerque, N.M. Sandia, along with other government and private research centers, designs 'microelectronic mechanical systems' – called MEMS or 'micromachines' for short."584

These incredibly small machines are already revolutionizing all areas of technology. As several separated innovations begin to merge, the merger of nanotechnology, biotechnology, communications technology and information technology creates a possibility of living, thinking and decision making objects devoid of a human soul. The decisions which drive the machines will be embedded in their programming by humans who define their missions and uses.

The perceptible size of the technology is mindboggling – "Packed full of sensors, lasers and communication transceivers, particles of 'smart dust' are being designed to communicate with one another. They could be used for a range of applications from weather monitoring to spying. The tiny 'motes' are being developed at the University of California, Berkeley, to produce the smallest possible devices that have a viable way of communicating with each other."585 "The Defense Department and several top engineering schools are developing tiny aircraft that could scout out enemy snipers, sniff poisonous chemicals or locate hostages in occupied buildings."586 The uses are endless and offer great possibilities for the benefit of humankind rather than its demise and destruction by warfare. The positive side of all of these developments are the things which can head us to a greater set of realities than the destruction of values, beliefs, human dignity, and the very planet which provides the sustenance for every living thing.

The power used to energize these tiny machines will either be present or sent in via directed energy sources such as microwaves. "'One of the biggest advantages of using microwave power is that you can make these UAVs smaller and smaller. With a battery, if you

583. *Sandia National Laboratory.* "About MEMS." http://www.mdl.sandia.gov/Micromachine/ EPI840
584. Boyd, Robert S. "Micromachines become reality." *Anchorage Daily News,* May 3, 1998. EPI763
585. Graham-Rowe, Duncan. "Dust Bugs." *New Scientist,* Aug. 28, 1999. http://www.newscientist.com/ns/19990828/newsstory2.html EPI878
586. Boyd, Robert S. "Small planes have big job." *Anchorage Daily News,* March 17, 1997. EPI703

continue making it smaller, you lose power,' Jenn explained."587 Remote power transmission lightens the remote by providing its energy supply. This will create a new generation of servants or warriors for mankind, as we so choose. Will these robots be programmed for peace or war? "The day when tactical mobile robots will serve as military 'point men,' surveying enemy terrain during combat operations, is one step closer to reality with the selection of NASA's Jet Propulsion Laboratory, Pasadena, CA, by the U.S. Defense Advanced Research Projects Agency (DARPA) to lead a consortium to create a miniature tactical mobile robot for urban operations."588

DARPA is also working on the smallest mechanical "bugs," in this case an insect more like a dragonfly."...(DARPA) has selected for negotiation six proposals to develop flight-enabling micro air vehicle technologies as part of the Micro Air Vehicle (MAV) program. ...Micro air vehicles are airborne vehicles that are no larger than six inches in either length, width or height and perform a useful military mission at an affordable cost."589

Macro Machines

The development of nanotechnology allows the larger weapons systems to advance in revolutionary ways. Troops as such are destined to be computer game operators directing remote units which allow the human soldier to remove himself from direct confrontation with an enemy. Immune from the emotion of battle, insulated from the stench of death, the soldier is immunized from the harsh reality of his actions. The horror of war can bring reflections of its awful waste. Will we lose this motivation for peace by reducing soldiers to remote-controlling technicians with power unprecedented on past battlefields? Who will direct the weapons of political control? Will these be weapons of the next revolution, or, better... the inspiration for an alternative revolution – a revolution which will lead toward a thousand years of peace.

"The day is not far off when the computer chip and the robot will be mobilized as the military's shock troops. The essence of this revolution is simple enough. It's the power of the microprocessor – the computer chip – applied to warfare. Which means that, for at least the next 15 years or so, the ongoing RMA greatly favors the world's dominant microchip manufacturer, the United States."590

The ability of larger military systems to overcome their technological limits changes the balance of power at all levels of conflict. "Basically, the Osprey converts from a helicopter into a

587. Kuska, Dale. "Micro-UAV's Possible in Near Future." American Forces Information Service. http://www.defenselink.mil/news/jan1998/n0107998_9801072.html EPI351
588. NASA. JPL To Develop Miniature Robots For Tomorrow's Soldiers. Sept. 8, 1998. http://www.robotbooks.com/war-robots.htm EPI1104
589. Office of Assistant Secretary of Defense (Public Affairs). "DARPA Selects Micro Air Vehicle Contractor." *DefenseLINK News* Press Release No. 676-97. EPI352
590. Watson, Russell and Barry, John. "Tomorrow's New Face of Battle." *Newsweek,* Special Issue. EPI862

speedy fixed-wing plane in 20 seconds, offering new capabilities for troop deployment, drug interdiction and search and rescue, including hostage rescue. 'This is the revolution in military affairs,' said Cohen, describing the aircraft as 'the epitome of what our forces will need and what they will become in the 21st century.'"591 These advances and their implications are understood by the U.S. Secretary of Defense and those at the top of military operations globally. The use of these systems must rest on the platform of human and democratic values and not just "platforms' hanging in space energized to dominate and control.

Urban conflict spilled into the streets of Seattle at the end of the millennium as demonstrations against the globalization of trade was met with a brutal use of "non-lethal" weapons. The globalization of economies through the World Trade Organization has created a counter force – a counterrevolution – which has united the cause of working people with the concept of planetary environmental stewardship. The coalition of human rights, labor, and the environment will propel the next revolution. The battle is engaged, and the masters have removed their velvet gloves. We believe that peaceful means are the route to change and preempts decision makers' ability to use new forms of force. Political control is the outgrowth of technology – political power inherently belongs to the people – all people. The cause of peace remains a noble cause. The models of economic homogeneity and domination of the earth and its diverse people is fundamentally wrong and works against the common interests of individual people and cultures. These cultural foundations are what underly our diversity and personal integrity. Urban conflict, police actions, and international conflicts are being driven by both economic and cultural conflicts which will either accelerate or be replaced with a new concept of freedom and peace which respects the individual's right to self actualization and expression.

Urban War

"While there will continue to be a role for traditional indirect fire systems in future urban operations, the vast majority of urban precision engagement capability in the future will come from robotic or unmanned systems operating within the city. These systems must be capable of semi-autonomous movement within the urban battlespace and be equipped with both high-speed and extended loitering capabilities. These systems must be capable of providing rapid, precise, lethal and non-lethal fires and be able to "occupy" areas of the city that have great military significance, but which do not require a physical human presence.

The 2025 urban warfighter must have an unprecedented level of lethality across a broad spectrum of capabilities. In addition to precision lethal fire provided by his next-generation individual assault

591. Briscoe, David. "Marines Introduce New Helicopter." *Newsday.com,* Sept. 8, 1999. http://www.newsday.com/ap/rnmpwh15.htm EPI1145

weapon, he should be able to access digital, voice, or possibly even thought-activated fires from a variety of robotic systems operating semi-autonomously within the urban battlespace.

He must also have direct access to precision fire support from distant platforms when larger engagement opportunities present themselves. Additionally, the future urban warfighter needs to possess nonlethal weapons to assure crowd control and minimize collateral damage in certain situations.

In addition to that provided by the advanced urban combat vehicle, the individual soldier will require enhanced individual mobility in both the horizontal and vertical dimensions. Some of this capability will be provided by UGVs (unmanned ground vehicles) that will carry much of the load currently borne by the soldier. Small, individual aerial assault systems can also be used to provide soldiers with a dramatically improved ability to move vertically and horizontally. With only small R&D investments, the 2025 Urban Warfighter System can include a Vertical Assault Urban Light Transporter (VAULT) that will give a soldier the ability to "leap" to the top of three- or four-story buildings and "jump" long distances over rubble at speeds of up to 30 knots."592

Geographical Positioning Systems

"Global positioning information comes from a network of 24 Department of Defense (DOD) satellites. Planes, boats, vehicles, and mapping and survey teams can determine their position on earth by using equipment that receives and interprets signals from these satellites. For civil applications, the satellites provide a signal that is accurate to about 100 meters without the use of DGPS.

DGPS is a technology for improving the accuracy of this positioning information. This greater accuracy is potentially useful in such ways as improving the accuracy of maps, enhancing search and rescue efforts, improving navigation in crowded waterways, and helping planes land in bad weather. DGPS increases the accuracy of the satellite signal through the use of earth-located 'base' or reference stations. The cost of these base stations varies from about $10,000 to $200,000 depending on the type of application and communication link needed to get the information to the user."593

The use of highly precision GPS systems allows for energy transfer to remote devices of every size, shape and description.

New Biological Weapons and the Food Supply

The advancement of genetic and biological technologies raises other questions. There is great concern about the use of crops which do not generate seed, for example. This technology is designed to

592. Hahn II, Robert F. et al. "Urban Warfare and the Urban Warfighter of 2025." *Parameters*, Summer 1999. EPI825
593. GAO. Global Positioning Technology. GAO/RCED-94-280. Sept 1994. EPI453

make farmers totally dependent in their food source, creating agraslaves where once there was a self-sustaining agricultural economics. Not every culture wants to become involved with western economic greed and its misplaced values.

There are also the suggestions of introducing biological material into natural environments without a full understanding of the impact on systems which are interdependent. "Deliberately infecting crops, even drug crops, starts us down a slippery slope. It tells the world that using disease to achieve an end is legitimate – and fairly easy. Whatever the UN might say, these fungi certainly look like biological weapons, which are supposed to be prohibited by an international treaty."594 The treaty intent is to forbid such violations of good sense.

EMP – Electromagnetic Pulse Weapons

"NATO has widened its target list to include power plants throughout Serbia. The action has had a serious effect on the general population. Although he was unable to give more details, the weapon is thought to be a so-called 'soft-bomb.' Unlike conventional explosives, these explode in the air above the power plants, showering splinters of graphite onto the electrical transformers below."595 These new systems are ten years old, having been used in Desert Storm. "During Desert Storm, the navy fired hundreds of sea-launched cruise missiles at targets in many parts of Iraq. Most were equipped with high explosives, but some carried small spools of carbon-fiber wire that were dispersed by a small 'burster' charge. These formed long strands, which rained down on electricity sub-stations and other switch-gear and short-circuited them – part of a program to destroy power supplies throughout Iraq."596 Was it really carbon or graphite filaments or was it something else? What was the rest of the story?

Electromagnetic pulse attacks on power plants will shut them down. "The introduction on non-nuclear electromagnetic bombs into the arsenal of a modern air force considerably broadens the options for conducting strategic campaigns. The massed use of such weapons would provide a decisive advantage to any nation with the capability to effectively target and deliver them."597 The military is not alone. Through technology transfers from the Department of Defense and our national research labs, a small version of these systems will soon be available throughout the country for police organizations. The National Institute of Justice issued the following solicitation which illustrates the points:

594. *New Scientist.* "A very unholy war." Sept. 11, 1999. http://www.newscientist.com/ns/19990911/editorial.html EPI559
595. *BBC News.* "Soft bombs' hit hard." May 3, 1999. http://news.bbc.co.uk/hi/english/world/europe/newsid_334000/334361.stm EPI328
596. Bennett, Felix. "New US weapon destroys electrics but not people." *New Statesman & Society,* June 1995. EPI1153
597. Kopp, Carlo. *The E-Bomb - A Weapon of Electrical Mass Destruction.* http://www.infowar.com/mil_c4/mil_c4i8.html-ssi EPI766

"The purpose of this solicitation is to request interested applicants offer their prototype electromagnetic (EM) devices for submission to NIJ's vehicle stopping Phase III testing (i.e., engineering field testing). Three types of EM devices will be considered for testing: electrostatic discharge, non-nuclear electromagnetic pulse, and high power microwave or radio frequency energy. NIJ is presently involved in a four-phase, multi-year effort to investigate the feasibility of EM technology to stop vehicles. Phases I and II are completed. In those tests, the EM devices were applied to targeted vehicles that were either idling or operating on a dynamometer with their engine and drive train engaged. The Phase II test results support the viewpoint that EM technologies have promise for eventual use in stopping vehicles."

ELF Impacts on Marine Life

Another strange and increasingly noticeable effect of our technologies is on other species. These effects are causing scientists to rethink the effects of these new technologies on humans as well. All biological systems are impacted in very different ways. However, the impacts are not well understood by those creating or consuming the technology. Mankind is on the threshold of either being reduced to total political domination or must initiate a cry for counterrevolution by acting on what we each know to be right and true. We can change the direction of our technological advances and insist that higher human values are applied to the decisions effecting them. We are responsible for the decisions our leaders and governments make. We can change the game from within by using existing structures to replace the leadership and activating those with the right heart for service.

Some of the impacts to animals have been noted. "Noise from supertankers, oil exploration and new military sonar equipment scrambles the communications systems of sea life, forcing changes in migration routes and breeding grounds, a new report warns."598 While, "In May 1996, twelve Cuvier whales washed up on the west coast of Greece, just days after NATO tested an LFAS system to detect nuclear submarines. The deep-diving species rarely gets stranded." 599 This was attributed to the NATO test of this new system. Similar technology has been tested off the coast of Alaska, California and Hawaii.

The idea that the brain is effected by magnetic fields is for some scientists a revolutionary discovery within their disciplines. "Caltech researchers have discovered tiny magnetic particles in the brains of humans, similar to those that have been found in other animals. In many other species, such as pigeons, salmon and whales,

598. AP. "Man-made underwater noise threatens sea life." *Anchorage Daily News*, June 28, 1999. EPI167

599. Gong Jr., E.J. "Sounding Out Humpbacks." *ABCNEWS.com*, March 4, 1998. http://www.abcnews.go.com/sections/science/Dailynews/whales0305.html EPI774

the magnetic particles allow the organisms to navigate in Earth's magnetic field, providing an inborn sense of direction."600 "The recent discovery of magnetite in the human brain could change the picture in a very significant way. The magnetic properties of this material, which are very different from those of typical cellular constituents, provide a physically plausible mechanism by which magnetic fields might perturb biological systems."601 "Scientists have discovered that about 60 species of animals can sense the Earth's magnetic field and may use an internal compass to help guide them on migrations that can span thousands of kilometers."602 The Earth's pulse, (7.83hz), is more than some disconnected energy – it is the pulse rate of the creative process in the human brain – the alpha rhythm – the point of beginning. This pulse rate is impacted by the fields we surround ourselves in and it does have an impact on all living things. The end is omega. What shall it be?

Owning the Weather

Nature can also create an electromagnetic pulse which can have the same effects as a weapon. "U.S. satellites and power grids could be disrupted for the next three years because of intense solar activity called Solar Maximum, scientists said Wednesday. Starting next year space storms, radiation showers and affects on power grids are all expected during the upcoming Solar Max period, which is expected to last about three years."603 The fact that nature creates its own space and terrestrial weather has led to a pursuit that one military article referred to as "owning the weather."

"Weather modification was another area of China Lake preeminence. Between 1949 and 1978 China Lake developed concepts, techniques, and hardware that were successfully used in hurricane abatement, fog control, and drought relief. Military application of this technology was demonstrated in 1966 when Project Popeye was conducted to enhance rainfall to help interdict traffic on the Ho Chi Minh Trail."604 Projections into the future are even more incredible:

> "In 2025, US aerospace forces can 'own the weather' by capitalizing on emerging technologies and focusing development of those technologies to war-fighting applications. Such a capability offers the war fighter tools to shape the battlespace in ways never before possible. It provides opportunities to impact operations across the full spectrum of conflict and pertinent to all possible futures.

600. Maugh II, Thomas H. "Caltech Scientists Find Particles In Human Brains." *Los Angeles Times*, May 12, 1992, pp. 3, A21. EPI50
601. *Microwave News*. "JASONs on EMFs and Health." Sept/oct 1993. EPI496
602. Woods, Michael. "Magnet in brain directs research on 'sixth sense.'" *Anchorage Daily News*, Dec. 12, 1992. EPI530
603. Reuters. "Solar Storms Endanger U.S. Satellites And Power Grid." Nov. 11, 1999. EPI1198
604. *The Rocketeer*. China Lake weapons Digest. http://www.nawcwpns.navy.mil/~bronkhor/clmf/ EPI845

Technology advancements in five major areas are necessary for an integrated weather-modification capability: (1) advanced nonlinear modeling techniques, (2) computational capability, (3) information gathering and transmission, (4) a global sensor array, and (5) weather intervention techniques.

As indicated, the technical hurdles for storm development in support of military operations are obviously greater than enhancing precipitation or dispersing fog as described earlier. One area of storm research that would significantly benefit military operations is lightning modification. Most research efforts are being conducted to develop techniques to lessen the occurrence or hazards associated with lightning. This is important research for military operations and resource protection, but some offensive military benefit could be obtained by doing research on increasing the basic efficiency of the thunderstorm, stimulating the triggering mechanism that initiates the bolt, and triggering lightning such as that which struck Apollo 12 in 1968. Possible mechanisms to investigate would be ways to modify the electropotential characteristics over certain targets to induce lightning strikes on the desired targets as the storm passes over their location.

Communications Dominance via Ionospheric Modification. A number of methods have been explored or proposed to modify the ionosphere, including injection of chemical vapors and heating or charging via electromagnetic radiation or particle beams (such as ions, neutral particles, x-rays, MeV particles, and energetic electrons). It is important to note that many techniques to modify the upper atmosphere have been successfully demonstrated experimentally. Ground-based modification techniques employed by the FSU include vertical HF heating, oblique HF heating, microwave heating, and magnetophospheric modification. Significant military applications of such operations include low frequency (LF) communication production, HF ducted communications, and creation of an artificial ionosphere...

An artificial ionospheric mirror (AIM) would serve as a precise mirror for electromagnetic radiation of a selected frequency or a range of frequencies. It would thereby be useful for both pinpoint control of friendly communications and interception of enemy transmissions.

This concept has been described in detail by Paul E. Kossey, et al. in a paper entitled "Artificial Ionospheric Mirrors (AIM). The authors describe how one could precisely control the location and height of the region of artificially produced ionization using crossed microwave (MW) beams, which produce atmospheric breakdown (ionization) of neutral species. The implications of such control are enormous: one would no longer be subject to the vagaries of the natural ionosphere but would instead have direct control of the propagation environment. Ideally, the AIM could be rapidly created and then would be maintained only for a brief operational period.

Artificial Weather. Nanotechnology also offers possibilities for creating simulated weather. A cloud, or several clouds, of microscopic computer particles, all communicating with each other and with a larger control system could provide tremendous capability. Interconnected, atmospherically buoyant, and having navigation capability in three dimensions, such clouds could be designed to have a wide range of properties. They might exclusively block optical sensors or could adjust to become impermeable to other surveillance methods. They could also provide an atmospheric electrical potential difference, which otherwise might not exist, to achieve precisely aimed and timed lightning strikes. Even if power levels achieved were insufficient to be an effective strike weapon, the potential for psychological operations in many situations could be fantastic.

One major advantage of using simulated weather to achieve a desired effect is that unlike other approaches, it makes what are otherwise the results of deliberate actions appear to be the consequences of natural weather phenomena."605

Welcome back to HAARP. Paul E. Kossey is one of the inventors of the original HAARP patents.606 He has also consulted and is credited by NATO France for his research into ionospheric modification for weapons applications. The above description is part of what HAARP is about.

The Russians also are advancing in this area and may even exceed the United States' technology. A Russian company created a plan which "calls for the use of new Russian technology to create cyclones – the giant storms known as typhoons and hurricanes – to cause torrential rains, washing the smoke out of the air. A Malaysian company, BloCure Sdn.Bhd, will sign a memorandum of understanding soon with a government-owned Russian party to produce the cyclone."607 What are the consequences of such manipulations? Does a weather change in one place shift the weather of the world?

605. House, Col. Tamzy J. et al. *Weather as a Force Multiplier: Owning the Weather in 2025.* Air Force 2025, Aug. 1996. EPI316
606. Begich and Manning, *Angels Don't Play This HAARP, Advances in Tesla Tchnology.* 1995, Earthpulse Press. EPI1229
607. *Wall Street Journal.* Malaysia to Battle Smog With Cyclones. Nov. 13, 1997. EPI322

The Politics of the Weather

"Resolved, That it is the sense of the House of Representatives that the United States Government should seek an agreement with other members of the United Nations on the prohibition of research, experimentation, or use of weather modification activity as a weapon of war.

I can see only two circumstances in which weather modification could be usefully employed by the United States as a weapon of war. First, some fundamental breakthrough might make weather modification a weapon of mass destruction.

The other possible use of weather modification is its employment in a covert war. In the future, it may be to a nation's advantage to engage in covert warfare rather than overt warfare to secure national advantages.

One important indicator of a nation's commitment to a ban on weather warfare is the nature of its weather modification research and the manner in which it is conducted. If weather modification research is open and aboveboard, if it is conducted domestically by domestic agencies, then a fairly high level of credibility can be established. If, on the other hand, weather modification research is conducted in secret, frequently outside the country by military or intelligence agencies, then I would suggest that any treaty would not be worth the paper it is printed on."[608]

"Weather Control – Scientists have experimented with weather control since the 1940's, but Spacecast 2020 states that 'using environmental modification techniques to destroy, damage or injure another state are prohibited.' Advances in technology, however, 'compels a reexamination of this sensitive and potentially risky topic.'"[609] It seems that we now have the answer about the treaty and what it means. The congressman was correct in terms of the 1975 hearings on the subject when he suggested that "any treaty would not be worth the paper it was printed on."

Weather War

The military has a great deal to say in this area and has published the following summary:

"POTENTIAL WEATHER MODIFICATION CAPABILITIES DEGRADE ENEMY FORCES

PRECIPITATION ENHANCEMENT

- Flood Lines of Communication
- Reduce PGM/Recce Effectiveness
- Decrease Comfort Level/Morale

608. Hearing Before The Subcommittee On International Organizations Of The Committee On International Relations, House of Representatives, July 29, 1975. Prohibition Of Weather Modification As A Weapon Of War. H. Res. 28. EPI408
609. Lawler, Andrew. "Visionary, Classified Plan Includes Nuclear propulsion." *Defense News*, Sept. 26-Oct. 2, 1994. EPI318

STORM ENHANCEMENT

- Deny Operations

PRECIPITATION DENIAL

- Deny Fresh Water
- Induce Drought

SPACE WEATHER

- Disrupt Communications/Radar
- Disable/Destroy Space Assets

FOG AND CLOUD REMOVAL

- Deny Concealment
- Increase Vulnerability to PGM/Recce

DETECT HOSTILE WEATHER ACTIVITIES

ENHANCE FRIENDLY FORCES PRECIPITATION AVOIDANCE

- Maintain/Improve LOC
- Maintain Visibility
- Maintain Comfort Level/morale

STORM MODIFICATION

- Choose Battlespace Environment

SPACE WEATHER

- Improve Communication Reliability
- Intercept Enemy Transmissions
- Revitalize Space Assets

FOG AND CLOUD GENERATION

- Increase Concealment

FOG AND CLOUD REMOVAL

- Maintain Airfield Operations
- Enhance PGM Effectiveness

DEFEND AGAINST ENEMY CAPABILITIES"[610]

These summary statements are revealing and give clear indications of where we are and where we are headed. The following is another summary which speaks to the present ability of existing technologies in these areas:

610. U.S. Army (Tecom).TTS '97 - Potential Weather Modification Capabilities AF 2025. http://www.tecom.army.mil/tts/1997/proceed/abarnes/sld004.htm EPI437

"SUMMARY

- Major Improvements in Short Term Forecasts by 2010
- 14 Day Forecasts by 2040

CURRENT CAPABILITIES

- Targeted Fog Dispersal
- Local Changes in Precipitation
- Cloud Modification - Surveillance/Coverage
- Hole Boring
- Create/Suppress Cirrus/Contrails
- Ionospheric Modification
- Energy Requirements Too Large For Major Storms
- Treaty Restrictions
- New Weapons Systems Push the Envelope
- The Environment Must Be Considered From the Start of the Concept/Design For All New Weapon Systems

As weapons and other systems become more sophisticated, the atmospheric environment will continue to be a major factor in the usefulness and operational effectiveness of these systems. For this reason it is imperative that atmospheric scientists be brought in at the beginning on any and all new proposed systems so as to avoid the costs of altering or abandoning the system at a later date."611 Are we waking up to the reality?

Stop the Tornadoes & Lightning

"Mr. Clinton said he would support money for a national weather center at the University of Oklahoma, where he hoped research could be done into how to 'deflate the strength of some of these very powerful tornadoes before they hit.' Earlier, on the helicopter tour, the president seemed stunned by the destruction."612

Dr. Bernard Eastlund was the lead scientist in the development of the HAARP technologies. He has completed additional research into these areas for the European Space Agency and others. The research took place at the University of Oklahoma some months before the President's statements. These technologies are already feasible. "Eastlund and Jenkins suggested Saturday that a coordinated system of ground – and/or space-based Doppler radars could warn of an impending storm in time to target and power up an orbiting microwave gun. 'The bottom line is that the elements to actually control tornadoes look like they might be there,' Eastlund says."613

In addition, it will be possible to control lightning for civilian advantage in the future. "If any of these approaches to sparking

611. U.S. Army (Tecom). Summary. http://www.tecom.army.mil/tts/1997/proceed/abarnes/sld021.htm EPI450
612. Rosenbaum, David E. "Clinton Orders Job Relief In Aftermath of Tornadoes." *New York Times*, May 9, 1999. EPI214
613. Discovery News Brief. "Tornados Zapped from Space." June 1, 1999. http://www.discovery.com/news/briefs/brief3.html EPI177

lightning with laser beams ultimately succeeds, application of the technique could be commonplace. Lasers might one day scan the skies over nuclear power plants, airports and space launch centers. And electric utilities of the 21st century, with their growing network of equipment at risk, may finally acquire the means to act on the threat of a gathering storm, instead of being destined to react only after the damage is done."614 It may not be lasers. Who really knows what conductive pathways can be created, but what are the risks of such controlled discharges? What if they begin to shift the energy from higher, more powerful, sources of energy like the ionosphere or magnetosphere? What type of discharges could occur and will they be able to be switched off once the connections are made?

What type of energy are we toying with? "In one day a hurricane would expend 2 x 1019 joules @ 5000 Megatons of TNT. The amount of energy available for man to use is small compared to the energy of these storms. In addition, the effects of adding large amounts of energy to these storms would have unpredictable and possibly undesirable side effects. CHAOS theory shows that small perturbations can end up having large effects, but available atmospheric data and model resolution are currently, and in the foreseeable future, not good enough to predict the outcome of adding energy to these storms."615

New Weapons of Mass Destruction

"There are some reports, for example, that some countries have been trying to construct something like an Ebola Virus, and that would be a very dangerous phenomena, to say the least. Alvin Toeffler has written about this in terms of some scientists in their laboratories trying to devise certain types of pathogens that would be ethnic specific so that they could just eliminate certain ethnic groups and races; and others are designing some sort of engineering, some sort of insects that can destroy specific crops. Others are engaging even in an eco-type terrorism whereby they can alter the climate, set off earthquakes, volcanoes remotely through the use of electromagnetic waves."616 United States Secretary of defense, William Cohen, when he said these words in 1997, put the world on notice of the beginning of a Revolution in Military Affairs.

"Unusual radio waves recorded hours before the Loma Prieta earthquake struck may someday provide a way to warn of an impending temblor, a Stanford University researcher said Thursday."617 These kinds of energy fields may be natural or could be enhanced by mankind either by design or by accident. Once the resonant or harmonic fields are recognized they can be modified by man. Herein is the danger and the light. How we use energy is relevant

614. Diels et al. "Lightning Control with Lasers." *Scientific American*, Aug. 1997. EPI411
615. U.S. Army (Tecom). Storm Modification.
http://www.tecom.army.mil/tts/1997/proceed/abarnes/sld018.htm EPI448
616. DoD News Briefing, Secretary of Defense William S. Cohen. Cohen's keystone address at the Conference on Terrorism, Weapons of mass Destruction, and U.S. Strategy. April 28, 1997. EPI317
617. Stober, Dan. "Strange radio waves preceded big quake." *San Jose Mercury News*, Dec. 8, 1989. EPI323

to each person. We can use our personal energy and the great powers that our creative capacities have provided us with to either breath life into the planet or snuff the remaining life from the face of the earth. We must each decide how we wish mankind to proceed.

Clean Up The Environment With What?

"In developing their high-powered microwave equipment, the Russian scientists discovered the beams destroyed chlorofluorocarbon molecules. The hope, Mr. Tulupov said, is to find a feasible way to remove the chemical before it floats high enough to combine with the ozone in the stratosphere."[618] This was also one of Dr. Eastlund's ideas in his original patents for the HAARP technology.

Natural Power Surges

"The eruptions produce shock waves and atomic radiation that disable satellites before their life cycle has been completed by crippling their onboard computers, rendering them useless. These storms can also cause disruptions of communications and electrical grid systems here on Earth. Forecasts from such an observatory will make space missions, manned and unmanned, safer and more dependable."[619]

"In the past solar activity has disrupted communications, navigation and surveillance systems, and has shut down electric power grids. Recent loss of a communication satellite, Telstar 401, has been attributed to solar emissions which passed the satellite at the time of the failure. Solar max will increase the IR background through which surveillance, threat warning and missile systems must operate. Heating of the upper atmosphere will change the orbits and cause the early demise of some satellites. PL/GP is developing methods to forecast and mitigate these changes."[620] For the last three hundred years, the waxing and waning of sunspots has been quite regular. If this cycle continues, the next solar maximum will be near the year 2000. Scientists at the Space Environment Center are getting prepared for this cycle which will be solar cycle 23."[621] "Damage from past solar storms has been impressive. One of the most notable events occurred on March 13, 1989, when electrical surges broke circuits across the grid of the Hydro-Quebec power company, blacking out the entire province of 6 million people in just 90 seconds. It cost the utility more than $10 million and customers tens or hundreds of millions more."[622]

618. Schmitz, Tom. "Calif.'s Livermore Weighs Plan to Use Russian Technology." Knight-Ridder Newspapers, April 8, 1993. EPI 320
619. Miller, David W. *The New Federalist.* "Mapping Magnetic Fields For Solar Forecasting." Oct. 24, 1999, p. 11. EPI484
620. U.S. Army (Tecom). Solar Activity - Next Maximum: Jan. 1999 March 2000 June 2001. http://www.tecom.army.mil/tts/1997/proceed/abarnes/sld005.htm EPI438
621. NOAA. "Solar Flare Knocks Out Radio Communications." Press Release No. NOAA 96-602, July 11, 1996. EPI404
622. Scripps Howard News Service. "Scientists: Big Storm is Coming." *Sunday Republican,* May 5, 1996. EPI406

Chapter 10

Earth Rising – The Revolution

"in the end they will lay their freedom at our feet and say to us, 'make us your slaves, but feed us.'"

Dostoevsky

The military has announced a new revolution – their revolution is a "Revolution in Military Affairs" (RMA). The first reference to the Revolution in Military Affairs in our research was from a military war college document we discovered five years ago.[623] The RMA encapsulates the idea that technology has changed to such a degree that the very nature of war is forever altered. It suggested that what was coming in new technology could be equated to the introduction of gun powder to Europe a few centuries ago or the discovery of the atom bomb in more recent history. The Revolution in Military Affairs states that these new systems are contrary to American values and that their introduction would be heatedly opposed in the United States. The authors of that paper proposed that in order to introduce these new weapon systems that American values would have to be changed!

It is particularly alarming when military "think tanks" begin to publish material in which they suggest that commonly held national and human values are insufficient to meet the demands of desired military objectives in introducing new technology. What is wrong with this picture? Do these institutions, and their extension to other public enterprises, *reflect popular values or should they be empowered to create popular values?* Are these public and quasi-public institutions, focused on defense and warfare, the right groups to determine values or should they mirror popular values so that a nation's foundational truths are expressed through their national institutions?

> "The buzzwords haunting the Pentagon today are 'revolution in military affairs'. The idea, simply put, is that the same technologies that have transformed the American

623. Metz, Steven and Kievit, James. "The Revolution in Military Affairs and Conflict Short of War." Strategic Studies Institute, U.S. Army War College, July 25, 1994. EPI516

workplace may have no less profound an effect on the American way of war."624

The Revolution in Military Affairs described a philosophy of "conflict short of war" ("terrorism, insurgency or violence associated with narcotrafficking") that requires new weapons and a change in public opinion. The RMA says that this change in opinion does not have to evolve naturally, but can be deliberately shaped by the government. The idea is that belief systems of Americans can be slowly altered to allow the military to introduce new weapons technology which, at this time, would be resisted by most Americans. What this paper puts forward is:

> "In its purest sense, revolution brings change that is ***permanent, fundamental, and rapid***. The basic premise of the revolution in military affairs (RMA) is simple: throughout history, warfare usually developed in an evolutionary fashion, but occasionally ideas and inventions combined to propel dramatic and decisive change. This not only affected the application of military force, but often altered the geopolitical balance in favor of those who mastered the new form of warfare."

The military's authors discuss emerging technologies which may go against Americans' beliefs in such things as the presumption of innocence, the right to disagree with the government, and the right to free expression and movement throughout the world. The examples of technology given include the disabling of aircraft while in flight, resulting in a crash which cannot be connected to the firing of a weapon – in other words "pilot error" or an "unexplained incident" likely would end up in official accident reports. Moving of funds out of "suspect" bank accounts is another technology discussed. Essentially the paper describes numerous ways an adversary could be attacked by these new technologies, how they would violate fundamental laws and how to change the attitude of Americans so that they could be used. What will those with the power to invade the privacy of individuals do and what would form the basis of their decisions? Based on what rationale? Will the holders of this power be trusted by the rest of the population? The military planners have anticipated that the population would answer with a resounding – "NO"! Therefore, they propose a series of events to shift the popular view to the opposite extreme. They propose a revolution in society, based on fear, which will allow for a Revolution in Military Affairs.

Terrorism is defined in one report as "a purposeful human activity primarily directed toward the creation of a general climate of fear designed to influence, in ways desired by the protagonists, other

624. Cohen, Eliot A. "Come the Revolution." Defense and Technology Issue, *National Review*, July 31, 1995, Vol XLVII, No. 14. EPI1260

human beings, and through them, some course of events."625 Under this definition the very condition the military intends to exploit in order to introduce their weapon concepts are the very things that they should be guarding against.

The United States Secretary of Defense, William Cohen, stated that:

> "The best deterrent that we have against acts of terrorism is to find out who is conspiring, who has the material, where they are getting it, who they are talking to, what are their plans. In order to do that, in order to interdict the terrorists before they set off their weapon, you have to have that kind of intelligence-gathering capability, but it runs smack into our constitutional protections of privacy. And it's a tension which will continue to exist in every free society – the reconciliation of the need for liberty and the need for law and order.
>
> And there's going to be a constant balance that we all have to engage in. Because once the bombs go off – this is a personal view, this is not a governmental view of the United States, but it's my personal view – that once these weapons start to be exploded people will say protect us. We're willing to give up some of our liberties and some of our freedoms, but you must protect us. And that is what will lead us into this 21st century, this kind of Constitutional tension of how much protection can we provide and still preserve essential liberties."626

This view reminds us of other periods in history occurring just before the general decline in civil liberties and basic concepts of freedom. This is the context of the current condition of our democracies and this is what must be reshaped.

The authors of an earlier U. S. Army War College paper lay out a fictional scenario where the illusion of the need for a trade-off of individual liberties in exchange for security is presented. In their scenario, a plan to desensitize the population to increasing control and introduction of the new technology, through systematic manipulation and disinformation by the government is described.627

In another section, The Revolution in Military Affairs discusses the reality of the RMA. Even with all the constraints, there is no question in the writers' minds that the new technology should be utilized. But then they present another alternative: "We could deliberately engineer a comprehensive revolution, seeking utter

625. Sloan, Stephen. "Technology & Terrorism: Privatizing Public Violence." *IEEE Technology and Society Magazine,* Summer 1991. EPI127
626. Cohen, Secretary of Defense William S. Hemispheric Cooperation In Combatting Terrorism, Defense Ministerial of the Americas III. *Defense Viewpoint,* Dec. 1, 1998. EPI660
627. Metz, Steven and Kievit, James. "The Revolution in Military Affairs and Conflict Short of War." Strategic Studies Institute, U.S. Army War College, July 25, 1994. EPI516

transformation rather than simply an expeditious use of new technology."

The implications, the tradeoffs and the direction of this technology shift are being developed outside of public scrutiny and behind veils of secrecy. The issue of global terrorism is even leading to the unilateral abandonment of the ABM Treaty "because of the threat posed by rogue states" deploying weapons of mass destruction.

The Revolution in Military Affairs describes "people's wars" as a shift to "spiritual" and "commercial" insurgencies, which they do not define well. They imply that these kinds of "insurgencies" represent national security risks to be defended against, which may be the case, but, who will decide what "people's wars" to fight and who will determine what is "spiritually" or "commercially" correct?

The writer Aldous Huxley, in one of his last essays published before his death, expressed great concern about the direction of technology and its impact on future generations. More importantly, his perspective included an assessment of the impact of technological change on the belief systems of young people. He concludes that, over time, young people would be taught to undervalue the basic elements of democracies and individual freedom mainly through neglect in education curriculums.

Huxley is perhaps best known for his novel *Brave New World,* which first appeared in 1932. This novel was about a scientific dictatorship which sought to control the thought of citizens beyond the wildest dreams of any Hitler, Stalin or modern dictator. By 1958, five years before his death, Huxley had assessed the technology of his day and wrote the essay *Brave New World Revisited.*628 In this work, he addressed the real concerns which were only fictionalized in his earlier book. When *Brave New World* was "revisited," Huxley considered the subject not in fictional form but on a platform of the state of the technologies available in 1958. At that time, he believed that it was forty years ahead of where he expected it to be when he wrote *Brave New World.* "Propaganda in a Democratic Society," "Propaganda Under a Dictatorship," "Brainwashing," "The Art of Selling," "Over-Organization," "Chemical Persuasion," "Subconscious Persuasion," "Hypnopaedia," were all chapters in his essay as were sections delivering solutions including the chapters "Quality, Quantity, Morality," "Education for Freedom," and "What Can Be Done?" The questions and points he raised in the concluding paragraphs included:

> "Does a majority of the population think it worth while to take a good deal of trouble, in order to halt and, if possible, reverse the current drift towards totalitarian control of everything? In the United States – and America is the prophetic image of the rest of the urban-industrial world as it will be a few years from now – recent public opinion

628. *Brave New World Revisited,* by Aldous Huxley, Harper and Row - Publishers 1958. EPI1258

polls have revealed that an actual majority of young people in their teens, the voters of tomorrow, have no faith in democratic institutions, see no objection to the censorship of unpopular ideas, do not believe that government of the people by the people is possible and would be perfectly content, if they can continue to live in the style to which the boom has accustomed them, to be ruled, from above, by an oligarchy of assorted experts."

Granted, Huxley didn't live to see the revolutions in thinking of the decades between 1960 and now. The ideas which concerned him then are the same as those which concern many of us now. The significant difference being Huxley "hadn't seen anything yet" in terms of the technology of the next four decades. Technology had been doubling about every fifteen or so years between 1932 and 1958.

The advances in science in the last 40 years had all been relegated to science fiction of the early 1900s through the 1950s – science fiction which eventually became reality, almost as if whatever the mind of mankind could envision, it could eventually translate into material reality. In fact, some modern and ancient orders of thinkers believe that this is precisely the case in terms of human creativity in the context of the creation itself – the idea that we each individually contribute to the outward expression of the whole of creation. Perhaps they are right? Perhaps they are correct in terms of what we create culturally and societally within all of our institutions.

Huxley goes on to describe the risks of a complacent society which is well fed with the trappings of comfort and how this really emerges as a subservient, almost slave-like society, where people are happy with some form of social order devoid of democratic principles. Huxley quotes Dostoevsky's parable in which the Grand Inquisitor declares, "in the end they will lay their freedom at our feet and say to us, 'make us your slaves, but feed us.'" Consider the state of technology when the Grand Inquisitor is depicted and the transformation of systems of control by the time Huxley completed his work in the late 1950s and the incredible leap in the four decades since. In *Brave New World,* Huxley empowers the new authority – the dictator – with the miracles of science which are used for "manipulating the bodies of embryos, the reflexes of infants and minds of children and adults." Unlike the Inquisitor of ages past merely talking about miracles and mysteries, the new scientific totalitarian uses science to create direct experience through drugs and other means to transform faith into experiential knowledge. Since Huxley sounded the alarm in 1958, the science he feared has become reality and gone even well beyond his imagination of what would emerge in coming decades. As we begin the next millennium we are challenged by the same risks magnified several times. In the last lines of *Brave New World Revisited* Huxley writes:

> "Under a scientific dictator education will really work – with the result that most men and women will grow up to love their servitude and will never dream of revolution. There seems to be no good reason why a thoroughly scientific dictatorship should ever be overthrown."
>
> "Meanwhile there is still some freedom left in the world. Many young people, it is true, do not seem to value freedom. But some of us still believe that, without freedom, human beings cannot become fully human and that freedom is therefore supremely valuable. Perhaps the forces that now menace freedom are too strong to be resisted for very long. It is still our duty to do whatever we can to resist them."

Other thinkers have continued to evaluate the trends in technology, some gaining the attention of mainstream media with their work. One of those not so well recognized by most readers is Zbigniew Brzezinski. For those who may not remember, Zbigniew Brzezinski was the National Security Advisor to President Jimmy Carter, a professor at Columbia University and a recognized global political and social strategist. His views have been reflected in public policy evolution around the world as an insider within the process of the social change which he writes about.

In Brzezinski's book, *Between Two Ages – America's Role in the Technotronic Era,*[629] he presents a rational projection of the future. This book was first published in 1970 and sets out a vision for the future based on the historic development of the present at that time. Brzezinski had a very different view, having personally seen the impact of totalitarian regimes of Eastern Europe and was able to consider the consequences of proliferating technologies on cultures, societies, politics, business and virtually every area of human interaction. His vision of the likely events proved to be true. In fact, as we count down the last few days of 1999 it is clear that much of his vision has been realized. His cautions have also been heard.

Several points are made which warrant even greater consideration today given the state of these technologies and the directions in which our governments are moving. He believed that our rapidly changing global reality would alter the very foundation of our existence. At the same time local social systems would be fragmenting as a result of globalization – and in the process losing the higher aspects of who we are individually. The result of both an explosion in technology globally and implosion of its effects in each locality can only end in insecurity, tension and the stress. He goes on to observe that "Life seems to lack cohesion as environment rapidly alters and human beings become increasingly manipulable and malleable.

629. *Between Two Ages - America's Role in the Technetronic Era.* by Zbigniew Brzezinski, The Viking Press, 1970. EPI787

Everything seems more transitory and temporary: external reality more fluid than solid, the human being more synthetic than authentic. Even our senses perceive an entirely novel 'reality.'" He continues to describe his increasing concern about the possibility of biological and chemical control of populations violating the very foundation of who we each are as individuals. He goes on to quote the ideas of an expert in intelligence control as follows:

> "I foresee the time when we shall have the means and therefore, inevitably, the temptation to manipulate the behavior and intellectual functioning of all the people through environmental and biochemical manipulation of the brain."630

The concerns which are raised by Brzezinski include the ideas of genetic manipulation, chemical control of populations, medical advances, scientific breakthroughs and media manipulation as contributors to a breakdown in self-perceptions. Extensive transplantation of global populations coupled with political, cultural and societal changes through the introduction of technology and access to global communications would create a highly controlled society. This controlled society would be the outgrowth, according to his predictions, of the insecurity and tension created by the impact of change resulting from technology. He analyses the impact of societal disruptions which cause mass migrations of populations and disruption of stable social and cultural structures. He describes the clash of values between young and old, the establishment and its order and new thinking, based on the changes, causing individuals to raise basic questions dealing with who we each are as people within the greater context of the whole presented by unavoidable integration and globalization.

Brzezinski raises the stakes in his presentation of what was theoretical then and is possible now when, in 1970, he quoted the work of Gordon J. F. MacDonald, a geophysicist specializing in problems of warfare. MacDonald believed that in the coming decades technology would advance so that artificially excited electronic bursts of energy could be created which would stimulate the electrified ionosphere above the earth in such a way as to create a returning electronic signal which could interact with the human body in a manner which would create behavioral changes in people over a large geographic area. He said, "...one could develop a system that would seriously impair the brain performance of very large populations in selected regions over an extended period...No matter how deeply disturbing the thought of using the environment to manipulate behavior for national advantages to some, the technology permitting such use will very probably develop within a few decades."631

630. Testimony by Dr. D. Krech, *Government Research Subcommittee of the Senate Government Operations Committee,* as reported by *The New York Times.* April 3, 1968, p. 32. EPI1259
631. *Unless Peace Comes, A Scientific Forecast of New Weapons*, Edited By Nigel Calder, "Geophysical Warfare, How to Wreck the Environment." by Gordon J. F. MacDonald, pp. 181-205, 1968. EPI573

This technology now exists, as the evidence has shown in this book. The other important observation that Brzezinski made was that only the most advanced countries would possess this new technology. For the most part this is true for the most sophisticated technology, but increasingly, over the last few years, the idea that others, including terrorist organizations, might posses this technology leaks to the media and surfaces in congressional inquiries, parliamentary investigations and academic literature, many of which are drawn upon in the context of this book.

We have presented a much different view of the world of possibilities in addressing the technological revolution which might otherwise overwhelm us. We are proposing a counterrevolution. A revolution which empowers individuals to assert increased control over the factors regionally unique to each area of the globe. We propose, that while the need to communicate increasingly in a global context is unavoidable, mutual respect for the differences in philosophy, religion, politics and every other area of human activity is required. What is "good" in the context of one's own region of the planet may ***not*** be "good" within another's cultural context. The global vision presented in the work of Brzezinski suggests a movement toward a global vision and religious philosophy – a dissolution of individual regional differences substituted by an increasing number of external factors which create a homogeneous society and global culture. Our individual experience also shows that since 1970 when Brzezinski's views were being expressed – the reality has begun to form and the resulting tensions are obvious to all. We take a different view, believing that regional belief systems and cultures have a right to their own evolution, or change, according to their own regionalized cultural context, even with increased international interactions. The need for mutual respect for very different groups must be translated into action. Values and beliefs must be respected as constituted regionally, with support from outside interests limited to providing resources when the external values and beliefs line up with those resident within a local population.

The vision of the modern globalist represents, to the greatest extent, the vision of economic interests. Commercial interests do not view humans as individuals but rather as "consumers" – a class of markets and market potentials not yet realized in the developing world or in the parts of the world previously inaccessible to multi-national economic interests. These economic interests eventually exploit the resources, labor and environments of the underdeveloped regions, creating another group of consumers in the process. The result of this shift has been devastating as crime rates increase; poverty rates rise; global hopelessness increases; and, illness and social tensions emerge coupled with the dissolution of local "norms" with the introduction of the conflicts presented by external influences.

The complexity of modern society also creates the need for increased transparency in governments and other institutions vested with public authority. These institutions become impersonal and reduce individuals' overall sense of power to reform or alter the direction of societal development. The sense of powerlessness is amplified in each impersonal interaction. What Brzezinski predicted was that if this feeling of isolation and loss of power were to pervade American culture, then the results would show themselves in withdrawal from political and social involvement and an individual inner retreat and outward conservatism. At the same time, an elite of "educated" people would emerge which is trained in technical areas, and believing that they possess "the answers." These educated technicians in many cases *are not* exposed, in the course of their educations, to the underlying, truly important questions explored in studies of the humanities, philosophy and religion. On the other side of the issue, the "doctrinaire liberal erred in assuming that economic progress would prompt social well-being," which underestimated the psychological and spiritual dimensions. It is these dimensions which we maintain are the foundation of human values and beliefs which when focused in regionally established goals yield culturally compatible results. Through democratic principles, increased transparency in governments and other authorities, open dialogue and mutual respect for the unique differences of each person and population, alternate possibilities are achievable. The old models have failed to create stability and are not balanced by underlying traditional values: they need to be abandoned. Those values begin with respect for personal freedom and individuality, access to justice and the desire to see each person achieve their highest and best potentials, which when expressed in action contribute to the whole of humanity.

The choices that this technology forces deal with the underlying ethical and moral questions regarding its use. Whether these technologies undermine our individual and collective rights will become increasingly clear as the state of the technology advances, presenting the greatest threat to the way of life of every person in every country of the world. The need to maintain traditional values in the front of our technological decisions must be made apparent to all.

Brzezinski goes on to suggest that a more directed and controlled society would be the outcome of our technological advances. He believed that the changes would be slow, gradual and continuous, becoming dominated by an "elite whose claim to political power would rest on allegedly superior scientific know-how. Unhindered by the restraints of traditional liberal values, this elite would not hesitate to achieve its political ends by using the latest modern techniques for influencing public behavior and keeping society under close surveillance and control. Technical and scientific momentum would then feed on the situation it exploits." He went on to suggest that in this condition a charismatic leader could emerge who could exploit the mass media to gain public confidence and would, in piecemeal fashion, transform the United States into a highly controlled society. He further suggests that the liberal left would justify the use of these new technologies as legitimate methods in achieving their version of social control and progress. On the other hand, he equally levels his concerns regarding the conservative right who he suggests are "preoccupied with public order and fascinated by gadgetry," and would be tempted to use these new technologies to effect social control and quell civil unrest, having failed to recognize that social control is not the only way to deal with the impacts of social change.

Chapter 11

Where is the End?

"There is only one thing which will really train the human mind and that is the voluntary use of the mind by the man himself. You may aid him, you may guide him, you may suggest to him, and above all you may inspire him; but the only thing worth having is that which he gets by his own exertions, and what he gets is propor-tionate to the effort he puts into it."

– A. Lawrence Lowell[632]

"As the United States military looks to the future, two themes dominate most projections. The first is advanced technology. Underwritten by the microchip, the technologies of war are changing rapidly. Weapons with microprecision accuracy, supercomputers linked by unlimited bandwidth, platforms providing continuous surveillance of practically any spot on the digitally mapped earth – all are coming into view. These emerging technologies are combining to produce orders-of-magnitude increases in military capabilities. Adm. William Owens, vice-chairman of the Joint Chiefs of Staff, calls this 'The Emerging System of Systems,' spawning a new revolution in military affairs. Understanding the ramifications of this revolution is an immense challenge for US military planners.[633]

Is this the Beginning of the Police State?

The major economic interests on the planet met with strong resistance in the United States when labor, environmentalists, civil libertarians and conservatives opposed to the deindustriatization of the United States and unrestricted exploitation of the rest of the world joined together in protest in Seattle, Washington. The World Trade Organization (WTO) was introduced to a different picture of America and the rest of the world had a chance to see that many of us do oppose the direction in which we are headed.

"A city that prides itself on tolerance and civility finds itself under a general curfew, shaken by violent demonstrations and a police

632. Barnett, Col. Jeffery R. (USAF). Defeating Insurgents With Technology. *Airpower Journal,* Summer, 1996. EPI231
633. Ibid.

crackdown. Arrests resumed shortly after daylight, and close to 300 people were taken into custody by early afternoon in addition to the 68 arrested yesterday. They were handcuffed, photographed and hauled away in buses to an old navy brig."634 "Seattle-King County Public Health Director Dr. Alonzo Plough refuted rumors circulated by some anti-WTO activists that protesters had been beaten while in custody at the King County Jail. He reports that his staff has identified injuries which occured prior to their entry into the jail and documented those conditions with Polaroid photographs."635 "Earlier this evening several protestors migrated from the downtown corridor to the Capitol Hill area. The group of approximately 200-300 blocked the intersection at Broadway and E. Olive Way and Pine Street. Agitators in the group had molotov cocktails and began to incite the crowd, which refused to move after they were given the order to disperse. Some of the agitators threw rocks at the police, compelling officers to utilize crowd control measures. CS gas and OC paint balls have been deployed to help disperse the crowd. The crowd continues to grow and has not dispersed at this time. Reports of property damage and looting by agitators are currently coming in."636

The protests quickly expanded and became unruly and additional orders were issued when the Mayor "expanded his civil-emergency order Wednesday to ban the purchase, sale and carrying of gas masks within city limits. In one area, police waved batons at television cameras and pointed for the cameramen to get away. Officers also were seen spraying pepper spray in people's faces as they ran by."637 "A local gas mask manufacturer is suing the city of Seattle. Gas Mask Incorporated is taking the city to court over its ban on the possession of gas masks this week. It is against the law for citizens to use the masks while police are using chemical agents during riot situations. The company sells its products over the Internet."638 Meanwhile, "The National Lawyer's Guild faxed an angry letter to Schell yesterday afternoon, complaining of excessive police force. The guild...claimed rubber bullets were fired by police at least twice and said officers beat a television cameraman and chased down a woman to dose her with pepper spray. Stamper refused to confirm that rubber bullets were used..."639 While everyone declined to comment on the weapons being used, primarily against peaceful civilians, the TV images were shocking and reminded many of the Martin Luther King riots, riots at the Democratic National Convention and major demonstrations of the 1960s and early 1970s in opposition to the Vietnam War, Watergate and demonstrations for Richard

634. AP. "Seattle 'war zone' filled with riot police, pockets of protest." *CNN.com*, Dec. 1, 1999. EPI1237
635. City of Seattle News Advisory. "Public Health Director Refutes Rumors From Jailed Anti-WTO Protestors." Dec. 3, 1999. EPI1238
636. City of Seattle News Advisory. "Disturbances On Capitol Hill." Dec. 2, 1999. EPI1239
637. CNN. "Seattle cracks down on protestors with second night of tear gas." *CNN.com*, Dec. 1, 1999. EPI1240
638. Yahoo. "Gas Mask Company Sues." *Yahoo.com*, Washington Headlines. Dec. 3, 1999. EPI1257
639. Sunde, Scott. "Chaos closes downtown." *Seattle Post-Intelligencer*, Dec. 1, 1999. EPI1241

Nixon's impeachment. How numb have we all become to civil violence in the past 30 years?

Modern Technical Gladiators

"After a rough, sad week, headlines about 500-plus arrests stir mixed feelings. Their beliefs inspired acts of civil disobedience, and they paid a painful price. They hunkered down to block streets under clouds of tear gas and pepper spray and rubber bullets in the legs. They were not off breaking windows or setting fires."640 "Tear gas, also known as CS spray, or orthochlorobenzalmalononitrile, is a complex organic salt that is quickly absorbed by the body. It affects muscles and glands, and causes watery eyes, runny noses, stinging skin and coughing. In some cases, people suffer from vomiting and diarrhea."641 "Police then made their move to begin clearing downtown. Block by block, officers fired canisters of gas into the crowds with a terrifying boom. Then they shot rubber bullets into the backs of protesters even as they ran away."642 "We thought that the police got a little aggressive," said Rich Lang, Capitol Hill Stewardship Council Chairman. "They seemed to not stop the people who were doing the looting...and turned on the regular citizens."643 The Mayor was hammered. "As bad as last week was for Seattle, it was devastating for Schell. The police-state image he so desperately wanted to avoid when the World Trade Organization came to town had been transmitted around the globe in photographs and video of police in riot gear lobbing tear gas into crowds of demonstrators."644 Welcome to the *Brave New World.*

"A news conference Thursday in which Schell apologized to the innocent victims hit by tear gas and rubber bullets and Stamper admitted being troubled by the behavior by some officers did not sit well with many in the police force."645 "We're not marching for workers' power," a man in a bikers jacket hollered back. "I'm here tonight to defend my neighborhood. I refuse to live in a police state, and I will not back down. We will not be silenced, and we will not be beaten like dogs in our own streets."646 "Late Tuesday, an ultimatum is delivered by a senior Clinton administration official to Schell: Clear out the protestors or the World Trade Organization conference might have to be called off."647

640. Dickie, Lance. "Sideshow punks shouldn't tarnish principled protest." *Seattle Times,* Dec. 3, 1999. EPI1242
641. Hodson, Jeff and Sorensen, Eric. "Police restock depleted tear gas supplies." *Seattle Times,* Dec. 2, 1999. EPI1243
642. Seattle Times staff reporters. "Police haul hundreds to jail." *Seattle Times,* Dec. 1. 1999. EPI1244
643. Robin, Joshua and McOmber, J. Martin. "Capitol Hill residents want probe." *Seattle Times,* Dec. 6, 1999. EPI1249
644. Cameron, Mindy. "Don't expect 'gentle Seattle' to be gentle with its mayor." *Seattle Times,* Dec. 5, 1999. EPI1245
645. *Seattle Times* staff. "Conference ends, protests don't." Dec. 4, 1999. EPI1246
646. Ith, Ian and Burkitt, Janet. "Capitol Hill rally protests police actions." *Seattle Times,* Dec. 3, 1999. EPI1247
647. Ith, Ian and Burkitt, Janet. "Capitol Hill rally protests police actions." *Seattle Times,* Dec. 3, 1999. EPI1247

"On top of allegations that police overreacted to protesters last week with tear gas, rubber bullets and concussion grenades, Siegel said protesters reported abuse at the hands of jail personnel – beatings and denial of medication, food and legal counsel."648 The abuses were incredible and widespread throughout the crackdown. Had military personnel been fully utilized the abuses would have been even worse, based on the lack of police training by military and their conflict of mission. "But a senior King County police official, speaking on condition of anonymity, said even though every deputy on duty in Seattle has been trained to deal with riots, some are going to lose their tempers and act unprofessionally at times. They represent a minority, just as only a relatively small number of protesters are responsible for the worst conduct, the official said."649 "The kick to the groin. The TV replay of a riot-clad police officer pursuing a young protester up a sidewalk and the painful-looking kick has become a symbol for an ugliness many thought they'd never see in Seattle."650 "Downtown's no-protest zone also served to push demonstrators onto Capitol Hill. On Wednesday, residents who tried to get the protesters to leave were gassed and shot with rubber pellets. City officials defend the Capitol Hill confrontation, saying demonstrators had planned to try to seize the East precinct."651 "Angry residents and human rights groups on Thursday demanded a probe of claims that riot police brutalized peaceful protesters as city officials admitted they botched security for an international meeting on trade. Besides pummeling innocent bystanders with teargas and rubber bullets, critics said police... trampled demonstrator's Constitutional rights of assembly and free speech."652

In a hail of public outcry which was still ringing at the end of December 1999, the Seattle "police chief announced his resignation Tuesday, becoming the first political casualty of the violent protests that disrupted the World Trade Organization conference. Stamper said he will cooperate in any investigation of the police department's role in dealing with the demonstrations. However, he declined to answer several questions Tuesday about the rioting."653 The Chief said he would cooperate but much was left unsaid. "In addition to hundreds of very visible Army National Guard troops called-up because of the civil disturbances in Seattle, more than 160 active duty military personnel, including a small number of Special Forces troops, were sent to Seattle by the Defense Department for the meeting of the World Trade Organization."654

---

648. Rahner, Mark. "Most jailed WTO protesters are no longer behind bars." *Seattle Times*, Dec. 6, 1999. EPI1250
649. Ibid.
650. Flores, Michele Matassa. "Seattle left less naive as it counts physical, psychological costs." Dec. 5, 1999. EPI1252
651. *Seattle Times* staff. "How did the police do? Everyone has an opinion." Dec. 6, 1999. EPI1253
652. Stetiewicz, Chris. "Serene Seattle Faces Rights Abuse Allegations." Reuters, Dec. 3, 1999. EPI1254
653. Klahn, Jim. "Seattle Police Head Quits After WTO." AP, Dec. 7, 1999. EPI1255
654. CNN. "Troops sent to Seattle as part of terrorism contingency plan." CNN.com, Dec. 2, 1999. EPI1256

Vietnam

"The American effort in Vietnam was the best that modern military science could offer. The array of sophisticated weapons used against the enemy boggles the mind. Combat units applied massive firepower using the most advanced scientific methods. Military and civilian managers employed the best techniques of management science to support combat units in the field. The result was an almost unbroken series of American victories that somehow became irrelevant to the war. In the end, the best that military science could offer was not good enough.

The American military establishment considers low-intensity conflict to be manifested in four different ways: (1) counterterrorism (assuming there is a terrorist to counter); (2) peace keeping; (3) peacetime contingencies (quick, sharp, peacetime military actions like the air raid on Libya in 1986); and (4) insurgency/counterinsurgency.

The results of the intervention of France, the U.S. and the Russians were decidedly one-sided. The super powers of the first world overwhelmed their opponents with technology, equipment, and manpower and still lost the war. In the war of ideas its really true that the pen is mightier than the sword and ideas are harder than bullets and bombs. The wars were settled politically with the major powers leaving the country over which they fought – with the exception of Chechnya. Chechnya was unique in that it was surrounded by Russian territory: the rebels had nowhere to hide and no one to help them. While they made life difficult for the Russian military they would have a very difficult time in achieving their goals.

A common error in the cases studied here is readily apparent: the intervening power did not consider what it was that the people living in the country wanted for themselves. The major powers seemed to think that they already knew what was best for the common people of the country when in fact the politicians had no real idea what was wanted at the grassroots level.

In short, there is much to be learned by studying the history of insurgent warfare. Chief among them is that technology is not the answer to everything a country tries to do. In the end it is the human element that will persevere; it is the human element that makes the difference in winning and losing an insurgency. Without recognizing this, technology will be to no avail."655

"An obsession with expensive weapons systems is robbing the nation of a chance to recast its defense strategy. The United States will spend more than $265 on defense next year."656 "..the Army's top officials are launching what they describe as an aggressive initiative to make the force lighter, faster and more lethal. They will experiment to

655. Hain, Raymond F. "The Use and Abuse of Technology In Insurgent Warfare." http://www.airpower.maxwell.af.mil/airchronicles/cc/ EPI877
656. Peters, Katherine. "Overkill." *Gov.Exec.com*, Dec. 1997. http://www.govexec.com/features/129751.htm EPI823

develop new battlefield equipment and an organizational structure they hope will ultimately supplant the old division structure that dates back to the 19th century and Napoleon's Grand Armee."657

International Law

"For new weapons, however, even with respect to which no collective pronouncements have been made, each party to the Draft Additional Protocols to the 1949 Geneva Conventions will have to apply the principles of the law of war. This duty is emphasized in the proposed Article 34 of Protocol I: In the study, development, acquisition, or adoption of a new weapon, means, or method of warfare a High Contracting Party is under an obligation to determine whether its employment would, under some or all circumstances, be prohibited by this Protocol or by any other rule of international law applicable to the High Contracting Party."658 "By targeting highly dangerous technologies that could include directed energy, acoustic and microwave weapons, non-proliferation goals have a better chance of being at least partially met, the experts said."659 "The fact remains that the scientists whose livelihoods were created by the Cold War have yet to secure their future in its aftermath. For western governments, the dilemma is how to justify maintaining an expensive scientific resource devoted to the military at a time when no single, clear threat is apparent."660 In what direction shall these technologies go? How shall they be used and what standards will be applied to Americans and others? Should these weapons, prohibited against our external adversaries, continue to be used against our own citizens?

Within these concluding pages of Earth Rising – The Revolution, we propose a number of possible solutions – we note the following:

• We propose that the existing treaties which address chemical and biological weapons be revised to include bans on electromagnetic and acoustic weapons which effect human health or behavior.

• We propose that those treaties dealing with environmental, weather or geophysical manipulation be revised to consider the state of the technology and forbid both cross-border and domestic testing and development of these systems.

• We propose that the issue of privacy as it relates to new technology be considered in terms of the spirit and intent of the law to maintain

657. Richter, Paul. "Army sets sights on war of Future." *Anchorage Daily News*, Oct. 12, 1999. EPI831

658. Roling, Prof. Bert A. et al. *The Law of War and Dubious Weapons*. SIPRI, 1976. Stockholm, Sweden. EPI474

659. Opall, Barbara. U.S. Experts: Focus Arms Control Goals. *Defense News*, Nov. 24-30, 1997. EPI420

660. Shukman, David. *The Sorcerer's Challenge*. Hodder & Stoughton, 1995. EPI1137

the right of people to a private life. We can not allow the invasion of hearth and home by either government or private persons.

• We further propose that an international "whistleblower" system be established. Such a system would include an international ombudsman's office. This office would be a place where any person could go and make any disclosure that the individual feels needs to be a matter of public policy review. The individual bringing the complaint would be protected in terms of confidentiality and supported, if need be, until the issues are resolved. This would allow people of conscience, whether they emerge from the government or private sectors, an opportunity to clear their conscience by speaking out on things that they have seen. This would allow people to bring forward the truth in the greater interests of humanity.

These things represent some of our ideas. We know that they are possible. It is time to expose to the light of truth those things hidden in darkness by institutions which have exempted themselves from public scrutiny for interests which no longer serve the cause of mankind. We call for the initiation of a counterrevolution based on truth, justice and freedom.

The adoption of new technological ideas and their potential impact on our lives is increasingly more important and relevant. The present age of discovery is offering both challenges and points of decision for our future directions. At the same time that we are presented great possibilities in science, we are also faced with the darker side of technology.

Historically, mankind has always coupled political structures with a method of influencing the will of the political leadership. Enforcement has usually come at the end of a gun barrel by developing and maintaining military organizations. Ultimately, the law of force and power determines who's political will shall prevail. In the last hundred years, politicians have increasingly recognized that technology and science offer new mechanisms for control. World War I and World War II brought a clear recognition of the role of science and technology in creating the most powerful war machines in recorded history. Science offers governments a seemingly endless supply of new innovations for maximizing their destructive capabilities.

Innovations based on very complex science have also led to the greatest concentration of wealth ever contemplated by industry. Trillions of dollars have been dedicated to developing the best systems for destroying people and property. Huge transfers of wealth have been made from taxpayers to corporate bottom lines.

At the end of the cold war people were relieved and hoped for a shift in resources to more productive ends. Many of us realized that although the largest risks were seemingly over, new threats to stability would emerge. The need for strong armies is far from over and will likely remain for decades to come. However, it was thought that there would be a "peace dividend" which would translate into tax relief for an overburdened middle-class and technology transfers to industry. It was the hope of many that the technology transfers might give access to new information in pursuit of more humane uses for our technologies. Eight years later, the results of the "peace dividend" are slow to materialize and those which have materialized are headed down the wrong path.

Some attempts have been made to transfer technology to private hands. Much of what is being released is flowing to large multi-national companies – the same companies who built much of their wealth in the defense industry. Now these firms are being given the opportunity to exploit those same technologies for even greater profits. They are perhaps in the best position to develop the science, but should they be the only organizations with such access? Shouldn't this publicly funded research be considered "in the public domain" – accessible to both small and large organizations?

We have all participated in debates about our "national interests" and have considered the distribution of our national resources. Those national resources have primarily been centered in land, minerals, water, energy and other physical property. At the same time, all national governments are discussing the value of the "information age" and the promise of great innovative potentials. The combined resources which the world's governments are investing into new technologies is staggering. Why are these public expenditures, which often result in new patents and innovations, held by private companies rather than held in common as national interest resources equally available to the public which paid for the technological developments? Many of these developments take place in national laboratories in cooperation with research university laboratories and private companies. These are primarily publicly funded activities and yet the greatest value of the innovations falls to the companies which are feeding on tax revenues. This would be no different than an employee of a private research facility having the opportunity to work in a lab, develop his ideas and then patent them in his own name and take the technology with him the moment he parts company with his employer. Would any private organization allow this as a practice in their normal business activity? We think not. Yet this is exactly what our various governments often have done. They have created an information welfare system where innovation and technological development are owned by private organizations and paid for with public funds.

Science and innovation have historically offered governments power. That power is now being shifted to private companies which emerge as organizations without national interests because of their overlaps around the world. Their interests may very well conflict with the very interests out of whose resources the science was developed. Many of these companies move their production into areas of the world where labor is cheap and regulation is low, then bring back the production to where the demand exists for these new products. Most important is the idea that the technology transfers are building huge private holdings at the public's expense in the same way that land and resource transfers did in the United States over the last two centuries. We are not suggesting that the transfer to private hands is entirely bad. We believe that private enterprise should be the institution that develops these ideas into commercial possibilities. However, the transfer of publicly funded science should be available to everyone as "public domain science." Business organizations should not be permitted to take the science into production outside of their country of origin in order to capitalize on slave labor wage rates, poor working conditions, lack of environmental controls or other factors which run against the ethical and moral fiber of the country of origin. Our public foreign policy should never travel absent our ethical base. Corporate interests are, in most cases, driven by the profit on the bottom line, neglecting the higher values which allowed those same companies the possibility of existing in the first place.

This is a time of transition and the shift in resources must be coupled with an understanding of what these transfers imply. We have a great opportunity to move these resources to projects which could develop solutions to the truly great problems we face. The "peace dividend" should be used to enhance life rather than continue to exploit life. Exploitation of third world countries and people ultimately leads to political instability and runs counter to the hopes of many of us in seeing a lasting peace created in coming decades.

Science in the USSR

The situation in the former Soviet Union is dismal for the scientific community. Efforts have been made by individuals and various governments to financially stabilize the scientific community in the new countries formed after the dissolution of the USSR. The problem has been in uniting the scientific community around a new set of objectives and goals. Throwing money into a group without a unified mission is a useless and ultimately a wasteful exercise. An incredible opportunity is also being missed.

There resides in Russia a great set of possibilities if the scientific community can be organized around working on solutions to problems in health, agriculture, the environment and other positive programs. Top notch scientists in Russia are virtually unemployed, with the lucky few receiving small amounts of income, usually less than $400 a month.

Several years ago there was a good deal of concern being expressed that many of these scientists would be attracted to work in the laboratories of countries seeking to advance their own weapons capability. In fact, our own government, in the same way that they attracted scientists from Germany after World War II, have brought several of these scientists over to work in our defense industry. The possibility of organizing teams in Russia for the purpose of exploring specific problems in science is a great opportunity and may represent a dimension to the "peace dividend" which is otherwise being missed. Just throwing money at the problem is not the answer. Developing missions which have goals of benefit to the rest of us would be much more productive. The western countries have the ability to organize projects which can utilize the skills of these scientists to contribute to all of our lives.

The situation in Russia has been poorly reported in the western media. The disintegration has not been well understood by other than high level policy planners. There is increased concern, even fear, that the lack of a structured society will result in an even more chaotic situation. It was surprising to us to find out the true limits of knowledge in terms of the functioning of capitalistic systems. There are no points of reference for citizens of the former Soviet Union in terms of understanding those systems. How capitalism really works, aside from theoretical descriptions, is a great mystery and the assumptions that they reach are far off the mark. The need for infusing entrepreneurial thinking into their society, coupled with specific goals and outcomes, could build solid bridges for the future stability of the region. Without stability, the worst is possible and most likely will come to pass.

In the West we hear about the political corruption, the Russian mafia, the disintegration of the economy and support systems and the increasing level of social and political unrest. However, perhaps the greatest problem is one of leadership and confidence in the leadership. Much of the system has been reduced to the modern equivalent of the "survival of the fittest," without rules or order – something between predatory capitalism and a police state run by competing thieves has been created.

Russia's contribution to culture, through its people, has been noted throughout history. In literature, art, sports, science and virtually every area of human interactions, exceptional performance has been seen. Over the last seventy years, many of the scientific innovations created in Russia have been hidden from the rest of the world just as many of our own innovations have been hidden by the military-intelligence community in the west. A paradox now exists in that the hidden science of the east is becoming known while much of ours remains hidden. The new contributions possible from the former Soviet Union are just now being discovered at the same time as the risk of disintegration presents itself.

In the west, we are on occasion thrown a "bone" of information for the public to chew on. Some of these recent "bones" include electromagnetic weapon systems, CIA involvement in remote viewing and physic research, and space related innovations, among others. Contrast this situation with Russia, where the knowledge of science is pouring out freely for the first time in seventy years.

Interesting issues are emerging in science, some very clear, while some are highly speculative. What is obvious is that a great deal can be either lost or preserved at this point in our history.

Future Possibilities

A historic view of what happened in Russia in terms of keeping up with the west was, to the greatest extent, the outgrowth of their ability to develop science. What could not be built in productivity, distribution systems and their economy was made up for in scientific innovation. What places modern governments on an equal footing is their knowledge and control of science and technology. Science and technology are the handmaidens of political organizations. These represent the power of modern military machines and systems of control. While the systems of control continue to disintegrate in Russia to the point of near economic and political anarchy, scientists ill equipped to deal with either are being lost in the confusion and reduced to the level of basic survival. What is the greatest loss in all of this? It is the ability of humankind to use the knowledge in a way which benefits society. Russian scientists are being picked up by some private companies, other governments and most likely by criminal elements. The intellectual investment of all of the Russian people which was the outgrowth of their suppression and sacrifice over seventy years is being lost. The science of Russia, freed from the military, in combination with the productivity of the west, could present a needed core of innovative solutions to many problems in science. Moreover, the services of such scientists could be oriented towards a common interest through a distribution system based on capitalistic principles. In other words, a system that for seventy years attempted to develop a view of collective thinking for the benefit of society can now be synthesized into an individually driven system operating on a foundation of capitalistic distribution and democratic principals. This combination of a "collective good" driven by individual effort could yield combinations that solve significant problems.

Existing systems do not have the right level of independence in either the west or the east. A new system which becomes the outgrowth of organizational innovation will likely be what brings about the greatest possibility. Time is of the essence as science continues to be lost.

Solutions in Innovative Science

A clearing house for innovative science should be undertaken by organizations big enough to fund such efforts. Such a clearing operation would be the "Intelligence Link" for new ideas in humane science. The idea is relatively simple. A mechanism for connecting new ideas, scientists, business talent and financial resources could be established. The funding for such a project could come from revenues first saved from our shift from war economies to peace economies. In the long run, funding would come from the industries developed as a result of the effort. Small companies would be given the same opportunities as large organizations, based on information availability.

A partial model for such an organization has existed in Alaska for over a decade. The Alaska Science Foundation was funded out of "excess oil earnings" received from the development of Alaska's publicly owned oil resources. Alaska, with a population of just over one half a million people, established an endowment initially capitalized at $100 million to promote innovations in science which have benefits to the region. The interest on the fund is available for small grants and cooperative grants in the early stages of innovative technology. The grants are funded from the interest earned on the original $100 million. The grants are also repaid out of the projects that they fund at a rate which ensures that the fund will grow over time, being available for future generations. On a much larger scale the same kind of funding mechanisms could be established linking independent scientists with the business management and other resources that they need to contribute for creating positive change.

From Control to Power

Political leadership has become a system of manipulation rather than a responsible transfer of authority from groups to individuals. In American government we are taught that all government flows from the people to the leadership. However, over time this concept has become corrupted by a failure to maintain this ideal. At the same time the concept of authority flowing from the populace has become a lost memory for many political leaders and government bureaucrats. High standards of performance when filling leadership roles is required.

I have always been struck by the effect of leadership styles on people and what happens as a consequence of those styles. I have had the opportunity of speaking in mixed forums with other recognized leaders in various fields. I have observed numerous approaches to leadership, many of which were highly objectionable. The most important observation is that there is a gap in leadership training. Yet, leadership is fundamental to much of what we are each about; so much so that it requires some discussion.

What motivates leaders to lead? What creates intelligent movements for change? What are the possibilities, and how are individuals motivated to take specific actions in following a cause? The goal of this essay is to increase awareness of leadership and to create positive change. We hope that greater consideration and awareness of the roles we are each in and how we act to create those roles will lead to more fully recognizing our individual potency as people. It is our sincere desire that in putting these ideas forward we will stimulate much discussion.

Leadership which rises out of the circumstances at hand can be quite powerful. This type of leadership is not appointed or elected. It just manifests based on circumstances. This is the spontaneous activity of seeing issues or problems which require action and individuals rising to meet the challenge. In so rising to the call of events, these individuals are in leadership positions even if there is nobody behind them. This happens when an individual recognizes his personal need to pursue an issue and then begins to do so. They act on what they know, with their focus on a solution or program of action. This kind of leadership happens to someone rather than necessarily happening by directed intention. Acting on an issue independent of anyone else often places a person in the position of leading on an issue where "leading" was never part of the original plan. Leading in this instance is one of the possible outcomes of pursuing issues and ideas.

Recognition of the responsibilities which flow from positions of leadership and the idea of being accountable for those actions is sometimes lost. In political organizations accountability is often deliberately avoided. In order to control events and outcomes, manipulation and "spin control" becomes more important than assuring that outcomes are representative of the ideals of those governed. All political organizations use the power available to them to control outcomes. Is this good or bad? Does this matter? Should we expect anything different?

Zbigniew Brzezinski, as noted earlier, described his expectations for the remaining thirty years of the 20th century. His words lay a foundation, and sound an alarm, just as Huxley had decades before. The most startling ideas in his work dealt with the ideas of power and control by political leaders. He expressed concern that the political leadership (no matter who they were) would seek to exploit all technologies available to them in order to maintain their advantage over others. This concept is not new. This thinking represents the way things have always been, with the primary difference being that this kind of leadership used to achieve its power with clubs and guns and now has the ability of gaining its power with sophisticated media and new technologies.

The ideas and warnings put forward by both Huxley and Brzezinski are legitimate and are even more disconcerting when laid over the backdrop of the 1990s. Regardless of the view an individual takes of these two writers, it is clear that they had a perspective on events which most individuals miss. The important point is that they shared a vision of the future and attempted to inform people with their warnings and cautions. The technology to influence outcomes continues to increase faster than most people's ability to evaluate it, much less object to it. The need to become more aware of leadership style and approaches is important in preserving freedom.

Leaders must consider the ethical use of all technology and new methods of communication. They must consider more carefully their impact on the decisions of groups of individuals while recognizing their responsibility for the outcomes of the stimulated group actions. A strong and well articulated ethic is the foundation of the leader. The ethical positions of leaders determines to a great extent the means which they will employ to reach their desired ends. From my own personal experiences I learned at an early age the importance of communicating messages with care. The idea that well articulated thoughts could create tremendous energy releases in the direction of change was recognized and was a bit frightening. The idea that we could influence outcomes to a very powerful extent caused me to withdraw from leadership positions for about four years in the early 1980s. During that time, effort was made to gain a better understanding of leadership, motivation, communication and self evaluation before I would seek new leadership positions. The foundation laid at that time was imperative to long-term growth and increased understanding. The point being that people should realize the responsibilities and the need to be sensitive to the impact which goes with leadership prior to assuming these positions.

Individual Empowerment

Empowerment is an overused word in certain circles. When used it is usually in the context of some kind of complaint or whining call for something not earned. Real empowerment is the outgrowth of applied knowledge and information. The idea of giving people power is often used in the context of empowerment discussions. In this conclusion I am proposing that empowerment is something we each already have and it does not require that anyone pass along anything for us to act on this inherent power. This is the power of self determination as it affects us individually in our lives. As a birthright, a prerequisite to being human, we have the free will to pursue our interests. This is empowerment. The only thing lacking is the conscious awareness that all we do is the outgrowth of our individual power and potency. The limits we hold are those imposed by ourselves, by our individual communities of interest, and by society at large. The power to act in our own interest is fundamental and we do not require anyone's permission.

A Philosophy of Change

Anything can be changed under the right circumstances. Political activity can always be changed by those with a will for change.

The belief system which has motivated this book and most other aspects of our lives' is relatively simple. It is the idea that we should strive to reach the highest level of physical, mental and spiritual development through holistic actualization of who we are individually, or, more specifically, by acting in accordance with our highest probability options at a given moment. To try to make an effort toward the good, to be helpful in all expressions is the goal. Within the idea of holistic actualization resides our responsibility to assist all of those we are in contact with in achieving their highest potentials also. The last point in this simple belief system is the most important in that in helping others we learn a great deal about our own potentials, both positive and negative.

The only standing rule: "Thou shall not violate." The rule means that we cannot trample, manipulate or seek control of other people. We are all in different places in our personal growth and are here to help one another. Leadership requires responsibility.

A Path To Freedom

Cooperative and persuasive leadership styles are more in keeping with modern thought and generally leads to more progressive outcomes. In these instances a very different type of leadership is needed and proven to be most effective. This kind of leadership builds up and draws out the strength and talents of individuals. The objective of this type of leader is to empower rather than direct and control, to guide rather than manipulate, and to help individuals to actualize their fullest potential within the activity of the group. Individual integrity and strengths are recognized and developed as a priority of the leader, not in some kind of "group think" or dictatorial response to authority, but through a sincere desire to build into others the ability to recognize their own strengths and ability to lead.

This type of leader does not use manipulative methods to garner results. Manipulative methods are those which remove power from individuals. These tend to prey on negative emotions of fear, hate, greed and base self-gratification. To use these to get others to act results in violations of other people. The leader in these cases attempts to motivate by external coercion whereas the opposite is possible through proper information and persuasion which releases people to follow their own path to the right outcomes. Internally motivated people are much more effective than externally motivated people pursuing issues. This is where foundational ethical philosophy interacts with leadership responsibility and accountability.

Leadership which subjugates and controls takes what otherwise belongs to individuals: their right to free will and self determination. These leaders place themselves in a superior position to their followers and maintain their position of superiority through coercive means. The structure of the leadership system is hierarchal with the leader at the top and all others below in a pyramid of power. Effective progressive leaders do not require this same structure. Rather than a pyramid, they operate along with their "followers" on an open plane where each person leads in his own way while building up others. Some rise above the crowd and then they work to lift others to their new level of effectiveness rather than maintain a superior position.

I once heard it said, "If you don't know who's in charge, take over. If they throw you out of town, get in line and act like it's a parade." It is kind of a funny statement but it is also a reflection of reality. This is "spin control" – redefining the truth to be something else or, more simply, its propaganda production. We can assert our individual will to shape the perceptions of others. A union organizer I once knew said, "We are in the business of creating new realities." Perception is reality. Usually, perception is all that is being manipulated by the professional sellers of politicians, soft drinks and used cars. The sad reality is that most people are asleep and tend to be deceived by these manipulations. Critical thinking is blocked by failure to recognize the impact of emotionally developed propaganda which feeds base emotions.

Manipulation is essentially the misapplication of information in order to perpetuate the interests of the leader. Manipulation is clearly wrong and should be replaced by persuasion. Persuasion involves disclosure of the facts to create thoughtful decision making on the part of participants. Effective persuasion awakens an individual's ability to intellectually make decisions rather than react to emotionalized manipulations. This is fundamental to positive long-lasting change agendas. Truth will ring in the hearts of most people. Emotion should follow well articulated ideas rather than propel weakly contrived thoughts. In other words, the emotional response of followers should come after the ideas of the leader have been thought through and not the other way around. If emotion precedes thinking, bad decisions are made. This idea can also be viewed in the context of control-vs-empowerment. Control results from withholding, spinning or manipulating information, whereas empowerment comes from delivering all information necessary to form an opinion and then providing a specific "action set" for use in actualizing change based on that information. Realization of the facts should elicit an emotional response to act in one's own self interest. Persuasion is preferable to manipulation.

A good deal has been made of over control by government. People are increasingly questioning the intention of leaders by considering who and what motivates them. This is a healthy shift for the followers and a death knell for the manipulative leader. Much of what is believed to be power, force, or direction, is contrived by a few people with the ability to articulate and express their point of view into the public forum. Questions must be asked. Is the leader convincing, based on the facts? Is the leader able to demonstrate a strong positive ethic which is adequately demonstrated by his/her life's work? Is the leader honest in presenting all of the material necessary to empower individuals or is the leader taking power from individuals?

Recognition of individual strength is an important attribute of true leadership. This is what is being dissipated by over control of government that is dominated by manipulative leadership techniques. An individual's ability to self actualize his or her full creative capacity, at a point in time, is hampered by over control. Asserting an individual's will into factors for change is possible. Individual recognition of one's ability to first lead oneself is a precursor to leading others into actions resulting in real, lasting changes. Limits are indeed, to the greatest extent, self-imposed.

In the last ten years, I have had the opportunity to speak to groups from many different political, social, economic, religious and scientific backgrounds both within and outside of the United States. In all of these organizations, I have found a common thread. The thread throughout these very diverse groups are people with a common agenda. They are the unorganized organized – the Organic Internet. They are a leaderless movement, a revolution, a new way of thinking which is steadily gaining momentum as the direction materializes in front of each person. It is the same revolution which has always been here on earth. It is the search for individual truth.

These people who comprise the thread woven through diversity question those in authority regardless of their credentials or position. They raise questions based on their own individual analysis of the truth. These are people who are self assured and confident in insisting on answers. These are people who are thinking, alert and willing to modify their positions when faced with evidence which is compelling. They are not the blind followers who unfortunately still represent the majority in our societies. These are the "seekers" – those actively engaged in evolving beyond their current conditions.

The Chaos Theory of Human Interaction

Have you ever worked in a large organization of incredible complexity where the end product most closely resembles a sort of oozing to the finish line in accomplishing its goal? Large organizations function in a structured chaos which is generally full of inefficiencies. In large organizations people seem to forever engage in activities to correct, hide, forget or avoid some event which arises from their activity. It is like creating "order-out-of-chaos." We accept that there are those people in organizations, who by virtue of their positions, can create change by force of action. What is also clear is that most of what matters in organizations is accomplished by the working individuals in the organization and not by the Board of Directors. The world is a place where order is a reflection of the collective actions of all participants. The order, or lack thereof, is the creation of individuals acting in their own interests on the basis of their values and beliefs.

Every once in a while the existing "order" is overturned quickly through the simultaneous individual actions of many (For example: The fall of the "Iron Curtain") or more often by the slower pace of individual actions. However, in all instances change comes from individual actions. The smoothness or the chaos is determined by the foundational values and the prominence that these values have in the individual participants. The need to focus on "values" first, then "mission" and then "means" is required.

A focus on the underlying values of each individual within a group will create a smoother ride to the finish line more efficiently than any coerced or manipulated activity. The recognition of the effort made by others, and tolerance for the varied approaches, will build unusual alliances for creating change realities. Every change rests with individuals deciding to do something different in the areas they can effect, based on where they are at the moment. Change does not require any other external motivator.

Tolerance

Tolerance. It is the recognition that each person has his/her own road to the truth. We are not required to convince anyone that our road is better than any other or to stand in judgment of another's path to the truth. We are only required to keep looking and stimulating others to look also. This is the root of tolerance – Each of us is faced with life experiences and circumstances which serve to open or close our vision. We choose how we interact within the experience – learning and changing or getting stuck along the way. Actualizing potential in ourselves and stimulating the greater potentials in others by our own example is all that is really required.

Honesty in self-reflection is the beginning of change. Seeing the need and making an effort to find solutions is required. There really needs to be only one simple rule – that we shall not violate another's right to his/her free will. Out of this rule emerges tolerance and the greatest respect for the other person – the recognition of their right to be...

Accountability and Transparency

The idea of transparency starts with the individual. There are at least two perspectives in exploring the idea of transparency:

> Looking Inward. First, look from the position you occupy in self reflection. We can look through our lives and see areas where improvements can be made. We can see those flaws that no one else might see. We can begin to correct those things that create problems and inefficiencies or not...it is generally our choice. It is usually a matter of privacy as well. We do have a right to deny the truth and wallow in the place we are. It is always our individual choice. Truth begins in the mirror.
>
> Looking Outward. Secondly, transparency is the capability of truly discerning individuals or organizations.

> The less accountable or transparent an organization or individual, the higher its level of inefficiency. Organizational transparency yields truth and usually demands some accountability. Individual transparency rests with the individual who determines his/her level of visibility. Allowing transparency is the right of individuals to accomplish or deny.

A common theme in much of our writing involves that which is hidden from common knowledge and has to do with creating a more transparent society. We are committed to free and informed dialogue which will increasingly prevail over the hollow words of communications rhetoric surrounding most human interactions. There seems to be more concern for maintaining existing power structures than for recognizing that the only real power moves from the masses of people and not by the will of a few. We believe that fundamental changes are possible but only if built on foundations of values that are common to all and recognized.

Organizational privacy is different in that it reflects some collective standard. Exposing organizational flaws also exposes individual flaws. All choices we make which impact another requires some kind of accountability based on what is right and true and what is fair and just.

The increase in societal complexity requires greater individual accountability, first to ourselves, next, to whatever higher sources of inspiration and guidance we access in our search for truth and finally, to each other. At the same time, the security of individual privacy must remain as the basis of all freedom. The right to self determination, in a non-harmful way, is our foundational right from which all others spring. Our right to self expression and full human actualization is our hope in expressing our individual creation. All meaningful political, social, cultural, religious, or other interactions always begin with the actions of individuals.

Walls Of Secrecy

Increasing the transparency of governments and organizations and exposing problems will allow for greater accountability. Walls of secrecy in government are allowing a few people to manipulate the outcomes for the many. The few have been able to hide behind what is right and what is true – what is fair and what is just – in order to further their own ends at the expense of the many. The rights of life, liberty and the pursuit of individual happiness are being crowded out by a more directed, controlled and manipulated society. This is the wrong direction for society to take.

When people individually make decisions to change any aspect of their immediate conditions, the effect of the decisions move through everything connected to that individual. When numbers of people are doing the same thing, without any external motivator, they all arrive at the same finish line in much finer form. Like the Internet, all are connected, but are not waiting for leadership. They recognize that waiting for a leader is not what changing the world is about. It is about individuals taking individual actions, based on their unique effort at any given time, on a foundation of similar values. The values are expressed in beliefs which vary widely and bring about division in society. The underlying values are more important than the road traveled to find and recognize them.

To Change

These new seekers are always willing to change in light of what is true. They are individuals who first have discovered something about their own potentials and how to develop themselves. They recognize their own strengths and weaknesses and strive to overcome all obstacles. They are willing to admit error and benefit by it in trying to seek and live a true life. Their hallmark is not necessarily success – their hallmark is "sustained effort" – a willingness to try and a willingness to continue to try even after failure. These individuals recognize that it is in honest effort that success is found.

Another outstanding characteristic of these individuals is their willingness to contribute to the improvement of human conditions. They work from a most powerful position in this regard because they work from the point they occupy...the present life condition they are in at the time. They are not waiting for "the right time," "the right place," "the right circumstances" or "the right people to help" – they recognize that all of this exists in the very moment they occupy and only requires effort on their part. Truth seekers make things happen by acting on what they believe is true and right in the instance and place they find themselves.

In the last several years, Earthpulse has noted occasional news stories about individual people, despite personal risks, taking on adversaries in their pursuit of what they believe is right and true. Truly free people must have the ability to move forward on issues which are important to them individually and to us collectively. It is imperative that individuals at all levels of organizations, whether local government, science, religion, business or politics begin to insist on transparency and protection for those individuals who take risks in the public interest. To bring higher orders of transparency and efficiency requires open discussion of important issues. Solutions rather than the excitation of our differences and resulting conflicts must be foremost in our minds. A focus on solutions and changes which benefit individuals without violating or exploiting others should be encouraged in every way. Honest disclosure should be the standard when such is not the case. Justice should be possible in such a system and people should not be left alone without support. We believe this is possible and know of many working toward the same end.

Justice

People in all areas of human interaction have had occasion to come up against adversaries who would crush them to hide the truth from the rest. Most people yield before the battle has begun. Most people turn a bind eye to what they know is right and for years after are haunted by their decision to look the other way. On those rare occasions where someone emerges through the clouds of conflict into the light of day with some fragment of truth, the world applauds and says thank you or never notices in the first place, as the individual again disappears into his daily routines.

Earthpulse receives occasional letters from individuals reporting on "whistle-blowers" who become the victims of their effort to expose the truth. Disclosure after disclosure brings to mind the dishonest nature of many who hold positions of trust in politics, business, religion and science. The denial of truth serves only to enslave us all to those who control knowledge. Safety nets which protect those who expose the truth can be created without undue organization. Allowing abuses always results in higher costs than exposing them and forcing change early. Hiding the truth and protecting the villains while watching the messengers get slain is not the answer. There are plenty of examples – nothing is exempt:

> Business hides the truth about the risks or problems associated with its activities. Examples in the business community most recently include the tobacco industry's denial of the negative health effects of their products. They even enhanced the level of addictive substances in cigarettes, resulting in more deaths in the United States then otherwise would have occurred. Every day there are disclosures made for which a hundred more go unannounced, undiscovered or hidden behind someone standing in front of the truth.
>
> In politics it is a continual disclosure of endless scandals dealing with campaign funding, special interest favors or the latest failure in personal ethics by some would-be-statesperson.
>
> In science there is always innovation and discovery. We think of science as a set of beliefs grounded on facts. Much of science is really "hope grounded on dogma" and represents just another regulation of belief systems. Religion means "to regulate beliefs." So, in the broad sense, science can be thought of as a religion grounded on a faith in a group of theories thought to be based on "insurmountable facts," which seem to change as our per-

spective on them changes. Much is routed through "the guardian of the truth" arm of the particular science discipline in a way which tends to inhibit change. These "guardians" are groups of scientists protecting "their" science but need the balance provided by openness. Most science innovation, when first proposed, was rejected by the "official keepers of the scientific truth." Nothing has changed historically. Moreover, it is very likely that today's science will seem childish in many ways in the next hundred years.

The work of independent researchers disclosing the latest in health related discoveries, whether they be in medicine or environmental science, are the source of numerous reports. Additionally, there are always rumors and in fact, suppression of much innovation in favor of maintaining the status quo for vested scientific and industrial interests. These suppressed or restricted disclosures may be denying citizens opportunities to solve many of the problems which we as a society face. They include energy, medicine, agriculture and virtually every area of human interaction of significance.

In religion, there is almost always some level of suppression. The suppression may start by crushing of the human spirit as it is manipulated by the self interests or insanity of cult leaders. It can also be from mainstream religious leaders who are frequently in the news. These violations are no greater than any other except that we otherwise vest in these leaders our most sacred trust – the welfare of our souls. The fact of the matter is that the welfare of our souls rests primarily in ourselves and the choices we make.

The Current Fallacy of Justice

I have often said that justice has a direct proportional relationship to the size of your wallet. It was said as a sarcastic attempt at humor and, like most humor, it is in part based on the truth. Dealing with the court system, regulatory or elected body requires resources. Those with the most resources have the greater likelihood of success. Justice has more to do with your ability to move resources into your arguments than what might be right or true. This is the case whether you are the victim or the perpetrator.

Society has created an upside down situation where the perpetrator becomes the victim and the victim is victimized again, only this time it is by the system which is supposed to support justice. The rights of presumed innocence, fair representation, justice and accountability for wrongdoing are all fundamental. These fundamentals apply to both victim and perpetrator. Public resources should serve the interests of both equally. Only when there is equal access to the systems which maintain and create law can there exist true justice. Individuals who assert that they have been wronged should be provided simple means to test their arguments. People need access to juries with the power to change, enforce and interpret law in the same way as the other branches of government are so empowered. Juries need to assert their right to determine the effect and reality of law.

Public interest litigation and the right to bring complaints are equally fundamental to justice. Methods must be found allowing for the least costly and least complicated opportunities to be heard by a jury in an appeal for justice. The system should be made less intrusive and more equally accessible not only when individuals have been harmed but also when some larger community of interest has been harmed including the "public (human) interests." The right of a victim to redress under the law has to do with individual access to those systems of justice. To get justice as an average person, you must, under the current rules of the game, risk all of your resources in every instance. In addition, there are also the personal emotional costs associated with solving the conflicts.

There have been attempts at professional oversight by such groups as the AMA, Bar Association, Teaching Practices Commission and other quasi-public or quasi-private oversight organizations. These are the club rooms of closed files where the public does not get to see the hearing, evidence or results. The veil of secrecy often evades justice. Open hearings are the way in which justice is reached. It is in the open light of day that the perpetrator and victim are on an equal footing. It is in the open that the truth gets heard and acted upon by the public. There are of course issues of privacy, due process, rules of evidence and fair dealing that must be considered. The opportunity for closing off public disclosure should be very limited and be focused on protecting victims and the accused equally.

What now happens at various junctures is that someone musters the courage of their convictions, rises above the fear and risks, and screams from the housetops their version of the truth. What happens then? Usually not what was expected. The powerful interest they challenge is inevitably in control of much greater resources. The brief battle crushes the challenger and denies the world the opportunity to hear the truth. The method is usually the same. The first step is to crush the economy of the individual so as to take away his/her means of economic support. The next step is to cut off his/her communication with others in their respective field by declaring the individual an intellectual heretic, a disgruntled employee, a person suffering from stress related disorders or even worse. Another ploy is to file litigation which ties the person into an economic noose from which there is little escape. There are few people up to the task of continuing and there is many a battle lost in the first volley.

The solution is simple and begins where this writing began – with each of us individually acting on what we know to be true and right in the instance we now occupy. A great deal is being accomplished in every effort and we are never alone. We must only make an effort – We must try!

We shall someday begin a Thousand Years of Peace – we should begin the journey today.

The Earthpulse Institute

Over the last ten years Earthpulse has grown from a quiet research group to a quickly expanding publishing and news organization. Our desire has been to create discussion and solutions to some of the problems we see around us. Publishing has given us a valuable tool for producing work we believe stimulates thinking and contributes to improving the human condition.

Much of the work we are doing these days has shifted from our original research and writing roots to those of managing the day-to-day activity of publishing. Over the last year we have reorganized many of our operations in order to get back to research and writing projects put on hold over the years. Some of these projects are not commercial but should be pursued because these projects could result in the development of new technologies, inproved public debate and other contributions to a better world.

We started Earthpulse with the intent that our customers would be our source of support for our work as opposed to traditional grant funding sources. We have maintained this position over the years and are glad to have done so as it has served to assure our independence as an organization. This approach has kept us focused on the work which we have become known for over the last ten years. We seek to serve the public through information distillation and, as a result, have succeeded in our efforts because of each reader's participation in this work.

In the last year, we have had to pass over a number of projects we believe were important because of limits on our research resources. In the year 2000 we are forming the Earthpulse Institute in order to better fund our research and technology areas. This will allow us the opportunity to greatly expand our work in weapon systems research, alternative health projects, politics and other areas. To this end we are asking for your help. Please consider supporting the Earthpulse Institute by contributing to our efforts.

As well, information is the lifeblood of Earthpulse. We are interested in hearing from you about topics that concern us.

We, personally, want to thank each of you for your continued support and help over the years. Together we can change the world.

Dr. Nick Begich, Founder Earthpulse Institute
Shelah Slade, Founder Earthpulse Institute
James Roderick, Founder Earthpulse Institute

The Earthpulse Institute
P.O. Box 201393
Anchorage, Alaska 99520 USA
Fax: 1-907-696-1277

Resource Guide

Earthpulse Press
P. O. Box 201393
Anchorage, Alaska 99520 USA

24 Hours a Day
VISA or Master Card Accepted
Voice Mail Ordering: 1-888-690-1277
or 1-907-249-9111.
Fax: 1-907-696-1277
http://www.earthpulse.com

A free catalog of all of our products and books is available on request.

1. ***Angels Don't Play this HAARP: Advances in Tesla Technology*** is a book about non-lethal weapons, mind control, weather warfare and the government's plan to control the environment or maybe even destroy it in the name of national defense. The book is $18.00 Air Mail in the U.S. or $22.00 internationally.

2. ***Angles Don't Play This HAARP – THE VIDEO.*** This is a video lecture by Dr. Nick Begich produced by L. L. Productions of Seattle, Washington. Approximate running time 1 hour 45 minutes. 32.00 in the U.S. or $35.00 international.

3. ***Earthpulse Flashpoints*** is a Microbook series edited by Dr. Nick Begich. Microbooks cover four major areas: government, frontier health sciences, earth science, and new technologies. The goal of the publication is to get hard-to-find information into the hands of individuals on their road to self empowerment and self discovery. Nine issues in print @ send $4.95/each plus shipping $2.00/each.

4. ***Earth Rising – The Revolution: Toward a Thousand Years of Peace,*** by Dr. Nick Begich and James Roderick. This book is a book about new technology and the impacts of new technology on humanity. The book is footnoted with over 650 source references spanning fifty years of innovations. The book includes advances in the Revolution in Military Affairs, mind control and munipulation of human health, non-lethal weapons, privacy erosion and numerous other areas where technology is impacting mankind. The conclusion of the book contains an overview and series of actions which could be initiated which would return us to a more civil and open society. The book is $22.00 Air Mail in the U.S. or $26.00 internationally.

5. ***Earth Rising II – The Betrayal of Science, Society and the Soul***, by Dr. Nick Begich and James Roderick. This book is a continuation of the series on new technology and the impacts on humanity. The book is footnoted with over 350 source references spanning fifty years of innovations. The book includes advances in the Revolution in Military Affairs, munipulation of human health, non-lethal weapons, privacy erosion and numerous other areas where technology is impacting mankind. The book is $22.00 Air Mail in the U.S. or $26.00 internationally.

6. ***Holes in Heaven – THE VIDEO.*** This documentary film narrated by Martin Sheen explains both sides of the debate surrounding the HAARP issue and includes the military, scientists and those opposed to the technology. This video is $33.00 Air Mail in the U.S. or $36.00 internationally.

7. ***Weapons of the New World Order.*** This in-studio video is a two hour and forty minute discussion of new weapons technology from a Christian perspective. The topics include the latest information on HAARP, non-lethal weapons, weather modification, mind control and other controversial issues. The video is $28.00 Air Mail in the U.S. or $31.00 internationally.

8. ***The Secret Lives of Plants***, by Christopher Bird and Peter Tompkins is a classic on plants and early work in sound stimulation of plant growth. This product is shipped air mail in the U.S. at $19.00 or internationally for $24.00.

9. ***Secrets of the Soil,*** by Christopher Bird and Peter Tompkins, is the most important book ever published on new ideas for more productive agriculture. Faster growing rates and greater yields without using petrochemical based fertilizers are detailed in this incredible work. The book is of great use to home gardeners and commercial growers alike, an important book for all private collections. Available shipped in the U.S. at $22.95 or internationally for $27.95.

10. ***Bringing the War Home***, by William Thomas, is a controversial book which explores and challenges the ideas surrounding the events in the Gulf War. Through careful research, and a new perspective, the author describes the problems leading to the conflict and the startling results. The U.S. government's cover-up and military personnel's exposure to chemical and biological weapons is detailed in the most incredible exposé ever written on the subject. But it doesn't stop there – in the 448 pages of this book a cure for some of the diseases is detailed. The book is $22.95 shipped airmail in the U.S. or $27.95 internationally.

11. ***Scorched Earth: The Military's Assault on the Environment***, by William Thomas, is a detailed account of how the military has exposed Americans and people around the world to toxic risks they never disclosed. Every page contains well-footnoted evidence of how our lives have been negatively affected by the lack of concern for the impact of military technology. The book is $19.95 Air Mail in the U.S. or $21.95 Air Mail internationally.

A free catalog of all of our products and books is available on request.

Earthpulse Soundwave™!

Available Now $775 + $20 Shipping & Handling!

The most advanced sound transfer technology available anywhere in the world is currently being beta-tested in Europe for introduction to the United States in December, 2003. This new technology allows for the transfer of sound directly through the body by-passing the normal hearing mechanism. For more information on this technology contact Earthpulse Press at the above address.

> "This technology has now been replaced with superior technology developed in Europe in 2001-2003. The technology far surpasses the performance of the Flanagan invention in all applications by integrating modern circuit designs with 21st century engineering. The new technology delivers significantly greater energy efficiency through a more biologically compatible signal. This technology will be made available in late 2003 for accelerated learning applications, body system balancing and trials in hearing enhancement applications." ***Angels Don't Play This HAARP*, Begich & Manning, Earthpulse Press, 1995.**